COMPUTER AUDIT, CONTROL, AND SECURITY

Robert R. Moeller

WILEY

JOHN WILEY & SONS
New York • Chichester • Brisbane • Toronto • Singapore

**THE WILEY/INSTITUTE OF INTERNAL AUDITORS
PROFESSIONAL BOOK SERIES**

Gil Courtemanche • The New Internal Auditing
Gil Courtemanche • Audit Management and Supervision
David S. Kowalczyk • Cadmus' Operational Auditing
Philip Kropatkin • Management and the Audit Network
Robert R. Moeller • Computer Audit, Control, and Security

Library of Congress Cataloging in Publication Data:

Moeller, Robert R.
    Computer audit, control, and security / Robert R. Moeller.
      p. cm.
    Includes bibliographical references.
    ISBN 0-471-85310-0
    1. Electronic data processing—Auditing. 2. Computers—Access control. I. Title
QA76.9.A93M64 1989
657′.458′028558—dc20

89-36410
CIP

Printed in the United States of America

10 9 8 7 6 5 4 3 2 1

Preface

This book is a result of my frustration with much of the current computer auditing guidance material. In my work with internal auditors as well as my past association with the public accounting firm of Grant Thornton, I have often been asked to recommend publications to help computer auditors understand new and evolving computer audit, control, and security issues. I have found many current books to be out of date or to provide information in only limited technical areas.

Computer Audit, Control, and Security is my attempt to provide current guidance to the internal auditor as well as other professionals interested in these computer auditing issues. Each chapter contains both background discussions and sets of control objectives and audit procedures which, I hope, will be useful to the auditor in performing actual reviews. However, since every organization is unique, these same sets of control objectives and audit procedures are also included in a diskette format so that the auditor can tailor them to unique audit projects.

The materials in this book are based upon my own professional work and experiences. However, I would like to acknowledge Professor Fred Davis of The University of North Texas who reviewed my original book outline and encouraged me in the project. A second acknowledgment must go to Angie Chin, internal auditing officer at the Federal Reserve Bank of Chicago. She suggested that I include the audit program diskette along with the book.

Finally, I must thank my wife, Lois, who put up with me during this book project. She went to sleep many nights with the clattering of my computer printer as background music. We also postponed a variety of much needed house improvement projects while I worked in front of my computer on the project. Now that the book is completed, I guess that I had better get back to that half-completed sun porch remodeling project!

ROBERT R. MOELLER

Chicago, Illinois
June 1989

Contents

List of Control Objectives and Audit Procedures

Note: The Control Objectives and Audit Procedures contained in this book and listed below are also provided on a diskette included with the book in an ASCII format for use on an IBM-PC or compatible computer. See the Appendix for instructions on the use of the diskette.

CHAPTER 1

Introduction to Computer Control Procedures

INTRODUCTION

Computer systems first became common as tools for business data processing and information management in the early 1960s. They replaced a variety of punched card, unit record, and manual systems. Since then, the growth of computer-based systems and of computer technology has been phenomenal. Large mainframe computers are increasingly high-powered. More powerful microcomputers are used throughout large and small businesses. Programs once written in time-consuming assembler or compiler languages can be generated automatically with the input of a few parameters. Users once in awe of data processing professionals are now developing their own applications using a wide variety of high-powered tools

Auditors have been involved with reviewing controls over computer applications since the earliest days of data processing. This involvement was a natural occurrence since auditors traditionally were interested in the correctness of financial statement balances, and many of the first business data processing applications were financial. Over time, auditors began to realize that the correctness of computer-generated financial statement balances depended upon more than just the correct addition of a set of data. Thus, the auditor became concerned with the entire controls environment surrounding the computer system, including general management controls and computer systems security.

In response to a need for an overall data processing controls awareness, auditors developed standards covering reviews of data processing controls. These standards served both external and internal auditors, and tended to cover technology as it existed at the time the standards were developed in the late 1970s and early 1980s. Auditors viewed data processing as largely a batch-oriented, centralized data processing environment.

Dramatic technological and procedural changes have occurred over the past several years in the manner data processing functions are organized and business applications are designed, developed, and controlled. The objective of this book is to introduce these modern computer security, audit, and control concepts to the auditor.

The word "modern" does not mean that classic data processing controls, such as physical access controls, are no longer applicable. Rather, "modern" describes the ongoing changes taking place and how they will impact the auditor. Because of the increased use of on-line applications, for example, the auditor must place a greater emphasis on information security controls rather than just physical security. Similarly, the auditor cannot follow one set of physical security rules or standards for all auditor reviews. Many of the earlier standards developed for mainframe, central computer installations are not generally applicable for smaller mini or microcomputer systems.

This book will introduce the auditor to the concepts of computer security, computer auditing, and computer control principles. Its coverage will increase the skills of the experienced computer audit specialist as well as educate the beginning auditor.

ORGANIZATION OF THIS BOOK

The book serves as a reference rather than simply a continuous text on computer auditing. The reader should be able to open to any chapter of interest rather than starting at the beginning and searching through the book to find pertinent information. Following the introduction, the book is organized into five sections:

Controls in the Modern Data Processing Center
Auditing Data Processing Applications

Security for the Modern Data Processing Center
Audit and Control of End User Computing
Future Trends for Computer Auditing

Brief summaries of the sections and their chapter topics follow.

Section I: Controls in the Modern Data Processing Center

This section covers general data processing controls. These are the general or interdependent controls that affect all applications in a total data processing system. One example is a procedure for limiting access to computer program libraries. If there are no controls over who can change or update a production program library, any applications controls will be potentially weakened because production programs can be changed without management authorization. This area of data processing has been subject to major changes through reorganization of data processing operations.

Chapter 2. General Controls in the Larger Computer Center

Chapter 2 discusses some characteristics of a larger mainframe-based computer center and introduces various reviews that the auditor might consider. The chapter contains tables of control objectives and audit procedures for the following larger computer system control areas:

General organization controls
Controls over access to data
Controls over access to programs
Systems development methodology controls
Data processing operations controls
Systems programming and technical support controls
Data quality assurance

These tables of objectives and procedures will allow the auditor to design custom programs for specialized reviews of data processing general or interdependent controls. While most of these procedures are applicable to the larger computer center, the auditor may find some material useful in smaller data processing systems as well.

Chapter 3. Mini and Microcomputer Controls

Increasing use of mini and microcomputers has caused major changes in the way auditors should be thinking about computer systems controls. This chapter discusses these smaller, more informal computer systems and the control environment that the auditor must consider. The chapter discusses control objectives and audit procedures when mini or microcomputers are used as free-standing business data processing computers as well as when used for specialized process control functions.

Chapter 4. Controls in the Distributed Network

Computers are increasingly linked together into networks of processors where each computer performs some systems function. Chapter 4 introduces concepts of distributed processing network systems and related telecommunications controls. Also discussed are control procedures for intra-organization Electronic Data Interchange (EDI) systems and local area networks (LANs).

Section II: Auditing Data Processing Applications

Automated applications represent a critical control area in the modern data processing center. Often, earlier computer applications were replications of manual procedures with extensive front- and back-end checking and controlling. Control personnel were able to batch-balance inputs to an application and also perform extensive balancing procedures after the computer processing. This seldom happens in today's complex, highly integrated computer applications.

This section deals with the auditor's role in reviewing data processing applications, beginning with a pre-implementation review of application controls and continuing through a detailed, extended control risk assessment or review of the operating application. It also deals with the auditor's role in developing detailed procedures to gather evidence for application testing.

Chapter 5. Auditing Data Processing Applications

While auditors often perform general controls reviews of procedures covering the data processing operation, it is equally important that they review controls covering specific data processing applications. This chapter explains how to select applications for review, how to develop an understanding of application controls, and how to evaluate those controls. Chapter 5 contains large system and minicomputer examples of application review candidates discussing review approaches for each.

Chapter 6. Evidence Gathering and Testing Applications

A fundamental task of auditing is to obtain evidence as to the validity of application accounting treatments and transaction balances. Often, this can be accomplished through testing techniques ranging from the use of generalized audit software to the implementation of embedded audit software. Chapter 6 introduces these techniques for evidence gathering and suggests the best approaches for different types of audit objectives.

Chapter 7. Reviewing New Applications Under Development

The best time for the auditor to review new applications for the existence of appropriate controls and to make suggestions for improvements is when the application is still being developed or implemented. Chapter 7 discusses how the auditor can systematically select application candidates for such pre-implementation review based on the relative risk and perform the reviews.

Section III: Security for the Modern Data Processing Center

Although many larger organizations today have a separate, specialized data security officer or function, auditors are often viewed by management as computer security "experts." While data processing physical and information security should be part of any control risk assessment, the auditor often performs much of the detailed data processing security work. Physical and information security plus effective disaster recovery planning will be explained here.

Chapter 8. Physical Security Strategies and Risk Evaluation

The modern data center requires physical security controls to protect equipment from a variety of hazards. However, the auditor is often faced with the question of how much protection is really necessary. For example, a large mainframe-based computer complex requires more physical security protection than a departmental microcomputer. Chapter 8 presents several approaches to evaluating security risks to determine areas where controls should be implemented. The chapter then discusses various physical security controls. The risk evaluation approaches discussed in this chapter can be applied to other data processing security areas.

Chapter 9. Ensuring Information Security and Integrity

Information security includes controls over access to computer data and programs plus overall policies and controls to prevent and detect unauthorized system access attempts. Information security is an area where new tools and techniques are becoming available as concerns with computer crime and personal privacy violations increase. Chapter 9 comments upon tools and controls that can be installed to establish effective information security in today's data processing environment.

Chapter 10. Effective Disaster Recovery Planning

A data processing disaster recovery plan requires a formal set of procedures to allow data processing functions to continue to operate in an extended emergency where normal data processing facilities are not available. Auditors are often asked to review or help develop these plans. This chapter discusses the development of a data processing disaster recovery strategy in addition to documenting a formal disaster recovery plan.

Section IV: Audit and Control of End User Computing

Computer audit specialists once focused efforts on controls and procedures within the data processing function while operational/financial auditors worked outside the data processing areas and left computer systems issues to computer audit specialists. Today, this is changing. Information centers and microcomputer-based tools give end users facilities to develop their own applications outside the formal data processing function.

End user computing growth is perhaps one of the most significant changes impacting the auditor in today's data processing environment. Audit impacts of the end user information center and the significance of fourth generation languages will be explained in this section.

Chapter 11. Auditing the End User Information Center

The term "Information Center" means a variety of things at different organizations. For some, it refers to a formal facility where end users have access to terminal and software tools to develop their own applications. For others, an "Information Center" is little more than a training facility for departmental microcomputer users. Chapter 11 addresses various forms of end user computing and provides objectives and procedures for performing audit control reviews of the end user computing function.

Chapter 12. The Auditor and Fourth Generation Languages

Fourth generation languages are powerful software retrieval and development tools increasingly being used by end users and system developers. Characteristics of these languages and some unique control characteristics encountered when applications are developed will be discussed in Chapter 12. In addition, the chapter describes how the auditor can use fourth generation languages as a substitute for traditional computer audit software.

Section V. Future Trends for Computer Auditing

The modern computer audit specialist must be aware of technological trends driving data processing into new fields and directions. These trends have significant control implications. This section addresses how the contemporary internal audit department might better organize to cope with highly automated applications and extensive use of end user computing. This section concludes the book with a "crystal ball" chapter on evolving technologies that may impact the auditor in future years.

Chapter 13. Integrating Financial and Computer Auditing

Many audit functions have separate computer and financial audit specialists who have little formal interface or contact. In the modern data processing environment, this is not an effective way to monitor controls and perform audits of financial records or operational areas. This chapter discusses several approaches to better integrating the financial and computer audit functions within a single organization. Integration of functions is necessary in modern data processing, encompassing end user developed applications, distributed processing, and departmental computers.

Chapter 14. Evolving Technologies and the Auditor

New computer hardware, such as highly parallel processors or Write Once, Read Many Times (WORM) discs, and new systems development approaches, such as the use of expert systems, will change the manner in which future computer applications are designed and controlled. Briefly introduced here are some evolving technologies and their potential audit implications. Not all of these trends may evolve in the manner presently anticipated, but the auditor must be aware of current trends and techniques to meet evolving technologies.

USING THIS BOOK

As previously mentioned, this book is not designed to be read sequentially. Rather, the reader can use the chapter descriptions to select a pertinent section. If one chapter covers a technical area new to the reader, there will be a brief description of the technical concepts within that chapter.

This book, however, is primarily intended to help the auditor perform audits. Extensive tables of control objectives and audit procedures are included to help the auditor develop specific audit programs and perform effective audits. These tables are different from the more traditional audit checklists because they require the auditor to tailor them to specific audit situations. The tables list specific control objectives followed by audit procedures that determine whether those control objectives are being achieved.

Diversity of modern data processing environments prevents the auditor from using one set of procedures for all areas. For example, the types of specific controls the auditor might expect to find in a large, mainframe-based computer operations area are different from those found at a departmental microcomputer. Nevertheless, some basic controls objectives are the same. The tables of control objectives and audit procedures will allow the auditor to tailor procedures to specific computer system environments while still satisfying overall risk assessment and internal control objectives.

The same tables of control objectives and audit procedures found throughout this book are also contained on the 5¼″ microcomputer diskette packaged with the book. The auditor can print the figures from this diskette for use as actual audit programs using any microcomputer equipped with MS-DOS 2.0 or greater. More importantly, the diskettes are in ASCII format and can be used with standard wordprocessing software for custom tailoring to the particular organization being audited. The Appendix describes the use of the audit program diskette.

Computer audit, computer security, and general computer control issues are all important in today's data processing environment. The prudent auditor must be aware of these issues to be an effective advisor to general organizational management.

Section I

CONTROLS IN THE MODERN DATA PROCESSING CENTER

CHAPTER 2

General Controls in the Larger Computer Center

INTRODUCTION

Auditors first became involved with computers and data processing controls when they found manual and punched card accounting applications increasingly installed on computer systems. Frequently, these early business data processing systems were impressively installed in glass-walled rooms within corporate lobbies. However, the business data processing applications were not particularly sophisticated, and the auditor could "audit around the computer" by reviewing inputs and then system output totals. In the early 1970s, a company called Equity Funding changed all of this.

Equity Funding Corporation was a California-based insurance company involved in a massive management fraud. Its computer systems were used to carry on a portion of that fraud. Through management direction, fictitious policy data was entered on computer files. Equity Funding's auditors did not use procedures to verify the correctness of those computer files. In the aftermath of the Equity Funding affair, professional audit organizations, such as the American Institute of Certified Public Accountants (AICPA) and The Institute of Internal Auditors (IIA), began to emphasize the importance of reviewing computer controls. A new professional specialty, computer auditing, was launched!

In these early days of business data processing, most computer systems were complex, expensive, and "large." Fairly standard sets of auditor control objectives and procedures were developed for reviewing controls in these computer systems. While many control procedures are still applicable for all computer systems, the auditor must scrutinize other controls from a different perspective when auditing in today's computer environment. This chapter will examine controls in the modern, larger data processing organization. Special control concerns more unique to a mini or microcomputer environment will be discussed in Chapter 3. Finally, Chapter 4 will discuss distributed processing and telecommunications controls.

THE "LARGER" COMPUTER CENTER

The UNIVAC I was the first successful business data processing computer. Introduced in 1951, it helped predict the results of the 1952 US presidential election. The UNIVAC I occupied a sizable floor area, weighed 15 tons, and cost $1.3 million. However, a $5,000 personal computer today has 20 times or more the memory and speed of the UNIVAC I. Was the UNIVAC I a "large" computer system? In today's terms, it certainly was *large* based on cost or floor space occupied; however, when measured by functional capabilities, it can not be considered to be particularly large.

The definition of what is a "larger" computer becomes more difficult with modern computer systems. Some systems will be described by their manufacturers as "minicomputers." However, these same systems may appear to be "large computers" to the auditor. Such computer systems, although packaged in a small box, will support a large variety of peripheral equipment such as multiple terminals, printers, and disc drives. The computer hardware may also be supported by a sizable operations staff and handle numerous processing tasks.

Organization members have their own definitions for a "larger" computer system. The computer engineer often defines a larger computer system by describing its central processor's internal design or architecture while management defines a computer system's relative size by equipment cost and the size of its support staff.

The auditor is also interested in the size of the computer system to be reviewed. Computer system size impacts the auditor's approach and the audit control procedures to be performed. Some control procedures the auditor expects to find in a very large computer center would not apply to a small business computer system. The auditor would expect to find a formal systems programming staff in a larger systems organization but not necessarily in a smaller center.

A larger computer system tends to have some common characteristics. Though all may not apply to every larger computer system, the following will help the auditor to understand the characteristics of a larger business data processing system:

Controlled Physical Environment

Larger computer systems usually require specialized electrical power systems as well as dedicated air conditioning or water cooling systems. Because of these special needs and because computer systems consist of multiple pieces of equipment connected by heavy cables, larger computer systems almost always are located in specialized rooms with false floors and dedicated environmental monitoring controls.

Because the larger computer system represents a major investment in terms of the equipment installed and data carried in its computer files, these larger system machine rooms usually are protected by physical security controls such as locks on the door to the computer facility. Although sometimes smaller computer systems are located in similar secure, dedicated physical facilities, this type of facility is more characteristic of the larger computer center.

Multi-Task Operating Systems

Virtually all computers use some master program or operating system to control the various application programs run by the computer as well as tasks such as reading a disc file or supplying report data to a printer. These complex operating systems can run many programs during the same time intervals, in addition to many other tasks. Many smaller computers can process only one or two programs or tasks at once; a personal computer usually runs a single program at a time.

Often, a multi-task operating system is a characteristic of a larger computer system. While this operating system can perform a large number of tasks at once, this resource requires systems programmers to maintain the system.

In-House Programming Capabilities

The larger computer system will typically be supported by its own applications programming staff. Of course, some data processing departments may have very limited in-house computer programming capabilities. They may purchase the majority of their applications as package products from software vendors. Similarly, in an operating unit of a larger organization, all major systems may be supplied by a headquarters staff. However, usually a larger computer hardware system is supported by an in-house systems and programming department.

A larger data processing organization with its own programming staff may have systems analysts to help design new applications. Fairly formal procedures will help develop and implement new applications. These will typically include a systems development methodology or SDM, specialized library files to control computer programs, and technical documentation covering the programmers' work.

Multiple On-Line Terminals

Virtually all modern, larger computer systems have the ability to support multiple on-line terminals. These terminals may be located in data processing and user areas, and in geographically remote facilities. Terminal users may simply retrieve data or update transactions to application system files.

Characteristically, a larger computer system will have many local or remote terminals tied into application systems, requiring specialized information security controls to prevent improper access attempts. Also, the network may require specialized technical personnel within the data processing organization, to manage telecommunications.

Very Large or Critical Files

Although a computer system may be small in many respects, it may have one or more applications that maintain critical data on very large files. For an older computer system, these critical files may consist of many reels of magnetic tape. In a newer system, disc oriented database management systems will be used. Because of the criticality of such large files, a computer system with smaller hardware may take on characteristics of a larger system.

The existence of backup copies and the integrity of such critical files become very important to the data processing department. The organization may require specialized file backup procedures and personnel with specialized database management skills to ensure the accuracy, integrity, and completeness of the database.

Specialized Staff Positions

A computer system may be considered smaller rather than larger by all of the above measures. However, management may staff the data processing organization with other specialized staff personnel such as data security officers, telecommunications analysts, or quality assurance analysts. In an organization of this type, the auditor may want to structure the review procedures as if the computer system were much larger.

While the above characteristics do not define a larger or complex computer system in a precise manner, they provide guidance for the auditor in determining the review procedures to follow. While the auditor's control objectives will remain essentially the same for large and small installations, audit procedures will differ. Techniques for auditing smaller systems are given in Chapter 3, while specialized techniques for distributed systems are reviewed in Chapter 4. If the auditor has doubts, the safest approach is to consider the system to be reviewed as a larger, more complex one.

INTERNAL VERSUS EXTERNAL REVIEWS OF LARGER SYSTEM CONTROLS

Chapter 1 discussed different auditing standards and approaches used by internal and external auditors. Each should be interested in the controls environment in the larger computer center. However, because the external auditor is usually concerned with the impact of data processing applications controls on financial statements, the external auditor's emphasis will be primarily on financial applications and on the relative risks associated with the data processing controls environment. To understand the controls within those applications, however, the external auditor needs to understand the overall controls environment.

The internal auditor performs an internal controls oriented review under management's overall direction. The emphasis of this review will be on management-related controls within the computer center and on compliance with management policies and procedures. An internal auditor's review typically will be much more detailed than that of the external auditor. Because of this level of detail, the internal auditor spends considerably more time in performing the review and investigates many operational controls and details that may be ignored by the external auditor.

GENERAL VERSUS APPLICATION CONTROLS

Both internal and external auditors review general or overall data processing environments as well as controls surrounding specific applications. Because these controls are often interrelated, the auditor should have a good understanding of the differences between general—or interdependent—controls and specific application controls in the larger data processing operation. These differences are discussed below.

General Controls

These are procedural, processing, and managerial controls that affect all data processing applications. These controls include:

Management controls such as the plan of organization

Controls over unauthorized access to computer data files

Controls over access to computer programs

Sometimes, general controls are called interdependent, since specific application controls may be dependent upon some general controls and general controls may be interdependent upon one another. An example of a general control is the control over access to and updating of computer programs. If programmers or others, for example, can make changes to production application programs without management's review and authorization, there is a risk that an application system could be run with an incorrect or unauthorized version of one of its programs. This is general control rather than a specific application control because program change procedures affect all applications.

Application Controls

Application controls refer to controls relating to a specific data processing application such as payroll or accounts payable. Even though general controls may be good, controls built into a specific application may be weak. The auditor may be satisfied that the general controls are good, but the auditor may find a specific weakness in one or more applications.

Another example might better explain this concept. Assume that the auditor finds the general controls within a data processing function adequate. However, in reviewing the accounts receivable system the auditor may find that there is not an adequate audit trail over the application of cash to accounts receivable balances. Adjustments can be made to balances without a record of the transaction making that adjustment. If there is no further testing to establish some compensating control, the auditor may not be able to rely upon the balances produced by that accounts receivable system.

Application controls are explained more in Chapters 5 and 6. This chapter considers general controls as they exist in the larger data processing center. Chapters 3 and 4 will consider general data processing controls for smaller computer systems and distributed computer networks.

There are circumstances when the auditor reviews only general data processing controls in a larger computer system and does not perform reviews of specific application controls. However, the auditor must understand general data processing controls before reviewing controls surrounding a specific application. A general controls review of data processing functions can take several forms, including:

A review of general data processing and computer center controls to support detailed reviews and testing of specific applications

A management/operational review of general controls only to understand data processing operations

An economy/efficiency review to identify potential cost savings

A review to assess compliance with laws or regulations

Materials in this chapter should supply the auditor with many steps necessary to perform general data processing controls reviews in any of the above areas. However, there are different ways to organize a data processing organization. The auditor must be creative to do an effective job in performing general controls reviews of larger data processing systems.

AUDITOR SKILL NEEDS

For many auditors, large scale data processing organizations can be complex and difficult to understand. Data processing is an area filled with technical jargon and rapidly changing products and technologies. The auditor, whether internal or external, faces an additional problem because many data processing professionals do not want to explain technical issues to an auditor who does not appear to have a basic understanding of data processing concepts. This problem is compounded because data processing professionals use technical jargon when talking to their peers. The following conversation will illustrate this point:

Auditor: I'd like to start my review by following up on a past audit comment. What have you done to improve information security in your computer center?

EDP Manager: We're actively working on this matter. Over the past year, we reviewed both ACF2 and RACF. As part of our transition to XA, we plan to install RACF and will use it to separate production from programmer test data sets. In addition, we may install DES capabilities as part of our SNA implementation when we run a T1 to the XYZ Division.

Auditor: Thank you. Now, my next question is . . .

If the auditor really understood the EDP manager in this mythical interview, the auditor would realize that no definite action had been taken on improving information security. The auditor was told two computer security software packages are being considererd, Computer Associates' product called ACF2 and IBM's product called RACF (pronounced Rack-F). However, the EDP people plan no action until they install an enhanced version of IBM's large-scale operating system called MVS/XA. Once MVS/XA is installed, they evidently do not plan to extend information security protection to individual, key application files. In addition, they are considering implementing data encryption using DES (the National Bureau of Standards' Digital Encryption Standard) once they install their network architecture following IBM's SNA protocols. Also they are establishing a high speed data transmission line called a "T1."

An auditor without a strong data processing background should not expect to understand the fine points of data processing products such as XA, RACF, SNA, T1, or DES. However, the auditor should always feel comfortable in asking general questions for clarification. In response to the EDP manager's answer, the auditor might have asked, "Last year, we found that user terminal access security was weak. Users could gain access to files where they were not authorized. There were no computer security policies or procedures. How has it improved over this past year?"

What skills and information does the auditor need to perform general controls reviews of larger computer centers? This chapter gives the auditor a good basic understanding for doing such reviews. Other sources of information include the following:

Professional Organization Publications. A variety of computer audit related publications from organizations such as The Institute of Internal Auditors (IIA), the American Institute of Certified Public Accountants (AICPA), or the EDP Auditors Association (EDPAA), are available.

Computer Audit Series Publications. FTP Publications' *Auditing Computing Systems* and Auerbach's *EDP Auditing* are notebook publications with supplements covering new topics issued several times a year. Another good source is *EDPACS*, an EDP audit newsletter also published by Auerbach.

Data Processing Magazines. Data processing management oriented publications such as *MIS Weekly* or *Datamation.* Many are sent complimentary to qualified persons. These can provide useful information on new data processing products and techniques.

There are numerous sources for the auditor to keep up with data processing technical changes and trends. The skilled auditor should develop a personal/professional education program to keep current by reading some of the above publications. Also, by attending some of the computer audit-related seminars and conferences such as those sponsored by the IIA and EDPAA or by private organizations such as the MIS Training Institute, the auditor can add to an education program.

CONTROL CHARACTERISTICS OF THE LARGER DATA PROCESSING CENTER

General characteristics that define a larger data processing center as compared to a smaller one were discussed previously. In this section, we will consider some organizational and control characteristics of that modern larger data processing center. While there are variations in how a modern larger center is organized, these characteristics should help the auditor to understand the general controls environment in such a computer center.

Although much has been published about the "microcomputer revolution" and the growth of small business systems, there have also been significant changes in the larger data processing center. For the auditor, many early general controls issues have now evolved into accepted data processing operating procedures. However, there are other functions and control concerns previously not included in the auditor's list of general controls concerns.

Organization Structure

In the early days of computer auditing, checklists and audit programs were published that directed the auditor to ascertain that computer operators did not program and programmers did not operate the equipment. Because of the complexities of modern, large data processing operating systems and high production demands within the typical larger computer operations machine room, this separation of duties generally exists in the modern larger data processing organization. However, the auditor must determine that there is an adequate separation of functions within the data processing function.

Figure 2.1 shows a typical organization structure in a larger data processing shop. While there can be variations, the auditor should look for a separation of functions between Systems and Programming, Computer Operations, and Technical Support. Often the technical support function will include database administration, telecommunications control, and systems programming. However, sometimes these groups are set up as separate organizations reporting directly to the data processing director.

End user computing is another function that did not appear on earlier organization charts. This function may be responsible for information centers where users develop their own applications. Also, it may be responsible for personal computers located throughout the organization. End user computing is sometimes attached to the systems and programming function but is often a separate organization. The auditor should be aware of how end user computing is managed and controlled even if it happens to be outside the classic data processing organization. End user data processing controls are discussed in Chapter 11.

The Data Processing Director should report at an appropriate management level. Directors of the data processing function have historically reported to financial controllers since most data processing applications were originally accounting-oriented. The modern large data processing organization now supports manufacturing, purchasing, marketing, and many other applications. Today's data processing function should report at a more significant level in the organization to support these various users' needs. In some organizations, the head of data processing is given the title of Chief Information Officer (CIO) reporting at an executive level.

A variety of departments may be attached to the modern data processing department. Some data processing departments support office automation equipment and related functions throughout the organization. Others are responsible for voice as well as data telecommunications. Typically, these functions are separate organizations reporting to the director. They could be part of the technical support or systems and programming function. Modern data processing departments are taking on these additional responsibilities because of their overall responsibilities for information. Information management, data security, and integrity controls and procedures need to be estab-

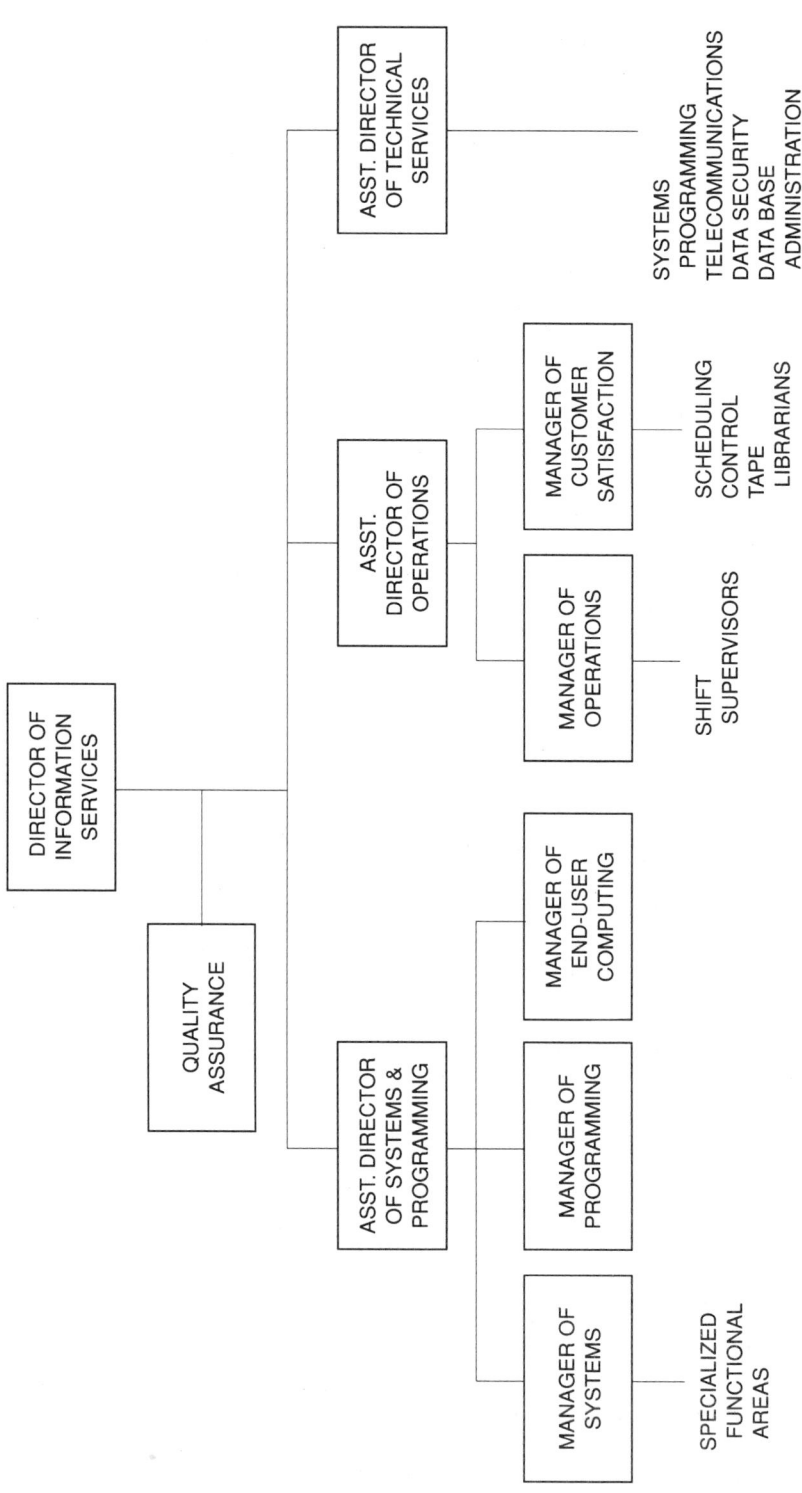

Figure 2.1. Typical larger data processing organization structure

lished whether a critical document is on the mainframe computer or on a departmental word processing machine.

Another function appearing frequently in modern, larger data processing organization charts is quality assurance. Such a function acts as an internal computer audit function within data processing; it may be a staff function attached to the director or part of the Systems and Programming Department. The auditor's relationship to a quality assurance function will be discussed in Chapter 7.

The modern data processing department should have some well-established management controls and procedures such as:

Departmental procedures and standards manuals

Documented position descriptions

Policies for staff training and career planning

These procedures should be equal to those elsewhere in the organization. If the organization has a formal policy for other departments, the auditor should see that it is followed within data processing. However, the auditor should not attempt to reorganize the entire organization's policies by pointing out a data processing-related deficiency.

Computer Hardware Configurations

Miniaturized electronic components in today's computer center take less space than at an earlier time. In addition, there have been significant changes in the design of some peripheral computer components. The traditional magnetic tape drives are gradually being replaced, for example, by cartridge-type drives with a much higher capacity. Similarly, removable disc pack drives have been generally replaced by fixed-pack devices with considerably greater capacity. Laser driven printers can replicate traditional paper forms and operate at high speeds.

There is no typical hardware configuration for the modern larger data processing organization. Often, the inexperienced auditor will be given a tour through a room filled with central processors, disc drives, printers, and other equipment with little recognition of what is there. Although IBM equipment is currently found in the majority of larger business data processing computer centers, the auditor may also find other large systems vendors. The basic components of a large computer system do not vary all that much from one vendor to another.

The auditor can learn the types of equipment in a computer center by requesting a configuration chart of the hardware. Figure 2.2 shows equipment for a larger computer installation. While the auditor may not be able to determine whether the computer center has the correct models of disc drives or other equipment, such a chart will indicate that management has done some planning in their computer hardware configuration.

While the number and type of tape drives, printers, and other equipment will vary, the auditor can expect to find similar characteristics in larger computer system facilities. These characteristics may include the following.

Physical Security Controls

The larger computer system should be located in a secure room or facility. Because the larger business data processing computer center maintains significant data files and represents a major investment in equipment, the computer equipment should be located in a room with locked access controls and no windows to the outside.

Computer room physical security helps to protect equipment, programs, and data from unintentional or malicious damage. Locked doors to the computer room prevent unauthorized persons from entering the area to pick up reports, ask distracting questions of the operators, or cause malicious damage.

While all business operations are subject to fires, floods, or vandalism, a computer center is particularly vulnerable because the equipment cannot handle easily stress of any sort. Because of the type and extent of data processed in the modern larger computer system, data processing operations centers should be located in unobtrusive locations and built to minimize exposure to fires, floods, or other hazards.

In an earlier era, it often was the responsibility of the auditor to convince data processing management of the importance of these physical security controls. The modern data processing executive, with responsibilities for one or more large computer centers, is aware of the need for computer physical security. However, the auditor will be asked to review and recommend improvements to data processing physical security controls.

Environmental Controls

Many modern larger computer systems require water chilling systems for cooling mainframe processors. Also, there will often be electrical power, air conditioning, and environmental monitoring systems independent from the rest of the building.

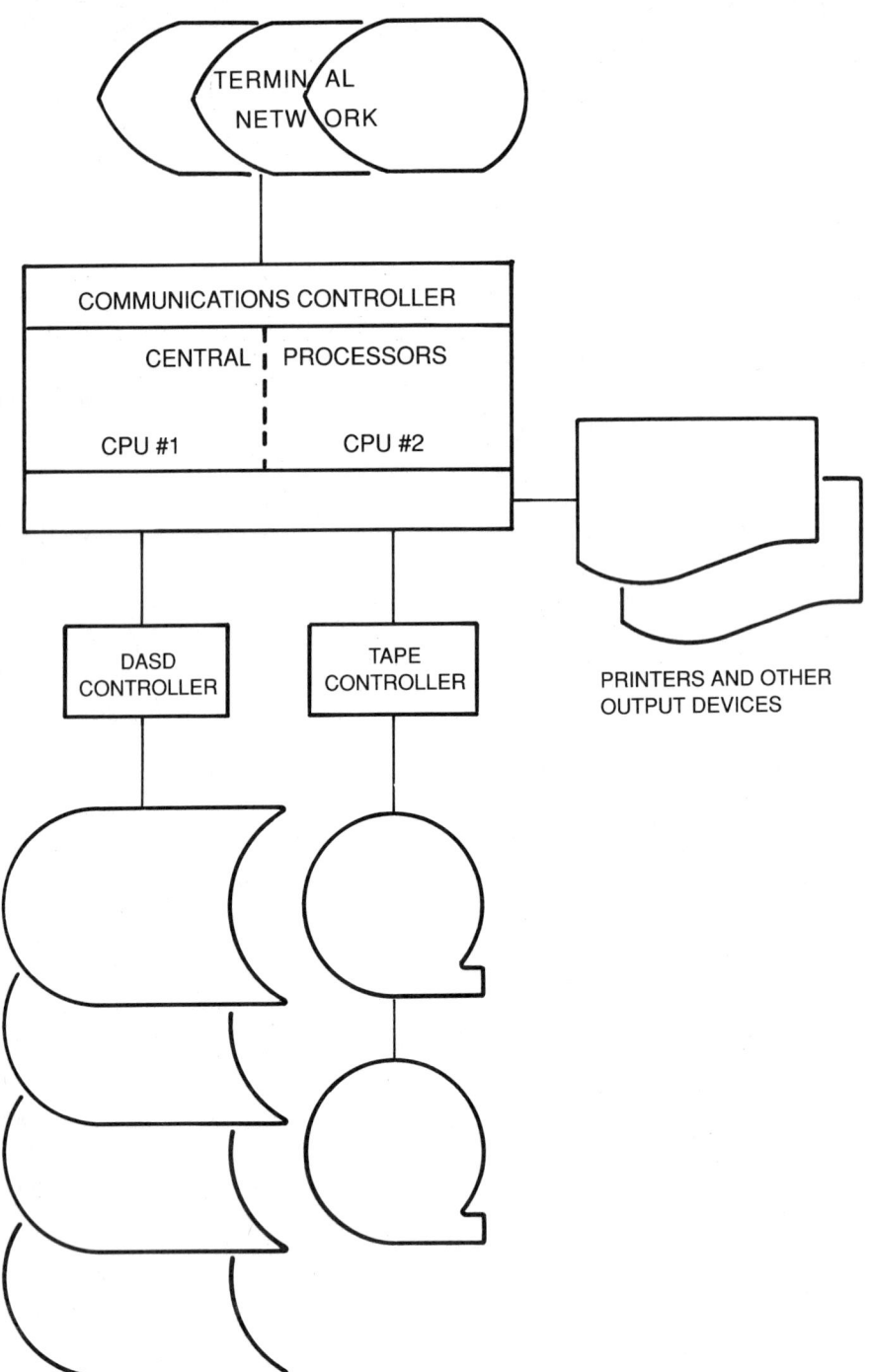

Figure 2.2. Larger computer equipment for a typical installation

Because the larger computer system and its applications are quite vulnerable to electrical outages or fluctuations in the power supply, many are equipped with emergency power sources. These may smooth out power fluctuations or provide emergency power to allow the computer system an orderly shutdown. Some larger systems may be supported by power generators to provide power for an extended power outage.

Weaknesses in the system of data processing environmental controls can cause failures in the operation of key data processing applications. The auditor should be aware of control procedures in the area, as noted in Chapter 8, and make recommendations for improvement where appropriate.

Separate Tape Libraries

Often, libraries or storage areas for magnetic tapes will be located in a separate room adjacent to the equipment room rather than on racks in the machine area. This provides extended physical protection for tapes and makes the mounting and backup of tapes a more efficient process. Some automated tools are used to schedule and call up tapes for mounting; others write internal labels on tape files and schedule them for backup rotation.

Usually, earlier generations of key tape files are rotated to another separate library storage facility, normally located in an area remote from the main computer room. In the event of a fire or other disaster within the computer room facility, these key tape files can be recovered from this remote facility for transfer to an emergency backup processing site.

Input–Output Control Sections

A larger data processing operation has some type of input–output control section to receive input data, distribute any outputs, and schedule and set up production jobs. When most production was run in a batch mode, such control functions often balanced input batches to system outputs and resolved any problems. In the more modern data processing operation, users generally take responsibility for their own data. Frequently, the data is submitted through terminals in user areas and outputs are transmitted back to the originating user areas.

Input–output sections exist to receive those manually submitted transactions that are still processed in batch mode. Such sections also take responsibility for distributing key reports produced on data center high-speed or laser printers.

Some input–output sections or related groups are also responsible for staging and scheduling production jobs in the computer room. This responsibility consists of setting any necessary parameters, assuring that all inputs are ready, and loading the job stream for processing. Automated tools have been designed to aid in this process.

The above characteristics do not define all aspects of the modern, larger data processing system, but they give the auditor an understanding of what to expect in the operation and organization of a larger computer system, no matter which model of computer is being used.

When auditors first reviewed general data processing controls, they looked for many of the characteristics discussed above, as well as for locked computer room doors, fire extinguishers, and proper batch controls. These controls are now in place as a matter of course in larger computer centers. While the auditor should keep them in mind, the general controls areas discussed later in this chapter are of more significance.

Operating Systems and Software

Typically, the larger computer system will use software with an advanced operating system that can process multiple jobs simultaneously. A larger computer system such as the IBM 30xx series might use the IBM MVS or the more advanced MVS/ESA operating system. Similar advanced operating systems are available for other vendors' equipment. The computer system may have a telecommunications capacity and an on-line monitor for better control of the local and remote terminals tied into the system.

The larger computer center typically has a COBOL language compiler as well as other specialized language processors. There may also be specialized data file software management tools, including a database management system.

Many larger computer systems have a fourth generation language (4GL), a powerful programming language often used by end users to speed the application development process. Some control issues surrounding 4GLs will be discussed in Chapters 7 and 11.

In addition, there will be numerous "utilities"—specialized programming tools to sort data, copy files, and correct file problems. Some of these utilities are quite powerful. An example is the "ZAP" or "Super-ZAP" program found in many IBM systems. It modifies data without leaving any sort of audit trail. The auditor should be aware of these utility programs and the controls over their usage.

A larger computer system will have some type of systems programming function to make changes to the operating system, maintain the other system programs such as operating languages, and "tune" the operating system to make it operate more efficiently. A systems programming function is staffed with highly technical personnel who are usually vital to the overall effectiveness of data processing operations. Some control issues surrounding systems programmers will be discussed later in this chapter.

Systems Development Procedures

The systems development function has the responsibility for designing and implementing the applications run at the larger computer center. This is an area that has undergone significant changes in recent years. Historically, the systems development function consisted of analysts who defined user applications requirements and programmers who wrote the code to produce them. In many organizations, this development process was slow and labor intensive. Analysts often designed systems that did not appreciate user requirements, and design changes were often necessary. Programmers wrote the programs in COBOL or other languages that took considerable time to debug and test.

There have been many changes in the way a systems development department operates in today's larger computer center. Formal systems development procedures help analysts design better applications; productivity tools such as code optimizer software packages help the programmer produce program code more efficiently; purchased software packages are used extensively; and users are much more involved in the new systems development effort. For an effective review of general controls, the auditor needs to understand these changes and their control implications.

Formal Systems Development Procedures

One of the recurring complaints of early computer systems or applications development projects was that the applications did not meet user needs. Applications were initiated through informal requests and when implemented often did not meet user expectations or requirements. To put controls into this development process and create better systems, data processing organizations implemented Systems Development Life Cycle (SDLC) procedures or Systems Development Methodologies (SDMs).

An SDM or SDLC is a formal set of procedures in which users document their requirements, and applications developers document their understanding of those needs for subsequent user review. Developers then document those requirements in a detailed fashion for program development. While this formal approach has resulted in better applications for many organizations, some early SDM approaches required such extensive user and application developer documentation that they were difficult to follow. For example, one SDM set of procedures offered by a vendor in the early 1980s specified some 70 different forms to be completed for implementing a new information system. Followed to the letter, the application development task became inflated to an almost bureaucratic process.

Modern data processing organizations have reduced SDMs to a more manageable level. They have used automated tools to document new systems and prototyping techniques to describe them. These automated tools are called Computer Aided Systems Engineering (CASE) tools. Prototyping refers to the creation of a test system or report to determine whether the needs of the user of an application have been met. If the user is not satisfied with test results, the test is revised.

The modern data processing systems development group should always have some type of SDM in use. Whether automated prototyping, or CASE, methods are adopted, the existence and use of a formal SDM is an important general control in a data processing department. The auditor should be familiar with the systems development procedure being used. In particular, this knowledge helps the auditor to review new systems under development.

Programmer Productivity Tools

For some time, programmers have recognized COBOL as the language of choice for business data processing applications on larger systems. Historically, programmers prepared detailed flowcharts of their programs and then coded them in COBOL with few standards or conventions. Using this approach, programs took a long time to write, were difficult to test and debug, and were particularly difficult to maintain.

Many modern data processing functions have adopted structured programming for writing program code. This is a COBOL coding technique that specifies a clear, straightforward, and easy-to-read manner for writing programs. If good structured techniques are used, programs are relatively easy to maintain, without the extensive hand-drawn flowcharts that were once required. Figure 2.3 illustrates the advantage of structured COBOL programming code over unstructured code.

An auditor should have a general understanding of the structured programming approach being used in an organization, but a review of the actual program code is typically not the auditor's function. An auditor probably would review program code only to determine if programming standards are followed, and then only on a limited test basis. In a larger data processing

Structured Versus Unstructured COBOL Programming Code

The following is an example of a *non-structured* COBOL procedure for a payroll calculation program. The code is filled with GO TO commands, making maintenance very difficult.

```
PROCEDURE DIVISION.
PARA-1.
      OPEN PAY-FILE, OUT-FILE.
      READ PAY-FILE AT END GO TO ...
      IF STAT-CODE = 'X' GO TO PARA-2.
      IF STAT-CODE = 'Y' GO TO PARA-7.
      IF STAT-CODE = ...

             .
             .
      GO TO PARA-1.
PARA-2.
      IF GL-CODE = '387-0093' GO TO EXCEPT-
GL.
      PERFORM TERM-EMP-CHK.
      IF EMP-NUMBER = ...
```

In structured COBOL code, GO TO commands are all but eliminated, and the code follows a consistent pattern. Maintenance is much easier because program logic does not jump around through the program code.

```
PROCEDURE DIVISION.
PARA-1.
      PERFORM PARA-2 THRU PARA-5.
      PERFORM PARA-6 THRU PARA-12.
      PERFORM PARA-13 THRU PARA-15.
      STOP RUN.
PARA-2.
         .
         .
         .
```

Figure 2.3. Structured versus unstructured COBOL programming code

organization, the quality assurance function normally would be responsible for reviewing program code for adherence to standards.

In addition to structured COBOL programming, many larger organizations have installed a variety of other automated programmer productivity tools. Some are on-line systems allowing the programmer to call up data names and definitions from a data dictionary and create specifications for programs to be written. Others are built around the fourth generation languages discussed earlier and are called programmers' on-line "workbenches" for efficient program development. These tools are changing the way programs are developed, documented, and controlled in the larger data processing organization.

Purchased Software

In the earlier days of data processing, most business applications were written in-house by the individual programming departments. In a large, multiple unit organization, it was not unusual to find that each business unit had written its own payroll system. Although payroll within a given country was not too different from one location to another, each unit's data processing function argued that its needs were "special." Each location then proceeded to develop its own payroll applications.

More recently, there has been a dramatic change from this reinvent-it and write-it-yourself approach. A large body of packaged software is available for purchase. These package applications have the flexibilities necessary for meeting unique organizational needs, and usually can be implemented faster and less expensively than traditional in-house programming methods. Numerous vendors offer many different types of software packages. Of course, extensive conversion, system interface, and tailoring efforts may be required to implement a given software package.

The auditor needs to be aware of the various system development productivity, or CASE, tools that may be used in the organizations to be reviewed. Auditor effectiveness can be increased if the auditor gains a hands-on familiarity with the various techniques being used. In particular, this familiarity should include a knowledge of the SDM and how it is used for both in-house and purchased software packages. Also, although not normally responsible for reviewing actual program code, the auditor should have an understanding of the programming standards used, including any stuctured programming techiques.

DETERMINING THE PURPOSE OF THE GENERAL CONTROLS REVIEW

We discussed in Chapter 1 how this book will take a control objectives approach to reviewing various aspects of modern data processing controls and security. That approach should begin by asking:

What is the purpose of the review?

Which specific controls and procedures should be in place?

How can evidence be gathered to determine whether controls work?

Using this approach, the auditor should develop an individual set of control objectives for each type of review rather than just using a standard set of internal control questions for all reviews. The objectives selected will depend upon the purpose of the review. If an economy or efficiency review is being planned, objectives will include such areas as the chargeback system and the job scheduling system. An outside auditor's review of the same data center might pay little attention to the chargeback system but would emphasize control procedures—for example, for program library updating.

Earlier in this chapter, some forms or types of general controls reviews in the modern, larger data processing organization were mentioned. Although a general controls review of a data processing organization has a variety of purposes, it often fits into one of the four types of general controls reviews described here.

Preliminary Review of Data Processing Controls

Outside auditors sometimes call this type of review a preliminary survey or an assessment of control risk. Its purpose is to gain a general understanding or an overview of the data processing controls environment. Internal auditors also perform this review to assess the internal controls environment. In a preliminary survey, the auditor asks questions, observes operations, and reviews documentation, but does only very limited testing. For example, the auditor may inquire about procedures for updating production program libraries and review the forms used for the approval process. However, the auditor would not select a sample of the programs on the production library to determine whether they had followed proper library update procedures.

A major objective of the preliminary review is to gain an overall understanding of the data processing controls environment for a more detailed general controls review or extended control risk assessment at a later date, or to gather preliminary controls information for a specific applications review. This preliminary review is limited in scope and may not cover all aspects of the data processing organization. However, the auditor will perform a preliminary review of data processing controls more often than other, more detailed review types. Examples of appropriate preliminary reviews include:

A preliminary controls review for data processing operations at a new acquisition

A follow-up review after a detailed controls review in an earlier period

An outside auditor's review to document a general understanding of data processing contols

Detailed Review of General Controls

This comprehensive and detailed review of data processing general or interdependent controls covers all aspects of the operation, including systems programming, telecommunications controls, and database administration. A detailed general controls review also includes testing. The auditor might test controls over program libraries by running specialized programs to compare source versions with production program versions.

A detailed general controls review requires good planning to make it effective, and the auditor must spend considerable time at the data processing operations and development departments. While the preliminary review can sometimes be performed by a less experienced auditor who is developing computer audit skills, a detailed general controls review is best performed by a more senior audit staff member with a good understanding of data processing controls and procedures.

Specialized or Limited-Scope Review

Auditors must often review specialized areas within the overall data processing functions. These specialized reviews can be limited to one department, such as database administration, or one function, such as output report distribution. Management will often request auditors to perform this type of review.

A specialized review takes considerable auditor creativity in planning the work. For example, if management is concerned about the equity of the computer chargeback system, the audit department may be asked to look at it. It will be necessary for the auditor to visit the data processing department to gain a general understanding of the system used, spend time planning the additional procedures and tests to be performed, and then return to the data processing department to do the actual review and testing.

As data processing departments grow in complexity and importance to the organization, auditors can expect to perform more of these specialized, limited reviews. When data processing was a relatively small function in the organization, comprehensive reviews of all data processing controls were appropriate. With the data processing/information systems function becoming a major resource in many organizations, it may be inappropriate to attempt to review *all* data processing general controls in *all* operational areas in one detailed review. This would be equivalent to a review of "manufacturing" in a major plant environment. It is accepted that not all manufacturing functions would be covered at once. For example, production control might be reviewed one year and receiving and inspection the next, and eventually most significant functions would be covered.

Legal Compliance Review

Auditors working with governmental agencies or in organizations that do extensive government contracting are required to perform compliance audits to determine whether the data processing department is following appropriate laws and regulations. These compliance laws and regulations differ from agency to agency and from one political division to another.

Often, a compliance-related data processing review can be combined with a preliminary or detailed general controls review. However, the auditor must be aware of the relevant auditing procedures and regulations published by the government agency requiring the audit. Most state bank examining agencies, for example, have published data processing controls guidelines. When operating in this type of environment, the auditor must be aware of the regulatory environment as well as any agency-published data processing control procedures.

Establishing the Review Purpose

The auditor must do some educating of management to get direction as to the type of general controls review to be performed. All too often, management tells an auditor, "Audit that computer center!" If the auditor does not explain the various types and scopes of general controls reviews, disappointments are certain.

When management expects a brief, preliminary survey of the data process-ing function and the auditor performs a detailed review, no one will be satisfied. The auditor will not be able to understand why management is exerting so much pressure to complete the work. Conversely, management will not be able to understand the relevance of the auditor's detailed comments—for example, on the lack of systems programming documentation.

A more common problem occurs when management requests a detailed review of data processing to assess perceived control or performance prob-lems. The audit function may not be able to deliver an in-depth review because of scheduling conflicts or unavailability of an experienced auditor to perform the review. The resulting report and findings may be quite general when management expected a detailed and comprehensive document.

This problem of expectations exists whether a computer audit specialist works for an internal audit department or on a public accounting engagement. The auditor should always endeavor to discuss and define the type of review to be performed, in advance of the actual work. This gives management an expectation of what might come out of the review and allows the auditor to define the necessary control objectives and audit procedures to be performed.

CONTROL OBJECTIVES AND AUDIT PROCEDURES

The control objectives approach to performing a data processing review requires the auditor to determine the type or scope of review to be performed, then to select control objectives that are appropriate to that type of review, and finally to use audit procedures to test whether the control objectives are being met. As was discussed in Chapter 1, generally, this approach will be used throughout the book to help the auditor define the audit.

In a larger data processing environment, the auditor is faced with a large variety of potential control objectives of concern. Further, depending upon the type or scope of audit, the audit procedures to be performed vary.

In the figures following, data processing control objectives will be defined for some major areas of control concerns within the larger data processing organization. Often the objectives statements will provide a brief explanation of why a given objective is significant. The first controls objective in Figure 2.4 can be used as an example to illustrate this review concept. That objective is:

> The information systems function should be organizationally independent from other departments or functions in order to allow it to properly serve all units of the organization.

The objective states the control concern—organizational independence—and also provides a reason for its importance.

Following each data processing control objective there is a series of audit procedures which may be performed to test for the objective. The auditor should select appropriate ones to perform. Continuing with the above example, Figure 2.4 lists three audit procedures to test for the organizational independence objective:

1. Review organization charts to determine whether the information systems function reports at a level high enough to allow it to act independently.
2. If the function reports to another operating unit, such as the Controller, determine through inquiry whether it appears to be adequately serving other units of the organization.
3. Review policy statements or other communications from top management which would support the independence of the information systems function.

For a limited or preliminary data processing controls review, the auditor would find it necessary to perform only the first of the three audit procedures. As the audit scope increases the second or third should be included.

The auditor should build these various controls objectives and procedures into a series of custom audit programs. Of course, each of these programs should provide flexibility to strike out or add procedures as appropriate.

In the following sections, control objectives and audit procedures for larger data processing organizations will be discussed for the following areas:

General organization
Access to data
Access to programs
Systems development methodology
Data processing operations
Systems programming and technical support
Data quality assurance

Control objectives and related audit procedures for these areas will allow the auditor to construct a general audit program for a larger data processing organization. Also, the figures contain sufficient procedures to allow an auditor to establish an audit program for a more detailed general controls review. However, for additional, more specialized control objectives and procedures that could be helpful for performing a general controls review of a larger data processing organization, the auditor should refer to the following chapters.

Chapter 4. Controls in the Distributed Network. Many larger data processing centers support a telecommunications network. The network becomes particularly complex in a distributed computing environment where processing takes place at central and local computer systems. Telecommunications general controls, as they apply to the single, larger computer center, are discussed in this chapter.

Chapter 7. Reviewing New Applications Under Development. This chapter explains procedures for performing pre-implementation application reviews, including control objectives and audit procedures for the applications development process. Even though the auditor may not be reviewing currently an application under development, this chapter may offer useful information in its discussion of applications development controls procedures in the larger data processing organization.

Chapter 8. Physical Security Strategies and Risk Evaluation. This chapter discusses an approach to evaluating physical and environmental risks in the data processing computer center. Many controls procedures discussed here will be appropriate to general controls reviews of larger computer operations.

Chapter 9. Ensuring Information Security and Integrity. This chapter discusses information security controls in detail. The auditor should be aware of information security control procedures as part of any general controls review of the larger data processing organization.

Chapter 10. Effective Disaster Recovery Planning. An effective disaster recovery or data processing contingency plan is an important control in virtually any data processing organization. Even when performing a limited review of a larger data processing center, the auditor should inquire about such planning. This chapter should provide the auditor with procedures for evaluating such disaster recovery plans.

Chapter 11. Auditing the End User Information Center. Information centers or other end user computing activities are increasingly common in the modern larger data processing organization. The unique controls environment surrounding end user computing controls are discussed in this chapter. Although frequently the subject of a separate audit, the auditor's general controls review should make some inquiries as to the end user computing controls environment.

Many subjects covered in these chapters include more advanced topics than the auditor may initially require when performing a limited general controls review in a larger data processing environment. However, the auditor should study these chapters and incorporate appropriate control procedures into overall general controls reviews as necessary.

GENERAL ORGANIZATION CONTROLS

A review of organization controls is often the starting point for any general controls review of the data processing organization. In addition to determining some overall management policies and procedures, such a review helps guide the auditor through the remainder of the data processing organization.

Often, an overall management perception of the data processing controls and organizational environment is one reason why an internal audit will be requested. Some symptoms that may indicate control, efficiency, or management problems within the data processing function include:

Extremely long development times required for new systems coupled with user dissatisfaction with the completed projects

Frequent application program and processing errors

Data processing turn-around problems and delays

Excessive staff turnover coupled with a perception of unpromotable senior data processing management

Frequent requests for new computer hardware and software products without adequate justification

Budget overruns and a perception of poor cost control

Organizational controls in the larger data processing department refer to management procedures and controls that exist within the entire organization and within data processing in particular. Often, the data processing function has grown over time without adopting the controls expected in a modern organization. For example, systems programmers may be allowed free access to the machine room because they have historically been allowed such access. Although data processing management realizes this practice is a control weakness, it may be reluctant to change. Top management also may not have imposed these controls and procedures upon data processing because of a reluctance to become caught up in technical issues. When reviewing data processing organizational controls, the auditor should try to be aware of these management perception issues.

The auditor should also address efficiency issues. For example, most data processing organizations once had an input-output control desk or department with responsibility for balancing input data and output reports. Over time, many data processing departments have built these balancing controls into their applications or have given that responsibility to users through on-line terminals. However, it is still common to see an input-output control desk with little to control! The auditor should attempt to make efficiency recommendations as part of any controls-related audit report.

A set of control objectives and audit procedures for general organizational controls is shown in Figure 2.4. These will be the starting point for most general larger system control reviews. The auditor generally does less testing in this portion of the audit than in others. Unless there is a reason to doubt the information given by data processing management, the auditor can often rely on assertions given by data processing management with only minimal compliance testing.

The overall plan of organization and related management procedures are perhaps the strongest control in this portion of the review. If the auditor finds that there was a massive staff turnover over the past year, for example, the auditor will probably find other control problems in other portions of the review.

The auditor may also find there are sensitive issues which may be encountered when reviewing general organization controls, which may require a considerable degree of diplomacy. Consider the following example:

Company XYZ has a large, single vendor mainframe computer center. Data processing operations are considered to be well managed and have expanded to meet the needs of increasing business. XYZ is a well run company with documented management policies and procedures. Among these is a requirement that all internal purchases must be subjected to competitive bidding and must be approved by the purchasing department.

As part of the general organizational controls review, the auditor finds that data processing equipment purchases do not go through established purchasing department procedures. The data processing operations manager discusses needs with the vendor's sales representative, places the orders, and passes the bills to accounting for payment. The purchasing department is bypassed.

XYZ management is aware of this situation but has chosen to ignore the policy violation. The auditor recognizes a potential conflict of responsibilities.

However, the auditor also realizes that the purchasing department may not have sufficient technical skills to evaluate computer equipment purchases. When the auditor expresses these concerns to the data processing manager, the manager takes serious offense at being accused of "being dishonest." How should the auditor handle this finding and recommendtaion?

There is no one correct answer to the above case. Rather, it illustrates the need for the auditor to become familiar with the various management issues and personalities associated with the data processing organization. In addition, it emphasizes the need for diplomacy in reviewing and reporting management controls.

DATA ACCESS CONTROLS

When unauthorized persons are allowed to access data files or records, the system of general internal accounting controls is weakened. As part of a general controls review, the auditor should keep this overall objective in mind. Many control procedures covering general controls over access to data are discussed in Chapters 8 and 9 on physical and information security. Specific application control objectives and procedures are discussed in Chapter 5. Of course, if there are weaknesses in the general controls, the application controls cannot compensate for those weaknesses.

Figure 2.5 describes control objectives and audit procedures for reviewing controls over access to data in a preliminary general controls review. If these procedures indicate that access to data could be compromised, the auditor may consider the more detailed security-related procedures of Chapter 9.

Information security is a difficult control to evaluate. The auditor can easily determine that controls are in place, but it is more difficult to evaluate the risk associated with any potential weaknesses. Computer security experts can virtually always find some weakness in any security system, but there may be little risk to the organization, or the cost of protection may be far greater than the expected loss.

The effective auditor carefully evaluates the strengths and weaknesses in information security controls and always attempts to keep recommendations reasonable given the risks. For example, in reviewing password systems, the auditor often finds some way to make that password system more secure. If users are required to change their own passwords every thirty days, the auditor might recommend that the security software automatically make the changes for users by producing a random-generated password. While passwords composed of sequences of random characters generated by the system may improve

**Control Objectives and Audit Procedures
for Reviewing General Organizational Controls**

Objective 2.4.1. The information systems function should be organizationally independent from other departments or functions in order to allow it to properly serve all units of the organization.

Procedure 2.4.1.1. Review organization charts to determine whether the information systems function reports at a level high enough to allow it to act independently.

Procedure 2.4.1.2. If the function reports to another operating unit, such as the Controller, determine through inquiry whether it appears to be adequately serving other units of the organization.

Procedure 2.4.1.3. Review policy statements or other communications from top management which would support the independence of the information systems function.

Objective 2.4.2. The information systems function should be organized to allow for a proper separation of duties such that its personnel do not perform incompatible functions.

Procedure 2.4.2.1. Review published organization charts for the overall plan of organization to determine whether it allows for a proper separation of duties.

Procedure 2.4.2.2. Match published position descriptions to positions on information systems organization charts to determine that position descriptions exist for all significant positions.

Procedure 2.4.2.3. Interview selected members of the information systems organization to determine that their duties and responsibilities correspond both to the published position descriptions and to organization charts.

Procedure 2.4.2.4. Observe data processing operations to ascertain whether separation of duties policies are being violated by, for example, programmers updating production libraries using computer room facilities.

Procedure 2.4.2.5. Visit a non-daytime shift of computer

Figure 2.4. Control objectives and audit procedures for reviewing general organizational controls

operations to observe whether separation of duties functions are being followed.

Procedure 2.4.2.6. Interview information systems management to determine that a proper separation of duties exists in functions that may not be defined clearly on organization charts, such as data base administration, information security, and systems programming.

Procedure 2.4.2.7. Where information systems personnel hold potentially sensitive positions, such as key programmers or operators, determine that policies exist for such matters as immediate separation from work duties upon resignation and the requirement to take vacations.

Procedure 2.4.2.8. Review training policies as well as actual training records to determine that personnel are adequately trained in the use of computer technologies.

Objective 2.4.3. Information systems policies and procedures should exist and be consistent with overall organization policies and procedures.

Procedure 2.4.3.1. Briefly review overall organization policy manuals and note those areas that could impact the informations systems function.

Procedure 2.4.3.2. Review the information systems policy manual and determine whether it includes or reflects overall organization policy statements.

Procedure 2.4.3.3. Interview members of information systems management to determine whether they understand overall policies and procedures and whether the policies and procedures are being communicated to staff.

Objective 2.4.4. There should exist long and short range plans, approved by upper management, which define the future objectives of the information systems function.

Procedure 2.4.4.1. Review procedures for developing long range information systems plans and for securing upper management approval for such plans.

Figure 2.4. Control objectives and audit procedures for reviewing general organizational controls *(continued)*

Procedure 2.4.4.2. Review published long range plans over several recent years and discuss with management any reasons for any significant variances from plan to plan.

Procedure 2.4.4.3. Determine that the long range plans give adequate attention to hardware, personnel, and technical software as well as to future systems applications.

Procedure 2.4.4.4. Review procedures for preparing short range plans and determine how these plans relate to the overall long range information systems plan.

Procedure 2.4.4.5. Review procedures for reporting variances against the short range plans and determine whether these procedures are followed.

Objective 2.4.5. There should be an overall information systems disaster recovery contingency plan which will allow continued operation of data processing services in the event of an unexpected interruption.

Procedure 2.4.5.1. Determine that information systems management has assessed the risks and vulnerabilities of overall data processing operations and has formulated a disaster contingency plan.

Procedure 2.4.5.2. Review the published disaster recovery plan and determine that this plan is consistent with the components of such a plan as outlined in Chapter 10.

Procedure 2.4.5.3. Determine that the disaster recovery plan has been tested on a regular basis and review the results of a representative test.

Procedure 2.4.5.4. Inquire regarding the existence of other non-data processing disaster contingency plans implemented within the organization and determine that the information systems plan is consistent with other plans.

Objective 2.4.6. There should be an overall management commitment to both information systems physical and data security.

Procedure 2.4.6.1. Assess the overall understanding of and commitment to data processing security by both information systems and upper management.

Figure 2.4. Control objectives and audit procedures for reviewing general organizational controls *(continued)*

Procedure 2.4.6.2.   Review physical security controls following the steps outlined in Fig. 8.8.

Procedure 2.4.6.3.   Review information security controls as outlined in Fig. 9.7.

Objective 2.4.7.   Information systems should utilize a budgeting and cost accounting system consistent with management accounting systems used throughout the organization.

Procedure 2.4.7.1.   Briefly review overall management accounting systems and determine that they are being used, where applicable, by the information systems organization.

Procedure 2.4.7.2.   Review procedures for budgeting hardware, software, personnel, and indirect costs; determine whether these procedures appear to be consistent with the entire organization and are adequate.

Procedure 2.4.7.3.   Review procedures for pricing or otherwise reporting on the costs of data processing to the user community.

Procedure 2.4.7.4.   Trace a limited number of transactions through the computer pricing system to determine that charges are being recorded properly in other organizational accounting systems.

Objective 2.4.8.   Procedures for acquiring new hardware and software products should follow the capital budgeting and procurement policies of the overall organization and should include formal justifications.

Procedure 2.4.8.1.   Review capital budgeting procedures and determine that such procedures are being followed for new computer hardware.

Procedure 2.4.8.2.   Review procurement policies and determine whether such policies are being followed for purchases of new computer hardware and software products.

Procedure 2.4.8.3.   Select several new computer hardware or software product additions and determine that budgeting and procurement procedures were followed.

Figure 2.4. Control objectives and audit procedures for reviewing general organizational controls *(continued)*

Objective 2.4.9.   There should exist a performance reporting system, reflecting both processing and development activities, which allows information systems management to report the results of its operations to upper management.

Procedure 2.4.9.1.   Review information systems performance reports over several recent periods and determine whether they adequately report both planned and actual results regarding overall costs, systems development activities, and data processing production processing.

Procedure 2.4.9.2.   Review procedures for developing performance reports and trace a sample of actual transactions into those reports.

Procedure 2.4.9.3.   Assess the technical understandability of performance reports and consider suggestions for improvements where appropriate.

Objective 2.4.10.   There should be overall management policies to ascertain that information systems internal controls are in place and working.

Procedure 2.4.10.1.   Determine that policy statements exist expressing management's responsibility for ensuring that internal controls are in place and working.

Procedure 2.4.10.2.   Review whether corrective actions have been taken on a selected number of past audit report recommendations from either internal or external auditors.

Figure 2.4. Control objectives and audit procedures for reviewing general organizational controls *(continued)*

Control Objectives and Audit Procedures
for Reviewing Data Access Controls

Note: These objectives and procedures are designed to provide the auditor with guidance for a general review of access controls. More detailed procedures can be found in Chapter 9, Ensuring Information Security and Integrity.

Objective 2.5.1. Responsibility for physical and information security access controls should be assigned to an appropriate individual or function within the organization.

Procedure 2.5.1.1. Through a review of organization charts and discussions with information systems management, determine that an appropriate, independent individual or function is responsible for physical and information security.

Procedure 2.5.1.2. Observe physical security facilities at the organization's major computer centers to develop an understanding of the security controls in place.

Procedure 2.5.1.3. Interview persons responsible for computer physical security to determine that proper attention appears to be given to the subject.

Procedure 2.5.1.4. Interview persons responsible for information security to determine that the function has the proper authority to install, monitor, and enforce information security rules and procedures.

Objective 2.5.2. Security policies should be established for both information systems personnel and users of computer services.

Procedure 2.5.2.1. Review overall organization policies to determine that there is management guidance covering the importance of data security. Also, determine if there are policies regarding actions to be taken for violations of data security.

Procedure 2.5.2.2. Review policy or standards manuals within the information systems function to determine that proper attention appears to have been given to security.

Procedure 2.5.2.3. Interview selected users of computer services to determine their understanding of security policies.

Figure 2.5. Control objectives and audit procedures for reviewing data access controls

Procedure 2.5.2.4. Interview selected information systems staff members, including systems development and operations personnel, to determine their understanding of security policies and procedures.

Procedure 2.5.2.5. Interview management or the organization's legal department to determine policies for prosecuting or otherwise handling data security violations.

Objective 2.5.3. Software tools should be installed to assure that only authorized persons can access and update appropriate files or data.

Procedure 2.5.3.1. Interview information systems personnel to determine the type of information security software installed on major computer systems.

Procedure 2.5.3.2. Determine that information security software covers all major application areas including on-line processors and database systems.

Procedure 2.5.3.3. Determine that all significant features of the security software have been installed including such features as "full abort mode" which locks a violator out of the system after a specified number of improper access attempts.

Procedure 2.5.3.4. If multiple information security software tools are installed, determine that all possess comparable features.

Objective 2.5.4. Improper software access attempts should be monitored and reviewed for subsequent appropriate action.

Procedure 2.5.4.1. Review the tools available from the information security software to determine facilities and reports available to monitor improper access attempts.

Procedure 2.5.4.2. Interview personnel responsible for information security to determine procedures for monitoring and following up on improper access attempts.

Procedure 2.5.4.3. Select a sample of improper access attempt reports and follow up on the actions taken by the information security function for all reported violations.

Figure 2.5. Control objectives and audit procedures for reviewing data access controls *(continued)*

Objective 2.5.5. Controls should be established over input documents and the processing of sensitive applications.

Procedure 2.5.5.1. Through interviews with management and observation, determine which data processing applications might be considered critical to the organization.

Procedure 2.5.5.2. For several selected critical or sensitive applications, review the types of special controls that have been installed over processing, including:

Appropriate marking and control logs for input documents

Security controls over critical files

Controlled distribution of system output reports

Adequate disposal of all carbon sheets, waste paper, and other material that may contain sensitive data

Procedure 2.5.5.3. Determine whether standard run times have been established for sensitive applications and whether actual run times are compared to standards to detect unusual occurrences.

Objective 2.5.6. Access to centralized computer processing centers should be restricted to authorized individuals and monitored.

Procedure 2.5.6.1. Observe procedures for controlling access to computer room facilities, including the use of key cards or cipher locks.

Procedure 2.5.6.2. Review procedures for assigning physical access, through key cards or cipher combinations, and for terminating access rights.

Procedure 2.5.6.3. Match a list of current employees to physical access log lists to determine that only current employees in appropriate positions have access to the computer rooms.

Objective 2.5.7. Controls should exist over the telecommunications network to limit incoming access transmissions to only authorized individuals or devices.

Procedure 2.5.7.1. Interview appropriate personnel to understand and document the telecommunications network, including any dial-up lines.

Procedure 2.5.7.2. Determine the adequacy of controls over

Figure 2.5. Control objectives and audit procedures for reviewing data access controls *(continued)*

any dial-up telephone lines including the use of call back or similar devices for controlling incoming calls.

Procedure 2.5.7.3. Through interviews with information systems management and others, determine whether it is necessary to have dial-up lines to the computer system.

Procedure 2.5.7.4. Determine that any dial-up access protocols used by maintenance or software vendors are changed on a regular basis.

Objective 2.5.8. Appropriate controls should exist over remote terminals and other devices to limit access to authorized individuals and for authorized use.

Procedure 2.5.8.1. Determine that any remote job entry terminals are properly secured to prevent improper access attempts.

Procedure 2.5.8.2. Review procedures for canceling password and systems access rights for employees who terminate or change responsibilities.

Procedure 2.5.8.3. Select the names of several recently terminated employees and determine whether they have had access privileges canceled.

Procedure 2.5.8.4. Visit several selected user areas and observe whether good information security practices are being followed including avoiding posting passwords on terminals and signing off terminals when not in use.

Objective 2.5.9. Controls should exist over the distribution of output reports and source documents to ensure confidentiality and security.

Procedure 2.5.9.1. Review procedures for logging in batches of source documents including the use of authorizing signatures.

Procedure 2.5.9.2. Review distribution procedures to determine that only authorized persons receive output reports.

Procedure 2.5.9.3. Determine that a mechanism exists for labeling or marking output reports to better identify the correct users.

Figure 2.5. Control objectives and audit procedures for reviewing data access controls *(continued)*

Objective 2.5.10. Controls should exist over tape files and other media to prevent unauthorized use or removal from data centers.

> **Procedure 2.5.10.1.** Review procedures for requests to release tapes from libraries, to return them from users, and to mount tapes during nonproduction jobs.

> **Procedure 2.5.10.2.** Determine that tape labels, volumes, serial numbers, and expiration dates are used for all tape files.

Figure 2.5. Control objectives and audit procedures for reviewing data access controls *(continued)*

controls over access to data, the costs of implementing and administrating such a system may exceed the added control that will be achieved. In addition, users unable to remember these random strings of characters may write them down on notes on their terminals. Thus, security would not be improved by this recommendation. The auditor must consider the types of applications run on the system as well as other potential risk related factors. Chapter 8 discusses this concept of risk evaluation in greater detail.

In a larger computer installation, access to information can be effectively controlled with the following control elements:

A separate, effective data security function or officer

Installed computer security software

A strong user awareness of the need for information security.

The data security function should be separate from the applications development function in the data processing organization. It should administer the password function by helping to set access rules, monitoring security violations, and controlling the computer security software system. The security software establishes the rule-based system which prevents or allows access to various information elements.

While the auditor can recommend installing or improving a data security function or security software, these recommendations will accomplish little unless there is strong user awareness of the need for information security. If this awareness is not evident, the auditor should attempt to work closely with both general management and information systems management to develop an information security education program. Newspaper clippings on other computer crime violations coupled with comments on the vulnerability of the current system to virus attacks will help promote user awareness.

PROGRAM ACCESS CONTROLS

In many respects, controls over access to programs are the most critical of the general controls in the modern data processing center, because it is often difficult to detect unauthorized or improper program changes. A modern, larger computer system often has many users who review various aspects of the data in output reports, and often good controls are built into the applications. If someone improperly manipulates computer data, therefore, there is some chance that the irregularity will be detected. However, it is very difficult to detect an unauthorized change to one of the application programs.

The auditor should be concerned regarding controls over access to computer programs for the following reasons:

1. Many important decision rules can be built into the source code of complex computer programs. Even a small change to a formula or decision rule in such programs can have major control implications.
2. The maintenance of important computer programs is often the responsibility of a small group of specialized programmers who may be the only persons who understand these programs.
3. While many data processing organizations place strong emphasis on their information access controls, often there is far less emphasis on access to computer programs.
4. Even while data processing organizations create program access controls, they allow them to be violated to respond to emergency requirements to fix program problems.
5. Many computer crime cases have been the result of an improper change to computer programs.

To review these controls in the modern, larger data processing organization, the auditor first must understand the overall procedure for authorizing program changes, for testing and approving changes, and for updating production program libraries. Although formal mechanisms for updating programs may not be needed in small data centers, the auditor should expect them to be in place in the larger centers.

Figure 2.6 contains objectives and procedures for reviewing controls over access to programs in the larger data processing organization. When reviewing program library controls, the auditor also should be aware that some key but not regularly used programs may be stored off-line. For example the annual

Control Objectives and Audit Procedures
for Reviewing Program Access Controls

Objective 2.6.1. Separate program libraries should exist for test and production purposes.

    Procedure 2.6.1.1. Discuss program library procedures with responsible information systems management to determine that a proper separation of test and production programs exists.

    Procedure 2.6.1.2. Develop an understanding of the software package or other procedure being used to control access to libraries and assess its adequacy.

    Procedure 2.6.1.3. Determine that proper levels of information systems management are reviewing the control and status reports produced by the library software and are following up on unusual reported events.

Objective 2.6.2. Access to production program libraries should be limited to authorized persons and production applications.

    Procedure 2.6.2.1. Review control procedures for limiting access to production libraries.

    Procedure 2.6.2.2. Select several applications being run in a "systems test" mode by the programming staff and determine whether only test program libraries are being used.

Objective 2.6.3. Changes to production program libraries should be approved by authorized persons.

    Procedure 2.6.3.1. Review procedures for updating production programs and determine that they require a proper level of review and approval.

    Procedure 2.6.3.2. Review procedures for transferring test library programs to production libraries and determine that appropriate management approval is required before transfer.

    Procedure 2.6.3.3. Select several production and maintenance programs recently placed into production and determine that the transfers were reviewed and approved.

    Procedure 2.6.3.4. Interview several programmers to deter-

Figure 2.6. Control objectives and audit procedures for reviewing program access controls

mine their understanding of the production library update process and follow up on any possible discrepancies.

Objective 2.6.4.   There should be naming conventions used in source and object version programs to improve program library integrity and control.

> **Procedure 2.6.4.1.**   Review information systems policies or standards to understand program naming conventions.

> **Procedure 2.6.4.2.**   Review production library source and object program listings to determine whether naming conventions are followed.

Objective 2.6.5.   Emergency procedures should exist for correcting problems with production applications during non-prime shift times and for temporarily running corrections from test libraries.

> **Procedures 2.6.5.1.**   Review procedures for emergency maintenance fixes of production processes where test library modules may be used temporarily.

> **Procedure 2.6.5.2.**   From production records, select several applications where nighttime emergency maintenance was performed and determine that these applications are now back in production libraries.

Objective 2.6.6.   For all load or object module changes, there should be corresponding changes to the source modules.

> **Procedure 2.6.6.1.**   Inquire into the use of SUPER ZAP-type programs which allow changes to object modules without corresponding changes to source modules.

> **Procedure 2.6.6.2.**   Through the use of naming conventions or program dates, attempt to identify any special object modules which lack corresponding source modules.

Objective 2.6.7.   Procedures should exist to prevent unauthorized theft or copying of programs from libraries.

> **Procedure 2.6.7.1.**   Review access control procedures over libraries which would prevent unauthorized copying.

> **Procedure 2.6.7.2.**   Select several programs now being modi-

Figure 2.6. Control objectives and audit procedures for reviewing program access controls *(continued)*

fied and determine whether correct procedures were followed for the transfers from production to test libraries.

Objective 2.6.8. Libraries should be purged periodically of unneeded or obsolete programs.

Procedure 2.6.8.1. Review procedures for purging libraries of unneeded programs and assess their adequacy.

Procedure 2.6.8.2. Select several applications where major revisions have gone into production and ascertain that obsolete programs have been purged or transferred to archive files.

Objective 2.6.9. When end users are using a fourth generation language or similar applications, a special library should be used to limit access to production libraries.

Procedure 2.6.9.1. Review procedures for end user or information center computing to determine that users operate from special libraries and not production libraries.

Procedure 2.6.9.2. Select an information center library listing and trace several program names to determine that all are authorized for end user purposes.

Objective 2.6.10. All production libraries should be backed up and stored in an off-site location on a periodic basis.

Procedure 2.6.10.1. Review procedures for backing up production files and libraries and assess their adequacy.

Procedure 2.6.10.2. Determine that backup copies of production libraries are included in off-site storage procedures.

Figure 2.6. Control objectives and audit procedures for reviewing program access controls *(continued)*

physical inventory compilation programs may be stored on tape during most of the year. Physical access to those programs should also be protected.

Controls over access to programs are interdependent with organizational controls, controls over access to data, and the other general controls discussed in this chapter. For example, while there may be proper segregation of duties within the overall data processing function, it may be lacking within the programming department. That lack of intra-department controls may allow programmers to update production program libraries without proper review and authorization. Similarly, a lack of data access controls may allow an unauthorized user to read a program source code from a source library and make unauthorized modifications in circumvention of the program's controls.

SYSTEMS DEVELOPMENT METHODOLOGY CONTROLS

A systems development methodology (SDM) is a formalized procedure for initializing, defining, developing, and implementing new information systems applications. A well-organized and functioning SDM is a key general control in the larger data processing organization. It helps the auditor to understand the control environment through which applications are developed, tested, and implemented.

Most SDMs follow a fairly similar set of procedures, requiring that users define their needs, developers document the systems under development, and users review the results prior to implementation. Some outside vendors publish formal SDM approaches complete with forms and training manuals. The older ones are very paper intensive, requiring numerous forms as part of the applications development process. Newer SDMs offered by vendors are less paper intensive and often are automated on a personal computer. Some data processing organizations have developed their own SDMs through the experience of data processing management and outside published sources. Figure 2.7 illustrates the steps in a typical SDM.

The auditor should not look for the implementation of one "correct" SDM in the data processing organization under review. Rather, the auditor should look for the existence and use of some formal and effective procedure.

Depending upon the type of review being performed, a review of SDM procedures is an area where the auditor may make control, compliance, and productivity recommendations. Some of the outside, vendor-offered SDMs require an extensive amount of paperwork for a new applications definition. If management has decided that such a procedure is to be followed, the auditor

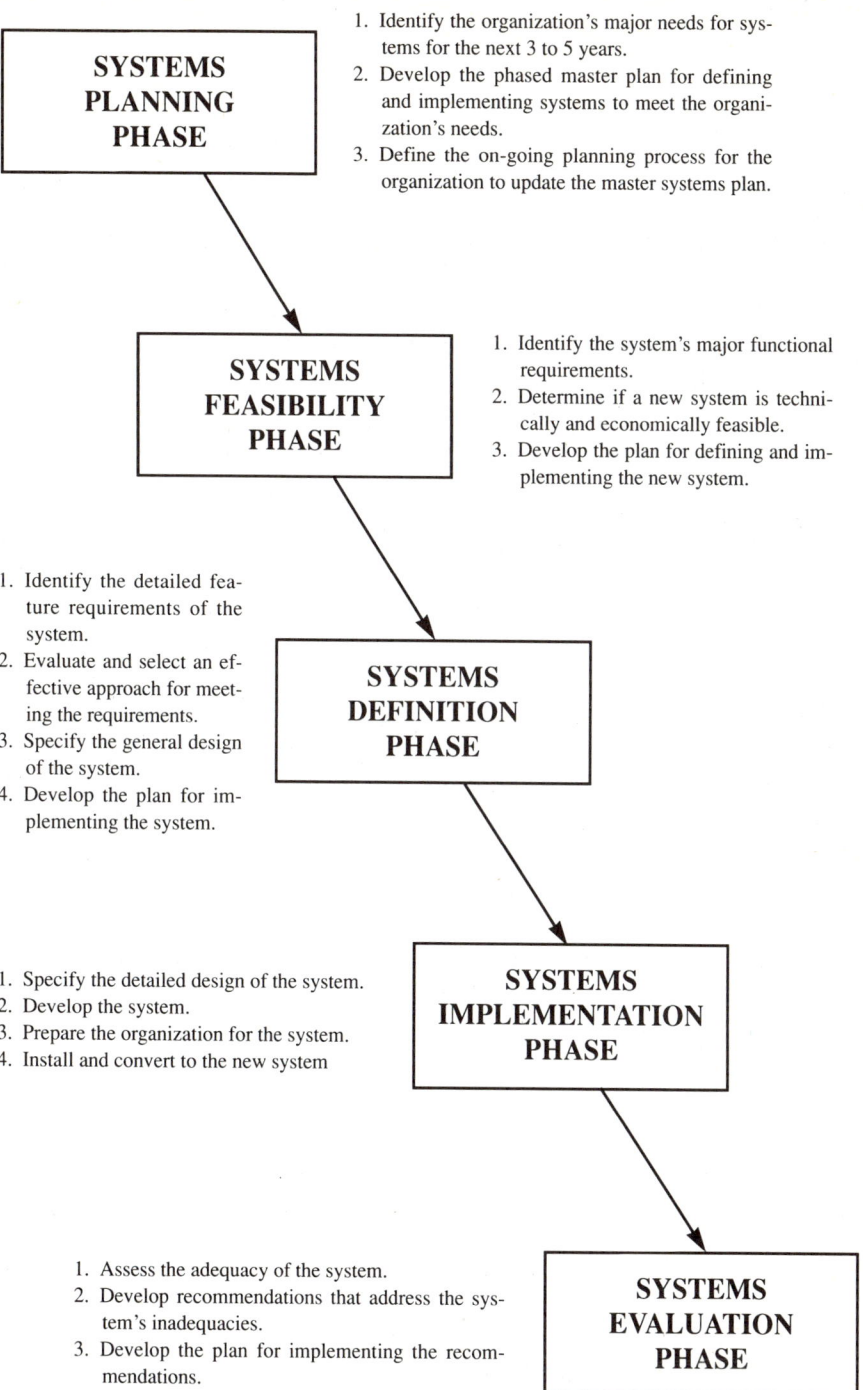

SYSTEMS PLANNING PHASE

1. Identify the organization's major needs for systems for the next 3 to 5 years.
2. Develop the phased master plan for defining and implementing systems to meet the organization's needs.
3. Define the on-going planning process for the organization to update the master systems plan.

SYSTEMS FEASIBILITY PHASE

1. Identify the system's major functional requirements.
2. Determine if a new system is technically and economically feasible.
3. Develop the plan for defining and implementing the new system.

1. Identify the detailed feature requirements of the system.
2. Evaluate and select an effective approach for meeting the requirements.
3. Specify the general design of the system.
4. Develop the plan for implementing the system.

SYSTEMS DEFINITION PHASE

1. Specify the detailed design of the system.
2. Develop the system.
3. Prepare the organization for the system.
4. Install and convert to the new system

SYSTEMS IMPLEMENTATION PHASE

1. Assess the adequacy of the system.
2. Develop recommendations that address the system's inadequacies.
3. Develop the plan for implementing the recommendations.

SYSTEMS EVALUATION PHASE

Figure 2.7. Example of a systems development methodology

would look for the existence and completion of this paperwork during a controls oriented review. However, in a management efficiency review, the auditor may consider recommending that extensive SDM paperwork be reduced to a more reasonable level.

Figure 2.8 contains a set of objectives and audit procedures for auditing SDM procedures. An understanding of the development controls in this area is important in reviewing applications controls, as discussed in Chapter 5, or in reviewing new applications under development, as discussed in Chapter 7. Depending upon the auditor's overall activities, work in this area during the general controls portion of the audit could be limited to obtaining a general understanding. Compliance with these SDM procedures could later be tested in conjunction with a specific applications review.

DATA PROCESSING OPERATIONS CONTROLS

In the older, traditional data processing organization, the operations area was of prime control concern. The computer operator had considerable power to make changes or bypass system controls. For example, computer operators could bypass tape label controls, change program processing sequences, or insert patches into programs. While it is still possible for an operator to do some of these things, the complexity of large computer operating systems as well as the sheer volume of work passing through them makes these types of unauthorized changes more difficult.

Many early audit recommendations regarding operations controls now appear archaic in the modern, larger data processing center. For example, auditors traditionally recommended that someone review computer console logs "on a regular basis." In a larger business oriented data center, console data is automatically written onto log files and extracted only for special purposes. The sheer volume of data makes a periodic human review of the "console log" reports unrealistic. However, there are other tools and controls to help the auditor understand operations controls.

An important first step for the auditor in reviewing data processing operations controls is to understand some of the basic operations procedures. This information might include the following:

1. *How is work scheduled?* Many operations use an automated production control system with operators initiating jobs from a production queue. Other operations give operators considerable authority in deciding which jobs to run.

Control Objectives and Audit Procedures
for Reviewing Systems Development Methodology Controls

Note: This set of objectives and procedures is designed to help the auditor with a general review of systems development controls. More comprehensive procedures can be found in Chapter 7, Reviewing New Applications Under Development.

Objective 2.8.1. There should be an overall Information Systems Steering Committee, comprised of members of user management, to set overall priorities for systems development projects.

Procedure 2.8.1.1. Review the overall structure of any Steering Committee to determine that it represents all key users of informations systems.

Procedure 2.8.1.2. Inquire as to the frequency of steering committee meetings and assess whether it is adequate for the level of systems development activity.

Procedure 2.8.1.3. If the information systems function initially screens all requests for new systems projects, assess whether all appropriate requests are forwarded to the Steering Committee.

Procedure 2.8.1.4. Determine whether new project priorities established by the Steering Committee are consistent with the information systems function's long range plans.

Procedure 2.8.1.5. Review status reports of approved projects submitted by information systems to the Steering Committee to determine whether they are informative and adequate.

Objective 2.8.2. The organization should have a formal Systems Development Methodology (SDM) procedure which defines the steps necessary to initiate, develop, and implement new information systems.

Procedure 2.8.2.1. Interview members of information systems management to gain an understanding of the SDM used by the organization and determine the level of management commitment to its use.

Procedure 2.8.2.2. Review documentation covering the SDM

Figure 2.8. Control objectives and audit procedures for reviewing systems development methodology controls

and determine whether it adequately describes SDM procedures.

Procedure 2.8.2.3. Interview members of the systems development staff to determine whether they understand and use the organization's SDM.

Procedure 2.8.2.4. Briefly review documentation for several recently developed systems to determine that SDM procedures generally were followed.

Procedure 2.8.2.5. For multi-location or divisional organizations, determine whether the same SDM procedures apply for all operating units of the organization under review.

Objective 2.8.3. The SDM should be used for all significant new systems development projects, major systems enhancements, and purchased software implementation projects.

Procedure 2.8.3.1. Determine the appropriateness of policies defining any types of systems which would not be subject to general SDM procedures.

Procedure 2.8.3.2. Select a sample from a list of all new systems projects over the past year and determine that SDM procedures were followed for each.

Procedure 2.8.3.3. If new development techniques are being used, review, through interviews and documentation examinations, the SDM procedures used for:

    Purchased software applications
    Applications developed under prototyping
    Applications developed with fourth generation languages
    Applications developed with computer aided systems engineering (CASE) tools

Procedure 2.8.3.4. Review a sample of applications under development which are using new development techniques to determine that SDM guidelines are being followed.

Objective 2.8.4. Internal Audit should receive notification of or should participate in Steering Committee activities sufficient to set priorities for pre-implementation reviews.

Figure 2.8. Control objectives and audit procedures for reviewing systems development methodology controls *(continued)*

Procedure 2.8.4.1. Assess Internal Audit's awareness of Steering Committee actions by matching Internal Audit planned pre-implementation review activities with Steering Committee priorities.

Procedure 2.8.4.2. Determine whether Internal Audit needs to increase its participation in Steering Committee activities to better understand overall new systems priorities.

Objective 2.8.5. The first phase of any SDM should include a project initiation process to define the general objectives, scope, and anticipated benefits of the systems project.

Procedure 2.8.5.1. Review control procedures over project request forms to determine that forms are complete and approval by proper levels of user management is required.

Procedure 2.8.5.2 Review project initiation and feasibility study SDM documentation for several new applications currently under development or recently implemented to determine whether general project objectives were defined and whether SDM procedures were followed.

Procedure 2.8.5.3. Determine that the initial general design or feasibility study phases of the selected applications were approved by responsible users.

Objective 2.8.6. The design phase of any SDM should define the input, output, and processing requirements of the application as well as controls and audit trails.

Procedure 2.8.6.1. Review design phase documentation for several applications under development or recently implemented to determine that SDM procedures were followed.

Procedure 2.8.6.2. Review the documented design for systems controls and audit trails for the selected applications to determine whether adequate consideration was given to this area.

Procedure 2.8.6.3. Determine that information systems management gives adequate attention to the review and approval of design phase documentation.

Figure 2.8. Control objectives and audit procedures for reviewing systems development methodology controls *(continued)*

Objective 2.8.7. The development phases of the SDM should follow information systems programming standards and documentation procedures.

    Procedure 2.8.7.1. Briefly review programming standards and determine whether they are followed in new systems development projects.

    Procedure 2.8.7.2. For one or more of the systems selected under the audit procedures of Objective 2.8.6., review the detailed design documentation to determine that it is consistent with overall general design and that any significant differences have been approved.

    Procedure 2.8.7.3. For purchased software packages, review vendor documentation to determine that it generally follows overall organization standards.

    Procedure 2.8.7.4. If other new development technologies such as prototyping or CASE are used, determine that appropriate SDM design requirements are being followed.

Objective 2.8.8. Prior to implementation, the SDM should provide for comprehensive systems testing with active participation by key users of the system.

    Procedure 2.8.8.1. Determine that SDM procedures and other information systems policies adequately cover program and systems testing.

    Procedure 2.8.8.2. Select several information systems recently implemented or near implementation to determine if testing standards are being followed.

    Procedure 2.8.8.3. Determine that key users actively participate in testing applications through activities such as preparing input, reviewing output results, and approving test results.

    Procedure 2.8.8.4. Determine that all testing takes place against extract test files using nonproduction libraries.

Objective 2.8.9. The overall development of any systems project should be controlled through a project management system which

Figure 2.8. Control objectives and audit procedures for reviewing systems development methodology controls *(continued)*

reports planned milestones, actual accomplishments, and estimated times to complete for the various phases in the project.

Procedure 2.8.9.1. Review the project management system to determine whether it provides sufficient detail to record time and resource estimates by task, identify critical paths for project completion, and show estimates to complete.

Procedure 2.8.9.2. Select several systems projects under development and determine whether the established project management system is being used.

Procedure 2.8.9.3. Trace project management reporting results from the selected systems to information systems management reports to determine whether the project management system is supplying sufficient data for management.

Procedure 2.8.9.4. Select a recent systems project that was implemented late or over budget to determine whether the project management system reports provided an early warning.

Procedure 2.8.9.5. Trace time reported by programmers from actual time reports to project management status reports to determine whether time is being recorded properly.

Objective 2.8.10. The SDM should provide for ongoing reviews of significant systems, through the post-implementation phase, by Internal Audit or a quality assurance function.

Procedure 2.8.10.1. If a quality assurance function exists, review its procedures for determining that new applications meet information systems standards.

Procedure 2.8.10.2. Review all new systems implemented over a one year sample period and determine how many were subjected to a post-implementation review.

Procedure 2.8.10.3. Perform a limited post-implementation review of one or two recently implemented applications to determine whether original objectives were met.

Procedure 2.8.10.4. Determine the level of priority given to correcting newly implemented applications that do not meet original objectives.

Figure 2.8. Control objectives and audit procedures for reviewing systems development methodology controls *(continued)*

2. *How are tapes and files managed?* Automated tools are used often in this area. In addition, some operations have a separate facility in the tape library where production tapes are mounted. However, if automated tools are not used for controlling tape libraries, there is a greater chance that there will be mounting errors.

3. *What types of operator procedures or instructions are used?* Documentation can take a variety of formats; the auditor should have a general understanding of documentation format and content.

4. *How is work initiated and how does it flow through operations?* In some operations, production is initiated through remote user terminals. In others, the production control function funnels all necessary input data to operations.

The above are representative of questions the auditor might ask when attempting to gain an understanding of data processing operations. The basic idea is to gain an understanding of how data processing functions.

Figure 2.9 contains objectives and audit procedures for reviewing controls over data processing operations in the modern, larger data processing operations center. Attainment of a substantial number of these control objectives will assure an adequate operational control environment for reviewing individual applications, as discussed in Chapters 5 and 6.

SYSTEMS PROGRAMMING AND TECHNICAL SUPPORT CONTROLS

The larger, modern data processing operation typically has a variety of specialized support functions to manage the specialized software tools used by the computer center. These include a systems programming function to maintain the operating system and related software, telecommunications support to manage network special products, and database administration to maintain the database software, including the data dictionary.

Data processing operations typically have a variety of technical staff to perform these specialized functions. While these functions are essential to the overall control environment of the data processing organization, often they are ignored by auditors, perhaps because it is difficult to think of these functions in the same context as controls surrounding the implementation of a specific application. However, there are strong analogies between applications controls and control procedures which should be in place in these technical support areas.

Control Objectives and Audit Procedures
for Reviewing Data Processing Operations Controls

Objective 2.9.1.   Computer equipment should be located in a secure, environmentally controlled facility.

  Procedure 2.9.1.1.   Discuss physical and environmental control procedures with information systems management to determine current policies and future plans.

  Procedure 2.9.1.2.   Tour computer room facilities and observe physical security strengths and weaknesses including:

> The existence of locking mechanisms to limit computer room access to authorized individuals
>
> The placement of computer room perimeter walls and windows to limit access
>
> The location of power transformers and air conditioning units to provide proper protection
>
> The general location of the computer room facilities within the overall building to minimize traffic
>
> The existence of fire detection equipment including zone controlled heat and smoke detectors
>
> The existence of an overall fire protection system including a zone controlled, Halon or equivalent system and local extinguishers

  Procedure 2.9.1.3.   Review computer room temperature, humidity, and other environmental controls and assess their adequacy.

  Procedure 2.9.1.4.   Briefly review maintenance records to ascertain that physical and environmental controls are regularly inspected and maintained.

Objective 2.9.2.   Production processing should be scheduled to promote efficient use of computer equipment consistent with the requirements of systems users.

  Procedure 2.9.2.1.   Through interviews with operations management, develop an overall understanding of computer processing demands including on-line and batch production work as well as any end user computing.

Figure 2.9. Control objectives and audit procedures for reviewing data processing operations controls

Procedure 2.9.2.2. Review procedures for scheduling regular production jobs including the use of automated job scheduling tools.

Procedure 2.9.2.3. Match a limited number of scheduled production jobs against actual completion times to determine whether actual schedules are followed.

Procedure 2.9.2.4. Determine that operating system job classes or priority codes are used to give proper priority to critical production jobs.

Procedure 2.9.2.5. Evaluate procedures for running "rush" or rerun jobs to determine that they are consistent with overall production schedule requirements.

Objective 2.9.3. Operations instructions should exist to allow operators to process correctly normal production as well as to respond to errors.

Procedure 2.9.3.1. Review documentation standards for production applications to determine that they provide operators with information regarding:

Normal operations procedures including instructions for special forms, tape files, and report disposition
Application restart and recovery procedures
Responsible programmer/analyst and user contact names for resolving production application problems

Procedure 2.9.3.2. Review procedures for turning new applications or revisions over to computer operations to determine that they are reviewed by operations and that standards are followed.

Procedure 2.9.3.3. Select a sample of production applications and determine that the operating documentation is current by comparing run listings with published documentation.

Objective 2.9.4. Computer operators should not be allowed to change programs independently or initiate production jobs without authorization.

Procedure 2.9.4.1. Determine that information systems poli-

Figure 2.9. Control objectives and audit procedures for reviewing data processing operations controls *(continued)*

cies prohibit computer operations personnel from performing programming tasks or running unauthorized jobs.

Procedure 2.9.4.2. Determine that production source libraries can not be accessed by operations personnel.

Procedure 2.9.4.3. Assess information systems procedures for reviewing periodically the contents of log files or otherwise monitoring improper operator use of computer equipment.

Objective 2.9.5. Procedures should exist to allow for emergency program modifications when error conditions prevent the processing of critical production work.

Procedure 2.9.5.1. Review and document procedures for changing production programs or procedure libraries when emergency situations require special handling.

Procedure 2.9.5.2. Determine that all emergency processing activities are properly documented and are subject to subsequent management review.

Procedure 2.9.5.3. Select several documented emergency program fixes and determine that the necessary changes were added to production processing libraries and documented.

Objective 2.9.6. Logs or records of computer systems activity should exist to monitor both regular and abnormal computer operations.

Procedure 2.9.6.1. Determine that a procedure exists, preferably through an automated system, to log all computer systems activity including:

Jobs and programs run

Any reruns

Abnormal terminations of any jobs or programs

Operator commands and other data entered through system consoles

Procedure 2.9.6.2. Determine that computer activity logs are reviewed periodically, that exception situations are investigated, and that the results of investigations are documented.

Procedure 2.9.6.3. Determine that tapes produced from the

Figure 2.9. Control objectives and audit procedures for reviewing data processing operations controls *(continued)*

computer operating system's log monitor are retained long enough to allow investigation of unusual activities.

Procedure 2.9.6.4. Review procedures for logging problems to determine that all abnormal software and hardware operating conditions are documented.

Objective 2.9.7. When batch jobs are run, procedures should exist to determine that only authorized input data is submitted at scheduled times.

Procedure 2.9.7.1. Determine that schedules exist for the submission of critical input data batches and that procedures exist to followup on missing data.

Procedure 2.9.7.2. Determine that all input data batches are initialed or otherwise identified to assure they have been submitted by authorized individuals.

Procedure 2.9.7.3. Review a limited sample of batch applications to determine that manual or automated batch balancing techniques are used.

Objective 2.9.8. Controls should exist to determine that computer system outputs are correct and are distributed only to authorized users.

Procedure 2.9.8.1. Determine whether users or information systems personnel are responsible for reviewing output controls and assess whether those control reviews are being performed.

Procedure 2.9.8.2. Assess procedures for reviewing output reports to determine they are complete, there are no printer errors, and correct forms are used.

Procedure 2.9.8.3. If output reports are distributed directly to users, determine that only authorized persons receive their own reports.

Procedure 2.9.8.4. If users pick up their output reports, determine that reports are controlled, through a lock box system or control desk, so that only authorized users can pick up their own reports.

Figure 2.9. Control objectives and audit procedures for reviewing data processing operations controls *(continued)*

Procedure 2.9.8.5.   Determine that sensitive or confidential reports are placed in sealed envelopes prior to distribution.

Objective 2.9.9.   Controls should exist to ensure that only correct data files are used for production processing.

Procedure 2.9.9.1.   Determine that all files are internally labeled and that production applications will process only current data files.

Procedure 2.9.9.2.   Review procedures, including the use of automated tools, for inventorying and controlling computer files.

Procedure 2.9.9.3.   Determine that procedures exist to monitor tape usage and to schedule tapes for cleaning and recertification when appropriate.

Objective 2.9.10.   There should be a job accounting system to monitor the use of computer resources and to charge or report such use to benefiting users.

Procedure 2.9.10.1.   Review and document the computer job accounting system to determine how charges are collected and allocated.

Procedure 2.9.10.2.   Review computer processing rates to determine whether they are equitable and understandable.

Procedure 2.9.10.3.   Interview several selected systems users to determine whether each has an understanding of the job accounting system.

Figure 2.9. Control objectives and audit procedures for reviewing data processing operations controls *(continued)*

The auditor is most effective in reviewing these technical areas if not put off by the technical jargon and mystique associated with these functions. For example, the auditor does not need to know the technical issues associated with installing IBM's MVS/ESA operating system to ask questions about how the project is being managed, what are the testing procedures, and how are results and progress being reported to management.

Systems Programming Controls

A larger data processing department usually has a separate systems programming function. If data processing management placed that function within the applications programming department, the auditor should raise separation of duties concerns.

Typically, a systems programming function has responsibility for the following general areas:

Installing new systems software products and upgrades, such as operating system components

Adjusting systems parameters or "tuning" hardware and software components to improve overall performance

Establishing and testing backup and recovery procedures if there is not a separate data security function

Resolving system-level hardware and software problems.

Consulting with applications programmers concerning job control language problems

The nature of many systems programming projects requires "emergency" procedures to fix a system problem or bug. In addition, systems programmers find it necessary to interact more closely with computer operations than do normal applications programmers. In many respects, the systems programming function operates much like an engineering or scientific support function. It should probably be managed as such, and the auditor should look for procedural controls to properly manage such a creative, technical support function.

Figure 2.10 contains a set of objectives and audit procedures for reviewing controls over the systems programming function. Many controls are management procedures, such as reviewing project planning or documentation. Since an auditor, typically, does not come from a systems programming background, the auditor may not have the skills necessary to evaluate technical materials.

Control Objectives and Audit Procedures
for Reviewing Systems Programming Controls

Objective 2.10.1. The duties of the systems programming function should be organizationally separate from the systems development function and computer operations.

> **Procedure 2.10.1.1.** Review the organization of the systems programming function, including its reporting relationships, and assess the level of separation of duties.
>
> **Procedure 2.10.1.2.** Determine that systems programmers are prohibited from changing production files and programs.
>
> **Procedure 2.10.1.3.** Determine that systems programmers are prohibited from operating computer systems when production files or application programs are resident.
>
> **Procedure 2.10.1.4.** Determine that responsibilities for various processors or software products are rotated periodically among members of the systems programming staff.

Objective 2.10.2. Systems programming activities should be documented and approved by information systems management.

> **Procedure 2.10.2.1.** Determine that all systems programming activities are approved by proper levels of management and, when appropriate, that project budgets are established.
>
> **Procedure 2.10.2.2.** Determine that documentation standards exist for all systems programming activities including descriptions of:
> Software options selected
> Vendor supplied changes, fixes, or version enhancements
> Special user exit routines
>
> **Procedure 2.10.2.3.** Determine that systems programmers record time against authorized projects or other activities as specified by management.
>
> **Procedure 2.10.2.4.** Trace the activities logged on selected programmer time sheets to documentation records to determine that the activities were approved and documented.

Figure 2.10. Control objectives and audit procedures for reviewing systems programming controls

Procedure 2.10.2.5.   Determine that all systems software documentation, including documentation supplied by vendors, is restricted to persons with a need to know.

Procedure 2.10.2.6.   Review the level of system utilities in use and assess whether such usage is appropriate.

Objective 2.10.3.   The selection and acquisition of new operating systems software and utilities should follow the same purchase justification and approval standards as used for application software.

Procedure 2.10.3.1.   Describe procedures for selecting new systems software products and determine if the same standards are used as for application software.

Procedure 2.10.3.2.   Review procedures used to catalog systems software products and to control their versions or release levels.

Procedure 2.10.3.3.   When a systems software product is used by applications programmers, assess the adequacy of procedures to communicate the use of this product and to help with any programmer questions.

Procedure 2.10.3.4.   Select several systems software products and both review documentation and discuss product usage to ensure that:
> The software product is still being used
> Vendor licenses for the product are current
> Documentation for the product is current

Procedure 2.10.3.5.   Select several software vendor invoices and trace itemized software products to library and documentation records to determine that the billings reflect the products and versions in use.

Objective 2.10.4.   Procedures should exist to test systems software to ensure changes or new products do not impact either existing systems software or production programs.

Procedure 2.10.4.1.   Determine that new or modified systems software products are tested against a pre-established, comprehensive plan.

Figure 2.10.   Control objectives and audit procedures for reviewing systems programming controls *(continued)*

Procedure 2.10.4.2. Determine that systems software testing includes documentation of test results and appropriate retention of test results.

Procedure 2.10.4.3. Assess the level of involvement in the software testing process of computer operations, quality assurance, and other appropriate departments.

Procedure 2.10.4.4. Review procedures for installing emergency fixes to systems software and determine that appropriate testing and documentation procedures are completed.

Objective 2.10.5. The systems programming function should monitor the overall performance of computer systems to improve processing efficiencies.

Procedure 2.10.5.1. Review the computer systems performance reporting package prepared for information systems management and determine whether the reports are comprehensive and understandable.

Procedure 2.10.5.2. Review the procedures and software tools used to monitor system performance and assess their adequacy.

Procedure 2.10.5.3. Review procedures for making performance efficiency changes to operating software and determine whether the changes:

Receive appropriate management approval

Are documented

Are followed up for corrective action if adjustments do not achieve expected results

Procedure 2.10.5.4. Determine that performance adjustment changes due to system emergency situations are subject to the same level of follow-up approval and documentation.

Figure 2.10. Control objectives and audit procedures for reviewing systems programming controls *(continued)*

The audit procedures listed in Figure 2.10, however, can be performed by inquiry and do not require specialized operating system knowledge.

Systems programming is an important element in the overall controls environment of a larger data processing organization. If the total data processing organization is sufficiently large, internal audit management might consider adding an individual to its staff with specialized skills to review systems programming controls.

Telecommunications Support Controls

The telecommunications support function is a rather new department in the typical data processing organization. The larger computer center usually has telecommunication networks of terminals and distributed processors attached to the mainframe computer center. These telecommunication networks may stretch to remote facilities in other cities or even other countries. Because of the many telecommunication technologies in use as well as recent deregulation, many organizations have established a telecommunications and network support facility within the data processing department.

Depending upon the organization's needs, telecommunications may be part of systems programming or may be a separate department. However, it will usually be involved in some of the following functions:

Helping applications developers and users plan the most efficient data communications network

Working with hardware and telecommunications vendors to resolve problems

Establishing standards and certification procedures for new devices on the network

Helping develop local area, non-telecommunications networks

In some organizations, the telecommunications function has responsibility for voice or regular telephone communications as well as data applications. Sometimes this function is also responsible for security over the telecommunications lines. As a first step to reviewing this function, the auditor must develop an understanding of its responsibilities.

Control objectives and audit procedures for telecommunications support are outlined in Figure 2.11. The auditing procedures have been limited to inquiry and observation. A detailed review beyond asking general management con-

Control Objectives and Audit Procedures
for Reviewing Telecommunications Support

Objective 2.11.1.   The information systems telecommunications network should be controlled through a central administrative function.

Procedure 2.11.1.1.   Interview personnel responsible for the data processing telecommunications network, and assess their authority and responsibility for the function.

Procedure 2.11.1.2.   Determine if the telecommunications network management function has developed appropriate records documenting the existing network.

Procedure 2.11.1.3.   Review procedures for making changes to the telecommunications network and determine if there is an appropriate level of management approval.

Procedure 2.11.1.4.   Review telecommunications network management policies and procedures and assess the extent to which these are communicated to users of the network.

Objective 2.11.2.   Inventory and identification controls should be maintained for all terminals on the telecommunications network.

Procedure 2.11.2.1.   Obtain an inventory listing of all terminals attached to the communications network.

Procedure 2.11.2.2.   Determine that standards exist to assure that only authorized terminals and models are connected to the network.

Procedure 2.11.2.3.   Select several terminals from inventory listings and verify inventory information by checking the actual terminal installations.

Procedure 2.11.2.4.   Select a sample of work stations or terminals in user areas and trace their attributes back to telecommunications records to determine whether records are correct.

Objective 2.11.3.   All system access should be controlled through passwords and authorization codes that are validated by security software.

Procedure 2.11.3.1.   Review procedures for granting authoriza-

Figure 2.11. Control objectives and audit procedures for reviewing telecommunications support controls

tion codes to terminals, and determine that access authorizations are canceled when a terminal has been inactive for a specified length of time.

Procedure 2.11.3.2. Review procedures for granting access passwords, and determine that passwords are assigned to specific individuals rather than functions.

Procedure 2.11.3.3. Review procedures for canceling passwords when an individual leaves the organization or has a change in responsibilities, and assess whether these procedures are adequate.

Procedure 2.11.3.4. Select several former employees from personnel records and determine that their terminal access rights have been deleted.

Objective 2.11.4. All telecommunications accesses to the computer system should be logged, and logs should be reviewed periodically by data processing or telecommunications management.

Procedure 2.11.4.1. Develop an understanding of the software tools available to monitor telecommunications activity.

Procedure 2.11.4.2. Review telecommunications administration procedures for reviewing control logs and following up on exception situations.

Procedure 2.11.4.3. Review telecommunications logs for several periods and determine that any exceptions have been reviewed and resolved by appropriate levels of management.

Objective 2.11.5. The use of dial-up lines to the computer systems should be controlled to prevent unauthorized access attempts.

Procedure 2.11.5.1. Assess the number of dial-up lines that are available within the telecommunications network.

Procedure 2.11.5.2. Determine whether existing dial-up lines are necessary and have been approved by management.

Procedure 2.11.5.3. Determine if call back devices, see-through security units, and other controls have been installed on dial-up lines to prevent improper access attempts.

Figure 2.11. Control objectives and audit procedures for reviewing telecommunications support controls *(continued)*

Procedure 2.11.5.4. Determine that the security system logs a potential user off the system after a very limited number of unsuccessful password or authorization code access attempts.

Objective 2.11.6. Master terminals, which can change the access rights of other terminals or users, should only be located in secure locations.

Procedure 2.11.6.1. Identify the number of terminals designated as master terminals for various applications and assess whether that number is excessive.

Procedure 2.11.6.2. Verify that all authorized master terminals are located in secure locations so that unauthorized individuals are prevented from access.

Objective 2.11.7. Sensitive data should be encrypted, or otherwise protected, when transmitted to prevent message interception.

Procedure 2.11.7.1. Determine whether the computer center handles any sensitive data which should be encrypted to prevent detection.

Procedure 2.11.7.2. Determine whether procedures exist to identify risks where data should be encrypted, and review encryption procedures as discussed in Chapter 9.

Objective 2.11.8. Manufacturer, software vendor, and third party access lines to the computer system should be monitored with frequent changes to lines and access codes.

Procedure 2.11.8.1. When outside users have access to the computer system, determine whether those requirements appear to be justified.

Procedure 2.11.8.2. Determine that outside user authorization codes are changed periodically.

Objective 2.11.9. All requests for uploads or downloads of data or programs from distributed processors, such as microcomputers, should be subject to management review and approval.

Procedure 2.11.9.1. Determine who is responsible for approv-

Figure 2.11. Control objectives and audit procedures for reviewing telecommunications support controls *(continued)*

ing requests for uploads or downloads of data.

Procedure 2.11.9.2. Review a list of users receiving or transmitting data to the central computer, and determine whether requests have been properly approved and appear to be justified on a need to know basis.

Procedure 2.11.9.3. Determine that remote user access to program libraries is limited strictly to approved users and to read only capabilities.

Objective 2.11.10. Telecommunications billings should be monitored to identify potential efficiencies and to detect unauthorized users.

Procedure 2.11.10.1. Determine that telecommunications billings are compared against authorized users to detect unauthorized usage.

Procedure 2.11.10.2. Interview telecommunications management to determine procedures to identify network efficiencies and to modify the network configuration.

Figure 2.11. Control objectives and audit procedures for reviewing telecommunications support controls *(continued)*

trols-oriented questions often must be performed by someone with specialized telecommunications skills. Chapter 4 provides an introduction to telecommunications controls and an approach to performing audits in this area.

Database Administration Controls

The typical modern data processing organization is making increasing use of database technologies to better manage its information and design more efficient data structures which share data rather than retain it in redundant storage locations. Because database software is a specialized product which provides data to many applications, often database administration is organized as a separate function, located with the other technical support departments or part of applications development.

Before reviewing the database administration function, the auditor should have a general knowledge of the database product being used in the organization. This is often called the database management system or DBMS. Database software is offered by many of the hardware vendors as well as by independent software suppliers. All such products, however, store data in an independent, logical manner which can be defined in differing manners by differing applications. Figure 2.12 illustrates the structure of a typical database. The stored data is independent of specific applications and may be used for differing needs.

The database administration function typically interacts with systems and programming for new applications and also works with systems programming on database system technical problems. The functions of this department include:

Design of logical and physical database structures

Operation and maintenance of the database software

Monitoring and tuning the software to improve performance

Maintenance of a data dictionary to define data elements for applications developers

Maintaining procedures for the security and the recovery of the database

Because of the importance of database files to many large applications, database administration should be included in any general controls review of a modern, larger data processing organization. Figure 2.13 contains control objectives and audit procedures for review of the database administration function. An important element in these objectives is the control over access to specific data elements, because they may be shared by various users.

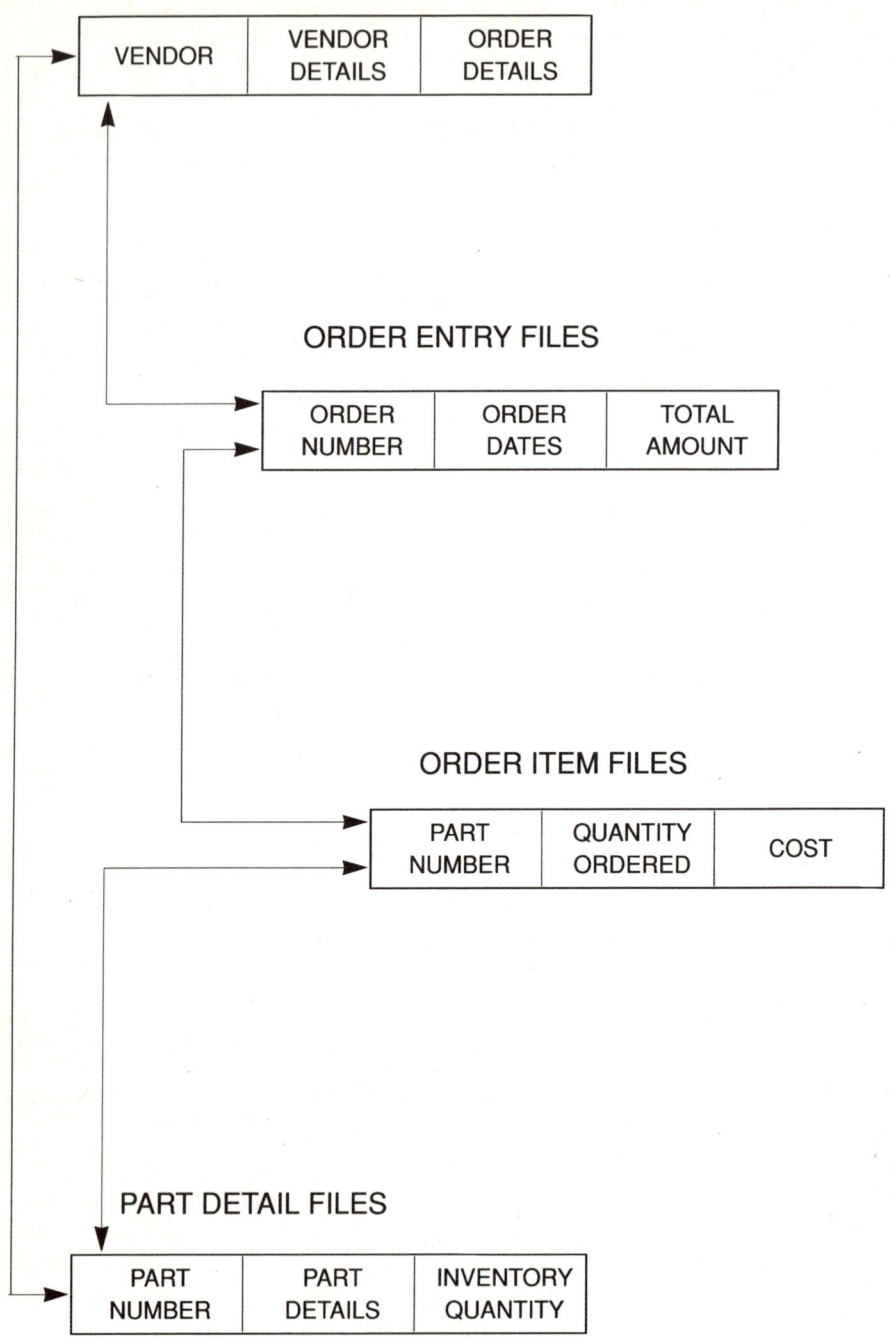

Figure 2.12. Typical database structure

**Control Objectives and Audit Procedures
for Reviewing Database Administration Controls**

Objective 2.13.1. When multiple applications share the same databases, there should be a separate information systems database administration function.

Procedure 2.13.1.1. Determine if a database administration function exists and is independent from the systems development and operations functions.

Procedure 2.13.1.2. Review the policies and procedures governing the database administration function to assess their adequacy.

Procedure 2.13.1.3. If there is not a separate database administration function, determine who is responsible for controlling shared databases and assess the adequacy of these arrangements.

Objective 2.13.2. A data dictionary should be in place to provide for common definitions of data items.

Procedure 2.13.2.1. Determine that a data dictionary is in place and assess its level of usage.

Procedure 2.13.2.2. Review the standards and procedures in place for entering data into the data dictionary as well as for its use.

Procedure 2.13.2.3. Review data dictionary standards and determine that the dictionary software provides:
Precise and unambiguous descriptions of data items
Restrictions on dictionary access or modification
Relationships of the data item to other data items or files

Procedure 2.13.2.4. Determine if the data dictionary is used for all production applications and, if not, assess whether the level of usage is appropriate.

Procedure 2.13.2.5. Trace the data definitions of selected items in a limited number of production source programs to the data dictionary and determine whether the dictionary is being used properly.

Figure 2.13. Control objectives and audit procedures for reviewing database administration controls

Objective 2.13.3. There should exist standards which govern the access and use of databases by applications programs.

Procedure 2.13.3.1. Determine that standards exist governing the access of databases by applications programs.

Procedure 2.13.3.2. Determine that the database management systems installed allow for the same level of information security as exists in other information systems applications.

Procedure 2.13.3.3. Review procedures for testing database applications as well as updating production libraries to determine the adequacy of controls.

Objective 2.13.4. The database administration function should be responsible for monitoring the use of any query language processors used for the database.

Procedure 2.13.4.1. Develop an understanding of the query language processors or retrieval languages used in conjunction with the database systems.

Procedure 2.13.4.2. Determine that use of the database retrieval language is limited to persons with an authorized need to access the database.

Procedure 2.13.4.3. Review procedures to train users in use of the database query language to prevent excess or inefficient retrieval requests.

Objective 2.13.5. The database administration function should be responsible for monitoring the overall integrity and performance of database systems.

Procedure 2.13.5.1. Review the availability and use of database log files to monitor access to the databases, to the program codes controlling such access, and to data dictionary "copybook" files.

Procedure 2.13.5.2. Review the procedures used to detect unauthorized access attempts against the database.

Procedure 2.13.5.3. Develop an understanding of the utility programs used in conjunction with database operations and determine whether documentation is maintained regarding such use.

Figure 2.13. Control objectives and audit procedures for reviewing database administration controls *(continued)*

> **Procedure 2.13.5.4.** Determine that the database software contains adequate controls to log input transactions so that files can be reconstructed in the event of a systems failure.

Figure 2.13. Control objectives and audit procedures for reviewing database administration controls *(continued)*

Database file structures are being used increasingly by larger data processing departments. The manner in which the data is organized and retrieved and the structure of such files are improving. An understanding of the data administration function helps the auditor in reviewing application controls and end user computing controls as well as general controls.

QUALITY ASSURANCE CONTROLS

Many progressive data processing organizations invite their internal auditors to review new applications under development and perform other related technical reviews. However, the size of many audit functions compared to the size of the data processing departments has prevented auditors from reviewing many areas of data processing management. To provide an ongoing review of the data processing applications development processing, some data processing organizations have established a quality assurance function. In many respects, the department acts as a computer audit function within data processing.

Where a quality assurance function exists, usually internal auditors work quite closely with it. Typically, quality assurance will be responsible for reviewing all applications for compliance with department standards before applications are placed into production. Internal auditors will review only selected applications, with a greater emphasis on internal controls. Internal audit may well rely on the procedures performed by quality assurance to limit the scope of its work in certain areas.

Because the auditor may work with and rely upon quality assurance on an ongoing basis, it is important to have an understanding of the general controls in the quality assurance function. Figure 2.14 provides control objectives and audit procedures for a general controls review of this function.

Quality assurance is a relatively new function which does not exist in all data processing organizations. Where it does, it is quite appropriate for the auditor to review general controls within this function. Where it does not, the auditor may recommend that data processing management form such a function.

Control Objectives and Audit Procedures
for Reviewing Quality Assurance Controls

Objective 2.14.1.   An organization or function should exist in larger information systems organizations to review the overall quality and integrity of new or revised computer applications.

Procedure 2.14.1.1.   Through interviews with information systems management, determine who is responsible for monitoring the quality and integrity of computer applications. Candidates include:

> A quality assurance function within the information systems organization
>
> An independent internal audit organization with responsibility to perform quality assurance functions
>
> An independent quality assurance function

Procedure 2.14.1.2.   If a quality assurance function does not exist, assess the adequacy of the information systems function or internal audit to perform such reviews.

Objective 2.14.2.   The quality assurance function (QA) should have sufficient independence to accept or reject new applications submitted for approval prior to production processing.

Procedure 2.14.2.1.   Review standards and procedures for turning new and revised systems projects over to operations.

Procedure 2.14.2.2.   Through discussions with information systems management and with individuals responsible for QA, assess the procedures used by QA to perform reviews.

Objective 2.14.3.   All systems and programming documentation should be subject to review by QA.

Procedure 2.14.3.1.   Determine that QA is involved in reviewing the following areas:

> New systems documentation including systems specifications
>
> Modifications or revisions to production programs including changes to operating procedures

Figure 2.14.   Control objectives and audit procedures for reviewing quality assurance controls

After the fact reviews for emergency program or systems fixes or corrections

Procedure 2.14.3.2. Determine that QA reviews purchased software as well as internally developed applications.

Procedure 2.14.3.3. Select a limited sample of new applications released to operations and assess the extent of any QA review, including a determination as to whether:

The new application should follow organization standards

QA review activities and systems improvement suggestions should be documented

Any disputes between the application development and QA functions should have been resolved by information systems management

Objective 2.14.4. QA should have established review standards that permit other functions to prepare materials for QA review.

Procedure 2.14.4.1. Interview members of the systems development staff to determine their understanding of QA procedures.

Procedure 2.14.4.2. Determine that QA reviews are formally scheduled and follow an overall plan.

Procedure 2.14.4.3. Determine that QA has the right to access all systems documentation.

Procedure 2.14.4.4. Participate in a selected QA review by attending key review meetings and reviewing selected documentation.

Objective 2.14.5. If separate functions, Internal Audit should both rely on the work of QA and should periodically review QA.

Procedure 2.14.5.1. Through interviews, determine that QA has an understanding of the role and responsibility of the internal audit function.

Procedure 2.14.5.2. Assess the emphasis of QA reviews on internal controls.

Procedure 2.14.5.3. Determine that QA has access to Internal Audit's findings and recommendations and that such comments are considered in subsequent QA reviews.

Figure 2.14. Control objectives and audit procedures for reviewing quality assurance controls *(continued)*

> **Procedure 2.14.5.4.**   As part of an overall information systems controls review, review QA for compliance with its own internal procedures.

Figure 2.14. Control objectives and audit procedures for reviewing quality assurance controls *(continued)*

CONCLUDING THE LARGER SYSTEM GENERAL CONTROLS REVIEW

A general controls review is a key element to understanding the controls and risks in a larger computer center. The information gathered in this review will prepare the auditor, whether internal or external, for more detailed reviews of selected data processing applications or functions. However, if significant control weaknesses are identified in the course of this general controls audit, the auditor should take care before evaluating controls within any application running in that data processing organization.

The auditor should use the control objectives and procedures set out in the figures in this chapter as a guide. Because there are variations in the way a modern larger data processing organization is organized, there is no "cookbook" approach to performing such a general controls review. Rather, the auditor must use a basic knowledge of the organization plus some creativity to perform the audit.

CHAPTER 3

Mini and Microcomputer Controls

INTRODUCTION

Auditors traditionally had problems evaluating controls in mini and microcomputer business data processing environments. These systems are typically installed with limited staffs in a more "user friendly" organization atmosphere. Auditors often reviewed mini and microcomputer controls in terms of more traditional, mainframe data processing controls, looking for such things as strong physical security controls and proper separation of duties among members of the data processing organization. These controls often do not exist or are only partially implemented in the smaller, business minicomputer environment.

When faced with evaluating controls in the mini or microcomputer setting, auditors sometimes recommended traditional "cookbook" types of controls. They advised that microcomputers be placed in locked rooms to provide physical security or that a small, two-person data processing staff be expanded to four to ensure proper separation of duties. While there may be situations where such controls are appropriate, they usually are not applicable in a typical business setting. The auditor can easily lose credibility if the control recommendations are not appropriate to the potential risk.

Organizations are implementing increasing numbers of mini or microcomputer systems to support smaller divisions or departments, or to provide data processing for entire organizations. Despite the smaller size of these computer systems, they present the auditor with many control concerns. This chapter discusses data processing controls in the smaller, mini and microcomputer business data processing environment, including unique control objectives and audit procedures for these systems.

THE MINI OR MICROCOMPUTER BUSINESS SYSTEM

In Chapter 2, the "larger" computer center was defined in terms of a variety of computer system and organizational attributes. The chapter advised that if the computer system was located in a secure facility, had a multi-task operating system, and had a large programming staff, the auditor probably should consider it to be a "larger" computer system and review for the appropriate control procedures. While not precise, this definition covered the typical major computer systems.

A similar attribute-based description is best used to define the mini and microcomputer environment. A strict computer hardware architecture definition does not help the auditor decide when to attempt to apply mini or microcomputer-based control procedures. For example, Digital Equipment Corporation (DEC) is, perhaps, the major manufacturer of minicomputers in the United States. DEC's larger VAX super minicomputers have virtually all the capabilities of a traditional mainframe computer. In addition, a series of these VAX machines can be connected to provide more computer power than some older traditional mainframe machines. When reviewing controls in this environment, the auditor should evaluate these linked or super minicomputers as the larger, mainframe systems discussed in Chapter 2.

The difficulty in using hardware to define a minicomputer is highlighted when the machine "looks" like a larger processor. For example, the IBM 43xx series of machines has been considered to be the lower end of IBM's mainframe series of computers. However, in spite of its appearance, the IBM 4361 is a minicomputer in terms of capacity and the way it is described by IBM literature. Many data processing organizations, however, implement controls for it as if it were a mainframe processor.

Minicomputers have been used for business data processing applications since the late 1960s. They are a product of the increased miniaturization of electronic components as well as different approaches used by computer engineers for packaging these components. Because they have been relatively inexpensive, are easy to use, and do not require elaborate power or air conditioning systems, minicomputers have been used for many business and specialized data processing applications.

Microcomputers or personal computers show a more rapid growth curve. During the middle 1970s, hobbyists began building their own microcomputers using the newly available integrated circuit electronic chips. Then, Apple Computer Corporation was formed in the late 1970s to produce the Apple II microcomputer. This computer was viewed as a curious toy by many until a spreadsheet software package, VISICALC, was introduced about a year later. Suddenly, many saw that an Apple II coupled with VISICALC could be used as a serious tool for business decision making. Several years later, in the early 1980s, IBM introduced its Personal Computer and legitimized the minicomputer as a serious business processing tool.

Today, mini and microcomputers are used for many business data processing applications. They may be used as the only computer system for a smaller company or one division of a larger organization. They may also be used for specialized departmental computing even though there may be a larger, mainframe computer capability within the organization. Sometimes, these specialized computers are used for applications such as laboratory research or manufacturing process control. These same machines may be used for some business processing applications in addition to intended specialized purposes.

The definition of a modern mini or microcomputer system often causes difficulties for the auditor because these computers are used for so many business related applications. An internal auditor may be directed by management to review general controls surrounding "all" computer systems in the organization. Clearly, this covers the mainframe computer center and freestanding divisional minicomputer systems. The directive may also cover the organization's departmental micro or personal computers. However, the auditor may wonder if it really covers the microcomputer in the engineering laboratory used for recording test results, or the microcomputer at the end of the distribution line that weighs and routes packages to the correct shipping dock.

While the above examples are truly computer systems, the auditor's review might best emphasize the computer systems used for business data processing purposes. To follow the previous example, the microcomputer at the end of the distribution line probably uses a standard set of software that cannot be modified by the local staff. It was probably purchased from an outside systems vendor and, after initial installation and testing, simply works with no programmer interaction. Such a machine generally has few business or control risk implications.

This chapter explains control objectives and audit procedures when mini or microcomputers are used for business data processing applications. Of course, these same control objectives and procedures may apply when mini or microcomputers are used for process control or other specialized applications. Where appropriate, those applications are discussed.

The auditor should use an attribute-based approach to determine whether to review a given data processing function as a larger or a smaller system. When the auditor has doubts, the controls review should be conducted as if it were a larger system as discussed in Chapter 2. The following should help the auditor to understand the characteristics of the typical mini or microcomputer business data processing system:

1. *Limited data processing staff.* The small business data processing computer system has a very limited staff. A microcomputer used to provide accounting reports for a smaller company may be maintained by a single person. A small business minicomputer system may have a manager, programmer, and one or two operators as the total data processing department. The nature of the equipment, applications processed, and size of the total organization limit data processing staff size. This small staff size should not automatically raise an organizational controls concern for the auditor. The auditor needs to look for compensating controls just as the financial auditor looks for compensating controls when reviewing a smaller accounting department where a classic separation of duties may not be feasible.

2. *Limited programming capability.* The typical modern small business computer system makes extensive use of purchased software packages. The few programmers in the department are responsible for updating purchased packages, maintaining program tables, and writing simple retrieval programs. There will be no specialized systems programming personnel who can make changes to the computer's operating system. If the auditor finds a larger programming staff or extensive in-house development activity, some of the control procedures discussed in Chapter 2 should be considered.

3. *Limited environmental controls.* Larger mainframe computer systems require specialized power, temperature, and other environmental controls. Some even require specialized systems to water-cool the computer's central processors. Small business computer systems tend to be just the opposite. Sometimes, they can be plugged into normal power systems and can operate within a wide range of temperatures. While some small business computer installations may be housed in formal, environmentally

controlled computer rooms, this is not a necessary attribute of a minicomputer system. In addition, a formal computer operations room does not necessarily create a large systems environment.

4. *Limited physical security controls.* Because there often is less need for environmental controls, many mini and microcomputer business data processing systems are installed directly in office areas. The level of auditor concern regarding physical security controls depends upon the type of computer equipment and the applications being processed. The auditor will sometimes recommend that physical security be improved if critical applications are being processed. In many other instances, this lack of physical security controls should not present significant control risks.

5. *Limited telecommunications network.* While many microcomputers are free-standing devices, microcomputers typically support from a few to many user terminals. The typical minicomputer system, however, will have a limited telecommunications network, with most terminals residing in the same building or facility as the computer. In a much larger network, the auditor may want to consider some larger systems control procedures discussed in Chapter 2. Some mini and microcomputer systems may be linked to a mainframe computer as part of a distributed processing network where files are both uploaded and downloaded. Similarly, microcomputers are increasingly linked together in local area networks (LANs) where they may share files or devices such as specialized printers. These features add different levels of control considerations to the mini or microcomputer system under review. These will be discussed in Chapter 4.

The above characteristics certainly do not define a mini or microcomputer system. They simply explain common system attributes. However, they should help the auditor decide upon the control procedures to be used. When in doubt the auditor should consider the computer system to be a larger, more complex version of a common system. However, the emphasis of this chapter is that it is not always necessary to use larger system control procedures when reviewing a smaller, mini or microcomputer business system.

REVIEWING MINI AND MICROCOMPUTER CONTROLS

Both the internal and external auditor will be concerned with controls over mini and microcomputer systems. However, the internal auditor will be more involved with reviewing general controls in small business computer systems than will the typical external auditor because often an extended control risk assessment review of a small computer environment is not warranted. However, both may find it necessary to review mini and microcomputer general controls if the applications are sufficiently complex and material to the organization's overall operations.

Chapter 2 briefly discussed the difference between general, interdependent controls and application controls. These differences are equally applicable for mini and microcomputer systems. First, the auditor must understand the general controls surrounding a mini or microcomputer system. If no significant weakness is encountered in this review, the auditor should then review specific applications (discussed in Chapter 5).

While the introductory comments to this chapter defined mini and microcomputer systems as one generic type of computer system, actual control procedures differ somewhat by the type, size, or usage of the minicomputer or microcomputer. Subsequent portions of this chapter will discuss controls in terms of:

Small business minicomputer system controls

Microcomputer business system controls

Process or non-business system controls

The auditor frequently encounters all three of the above types of small computer systems in a larger organization. Minicomputers provide total data processing support for a smaller business function or unit while microcomputers or personal computers support limited, departmental computing functions. Process or non-business system computers refer to numerous types of small computers used increasingly for manufacturing, distribution, and other operational control applications.

SMALL BUSINESS MINICOMPUTER SYSTEMS CONTROLS

A small business minicomputer system normally provides data processing support for a total organization or a separate operating unit of a larger organization. Such a system has many attributes of a larger, mainframe computer system, including a limited but formal data processing organization, production schedules, and procedures for implementing new applications. However, the smaller organization does not have other specialized functions such as systems programming. Examples of the types of computer hardware that the auditor encounters in such an organization might include an IBM AS/400, a Hewlett Packard HP3000, or any of several smaller DEC or Data General computer systems.

Numerous vendors supply such minicomputer systems in a competitive market where vendors are increasingly offering improved functionality and price performance. The auditor with an overall knowledge of some capabilities of various types of minicomputer systems hardware will be more effective in reviewing minicomputer system controls.

Organization Controls

The auditor often looks for a proper separation of duties in a data processing organization as a first procedure for evaluating general data processing controls. This organizational control often is lacking in a small business minicomputer data processing department. Good data processing control objectives call for a proper separation between those who run the computer applications, those who program the system, and those who operate the system. It is difficult to establish such strict organization controls in a small department.

When auditors first began reviewing general organizational controls in these smaller data processing departments, they often applied large mainframe system control remedies. An auditor looked at a two-person data processing organization and recommended that it be expanded to four to achieve a proper separation of duties! Such recommendations were hard to sell to cost conscious management. Probably the vendor that sold the minicomputer system had convinced management that it could be run with a two-person staff.

In such a small, minicomputer-based data processing department, the data processing manager may also be the principal programmer and operate the equipment when the need arises. Because much of the programming uses simple retrieval languages, operators often have some programming knowledge. The separation of duties that exists in a larger shop simply does not exist in this smaller environment. However, there can be some compensating controls, including:

1. *Purchased software.* Many minicomputer-based systems today operate with purchased software packages where programmers have little or no access to source code, greatly reducing concerns about improper modifications of programs due to a lack of organization controls.

2. *Increased management attention.* Although management personnel in an organization supported by a minicomputer may have little knowledge of data processing techniques, they often give considerable attention to the key computer generated reports. In a small company, it is not unusual for top management to review, for example, an accounts receivable aged trial balance in detail on a regular basis. This type of review reduces some auditor control concerns over improper modification of data.

3. *Separation of input and processing duties.* When minicomputer business systems were first installed, the data processing department typically was responsible for the keypunch-based data entry, computer processing, and output report distribution. In more modern minicomputer systems, users submit data inputs through terminals and receive outputs on remote printers or terminal screens. This distribution of responsibility becomes a compensating organizational control.

While the above points indicate there can be compensating organizational controls in small data processing departments, the auditor must be aware of potential risks and weaknesses in organization controls. Data processing departments continue to exist in which a data processing manager writes all applications, has responsibility for much of the processing, and is the only person in the organization who understands the data processing applications. While a limited staff may be quite acceptable in some circumstances, the organization faces a risk if all data processing knowledge is vested in one person.

There can be other symptoms of control weakness in the smaller data processing organization that typically do not exist in the larger department, including:

"Loyal" employees who do not take vacations or time off

Use of special, non-documented programs known only to the data processing manager

Direct input of special application transactions by data processing personnel, such as inventory adjustments

Control risk should be a major consideration when audit procedures identify potential weaknesses in these smaller business systems. A weakness may exist, but the relative risk may be minimal. For example, auditors often look in larger organizations for documented position descriptions as evidence of good management controls over the data processing department. However, many smaller organizations do not have position descriptions. While consideration might be given to suggesting that such descriptions be drafted for the entire organization, overall control risk may be minimal because of the size of the organization. However, the auditor will not be effective in making such suggestions solely for the data processing function while ignoring the rest of the organization.

Chapter 2 explained how the plan of organization and related management practices are among the strongest control procedures in a larger data processing organization. Often, this is not the case for the smaller organization. The small size of the organization and the informality typically associated with such a small group tend to weaken controls in this area. It is very important, however, that upper management understand the data processing function, its plans, and objectives. As mentioned in the objectives and procedures set out in Figure 3.1, the auditor should interview management at least one level above the data processing function as part of the data processing review.

A very important general organizational control for the smaller data processing organization is adequate documentation of systems and procedures. Management can be vulnerable if systems, programs, and operating procedures are not documented properly. There have been instances where both members of a two-person data processing organization suddenly resigned due to a disagreement or better employment offer. Without adequate documentation, it is difficult for someone else suddenly to take over. This is true even though the organization runs primarily packaged software since there may be many special procedures associated with those packages.

As outlined in Figure 3.1, the auditor should carefully review systems and programming documentation for adequacy. While the auditor often reviews documentation only to verify that it exists, selected portions should be reviewed in greater detail in the smaller organization. This is because the limited personnel in the data processing department may give in to management pressure to create documentation but not take time to provide much content.

Sometimes, the smaller data processing organization is located at an operating unit of a larger organization with central data processing facilities. Even though the smaller data processing organization is entirely freestanding, it may receive central directions as to standards and procedures. In order to audit compliance with these standards adequately, the auditor must have a general understanding of them, including an understanding of how strictly the central or corporate standards are followed by the smaller organization. When a smaller organization is expected to follow such procedures, the auditor may use the objectives and procedures from Chapter 2.

Despite the smaller and more informal nature of a typical minicomputer business system, the auditor should have the same general organizational control objectives as discussed for larger systems in Chapter 2. While some procedures can be the same, the auditor may want to modify others due to the more informal nature of these smaller systems. Figure 3.1 provides a set of control objectives and audit procedures for evaluating general organizational controls in the modern small business computer system.

**Control Objectives and Audit Procedures
for Reviewing Minicomputer General Organizational Controls**

Note: This table of objectives and procedures covers controls in a minicomputer or small business systems computer environment. Refer to Figure 2.4 for a larger computer systems environment.

Objective 3.1.1.   The data processing function should report at a level that allows it to serve all other units in the organization.

> **Procedure 3.1.1.1.**   Survey the types of applications processed by data processing to determine whether the function reports at a proper level to serve its users.
>
> **Procedure 3.1.1.2.**   If the data processing function reports to another operating unit, such as the Controller, determine through inquiry whether it appears to be serving adequately other units of the organization.

Objective 3.1.2.   The data processing function should be organized to allow for a separation of duties to the extent allowed by staff size constraints.

> **Procedure 3.1.2.1.**   Review the plan of organization and identify areas where there is a formal separation of duties as well as areas where limited staff prevents implementation of such a separation.
>
> **Procedure 3.1.2.2.**   Determine that there are published position descriptions, defining duties and responsibilities, for all members of the staff.
>
> **Procedure 3.1.2.3.**   In areas where limited staff causes a potential separation of duties conflict, determine what steps are taken to introduce compensating controls such as:
>
>> Independent management review and approval of various data processing functions.
>>
>> Controls over certain privileged functions, such as requiring that all program library updates are assigned to designated personnel.
>>
>> The assignment of certain functions, such as master terminal information security, to persons outside of the data processing function.

Figure 3.1. Control objectives and audit procedures for reviewing minicomputer general organizational controls

Procedure 3.1.2.4.   Interview persons responsible for the data processing function to determine whether they have an understanding of the need for a separation of duties.

Procedure 3.1.2.5.   Review training policies as well as records of participation in actual training programs to determine that personnel receive adequate technical training.

Objective 3.1.3.   Policies and procedures should exist for the data processing organization which are consistent with those in the overall organization.

Procedure 3.1.3.1.   Briefly review any overall organization policy manuals to determine those impacting the data processing function.

Procedure 3.1.3.2.   Determine if formal policies and procedures exist for the data processing function and whether they reflect overall organization policy.

Procedure 3.1.3.3.   If a formal, published policy and procedures manual does not exist within the data processing function, interview management to determine how these policies and procedures are communicated to the staff.

Objective 3.1.4.   There should exist overall data processing long and short range plans, approved by management, which define the future objectives of the function.

Procedure 3.1.4.1.   Review procedures for developing long and short range data processing plans and for securing upper management approval.

Procedure 3.1.4.2.   Review past long and short range plans from several recent periods and discuss with management reasons for any significant differences.

Procedure 3.1.4.3.   Review procedures for reporting to management variances against plans and determine whether these procedures are followed.

Objective 3.1.5.   There should be an overall data processing contingency or disaster recovery plan which will allow continued operation of data processing services in the event of an unexpected interruption.

Figure 3.1. Control objectives and audit procedures for reviewing mini-computer general organizational controls *(continued)*

Procedure 3.1.5.1. Determine that data processing management has assessed the risks and vulnerabilities of data processing operations and has formulated a disaster recovery plan.

Procedure 3.1.5.2. Review the published disaster recovery plan and determine whether the plan is consistent, given the size of minicomputer operations, with the steps outlined in Chapter 10, Effective Disaster Recovery Planning.

Procedure 3.1.5.3. Determine that a back-up processing site has been identified and review the results of any testing at that site.

Objective 3.1.6. There should be management commitment to both data processing physical and information security covering minicomputer operations.

Procedure 3.1.6.1. Assess the overall understanding of and commitment to data processing security by both data processing and upper management.

Procedure 3.1.6.2. Review physical security controls following the steps outlined in Figure 8.8, *Objectives and Audit Procedures for Small Computer Physical Security Reviews.*

Procedure 3.1.6.3. Review information security controls, consistent with the size of the minicomputer installation, as outlined in Figure 9.6, *Objectives and Audit Procedures for Reviewing the Information Security Function.*

Objective 3.1.7. Data processing should utilize a budgeting and cost accounting system consistent with the accounting systems used throughout the organization.

Procedure 3.1.7.1. Briefly review overall management accounting systems and determine that they are being used, where applicable, within the data processing function.

Procedure 3.1.7.2. Review procedures for both budgeting and tracking hardware, software, personnel, and related indirect costs to determine whether these are consistent with the overall organization.

Figure 3.1. Control objectives and audit procedures for reviewing minicomputer general organizational controls *(continued)*

Objective 3.1.8. Procedures for acquiring new hardware and software should follow overall organization approval and procurement policies and should include formal justifications.

Procedure 3.1.8.1. Review any plans for new hardware and software purchases and determine whether appropriate justifications and procurement approval procedures are being followed.

Procedure 3.1.8.2. Select several new computer hardware or software product additions and determine that proper budgeting and procurement procedures were followed.

Objective 3.1.9. There should be a data processing performance reporting system which reports both processing and development activities to upper management.

Procedure 3.1.9.1. Review performance reports over several recent periods and determine whether they appear to report adequately both planned and actual results regarding data processing systems development and production activities.

Procedure 3.1.9.2. Assess the technical understandability of performance reports and consider suggestions for improvements where appropriate.

Objective 3.1.10. There should be overall management policies to ascertain that data processing internal controls are in place and are working.

Procedure 3.1.10.1. Determine whether policy statements exist defining management responsibilities for ensuring that internal controls are in place and working.

Procedure 3.1.10.2. Review whether corrective actions have been taken on past, data processing recommendations from either internal or external auditors.

Figure 3.1. Control objectives and audit procedures for reviewing mini-computer general organizational controls *(continued)*

Data and Program Access Controls

When unauthorized persons are allowed to access and modify computer data files and programs, general controls are weakened. In the smaller data processing organization, it is more difficult to satisfy control objectives in this area. However, the auditor should consider access to data and programs to be a major general controls objective when reviewing the smaller data processing organization. This is true whether the software used on the system is purchased or is developed in-house.

Controls over access to data can be considered in terms of both specific applications and general controls. Specific application controls are discussed in Chapter 5. However, in smaller computer centers, general controls often have a greater importance than specific application data access controls, because applications operating on the same system typically operate under the same set of general data access controls.

In a modern minicomputer system, data can be improperly accessed and modified primarily in three ways:

Improper data access through user terminals

Unauthorized use of specialized utility programs

Improper data processing requests

Of course, there may be other ways to access data. In older minicomputer systems, decks of data cards could be manually accessed. In some newer systems with telecommunications capabilities, it is possible for outsiders to access files through dial-up telephone lines.

Improper Data Access through User Terminals

Small business minicomputer systems typically do not have the sophisticated information security systems available on many larger, mainframe systems. Rather, these smaller systems have a user log-on/password identification coupled with menu based information security. A systems user typically enters the assigned log-on identification code into the terminal and receives a display of a menu screen with the applications available to the log-on code. Only then can the user access the applications assigned to that menu.

For example, a minicomputer system may have a series of accounting applications that some users are allowed to update and others only to access. Two series of log-on codes should be assigned. The first would display a menu of programs that could update these various accounting applications as well as other menu selections to display the files. Persons limited to accessing data would log on and receive a menu with access-only selections. Persons allowed to update some applications and only access others, and precluded from even seeing others, would receive an appropriately limited menu upon submitting the assigned user identification.

This type of menu based security provides a fairly effective control against improper access attempts. However, these controls tend to break down due to the informality of smaller organizations. Often log-on IDs are not changed on a regular basis, one general menu is given to virtually all employees, or terminals with more privileged IDs are left on for all to use. Because users generally are unaware of potential data sensitivities and vulnerabilities, minimal management attention may be given to such security issues.

To review controls in this area, the auditor should first gain a general understanding of the vendor's data security system installed in the given minicomputer under review. Controls will range from some good password systems with structured menu systems to other more rudimentary sets of procedures. The second step is to gain an understanding of how the security system has been implemented in the system under audit. Third, the auditor should determine how that system is being used. This last step requires the auditor to spend some time reviewing use of the application and its controls in user areas.

A small business minicomputer system typically does not have log-on mechanisms to monitor invalid access attempts. Rather, the auditor should review the overall administrative procedures covering the security system. These include reviewing how often log-on codes are changed, who has access to the system administrator's menu, and the general appreciation of access controls among data processing personnel.

Unauthorized Use of Specialized Utility Programs

Modern minicomputer systems often are equipped with very powerful utility programs which easily can change any application data file. These programs are designed to be used for special problem-solving situations. They generally produce no audit trail report. All too often, however, smaller data processing functions treat these utility programs as substitutes for normal production update programs. They are sometimes used by the data processing manager for special functions, and sometimes by operating personnel for routine functions.

As an example of an inappropriate use of utility programs, consider an organization with an inventory status system on its minicomputer. While the system normally provides proper stockkeeping records, the inventory status can become misstated from time to time due to a variety of reasons. To help users correct these inventory recordkeeping problems, the data processing manager may develop the practice of correcting inventory balances through the use of a utility program with no audit trail.

Utility programs go by a variety of names depending upon the type of computer used. For example, on the IBM System/36 the utility program is called Data File Utility, or DFU. It can be used for a variety of special reports or file modifications. The auditor should gain an understanding of the types of standard utility programs available for the system under review.

Improper Data Processing Requests

The informality of smaller organizations often allows data to be improperly accessed through normal procedures. For example, someone known to the data processing function may request a "special" computer run which results in an improper access of data. In larger, more formal organizations such a request would often require some type of special management permission; smaller, informal organizations often waive such requirements. This type of access often presents a greater control risk than access through use of improper programs.

The auditor should look for controls to prevent such casual data processing requests. The best control is a formal "Request for Data Services" form, to be approved by management. In addition, logs should be maintained listing all production activities as well as the names of the persons making the requests and receiving the reports.

Controls over Access to Programs

Many of the auditor's concerns over improper access to data also apply to program libraries. Small business mini or microcomputers typically do not have such sophisticated software control tools over program libraries as are found in larger, mainframe systems. Many have menu based systems which can limit improper access, but without such a security system, it can be relatively easy for someone with a little knowledge to locate and potentially modify program library files. The control objectives and procedures are similar to controls over access to data as noted in Figure 3.2.

In many smaller data processing organizations, the auditor will find weak controls within the data processing department over program library updates. The one or two data processing personnel who act as programmers in a smaller data processing department typically update program libraries with little concern for documenting changes or obtaining upper management authorization. While some of these changes may be justified in order to respond to user emergency requests, others may not be proper.

The small size of these data processing organizations makes it difficult if not impossible to install separation of duties organization controls over the system program libraries. In addition, it probably will not work to require that upper management formally review and approve all program library updates; they will neither be interested nor have the technical skills to perform such reviews. The best control method might be to install procedures that rely upon manually logging all changes to the production program library, with the logs subjected to ongoing internal auditor reviews. This type of control takes advantage of the fact that minicomputer program compilers maintain a count of program size in bytes and also have the ability to retain some form of date or version number to identify modifications.

The auditor might suggest the following controls for a minicomputer program library:

1. Establish a program-naming convention that includes the date or version number within the program name.
2. Require that the version number, date, program size, and reasons for any program changes be logged in a formal notebook, and require a periodic management review of the logs. In addition, the source code should contain comments explaining the change.
3. Maintain at least one backup of the source listing, whether as a paper listing or on a machine readable file. If at all possible, mark changes in the source code listing so that changes can be traced back to the logged description.
4. Rotate a copy of the program library file to an off-site location at least once per week.
5. Strengthen access controls so that non-programming personnel cannot access the programming library files.
6. Perform an internal audit review of the library change log on a periodic basis. That review should match logged program versions, dates, and sizes with data reported on the program library file.

**Control Objectives and Audit Procedures
for Reviewing Minicomputer Data and Program Access Controls**

Objective 3.2.1.   Responsibility for information and physical security in a minicomputer environment should be assigned to one individual or function to administer the function independently.

> **Procedure 3.2.1.1.**   Through discussions with management, determine who has responsibility for information and physical security and assess whether that individual or function has the ability to act independently.

> **Procedure 3.2.1.2.**   Observe physical security arrangements for minicomputer operations and determine whether controls are appropriate given the size and purposes of the installation.

> **Procedure 3.2.1.3.**   Interview persons responsible for information security and determine that the function has proper authority to install, monitor, and enforce information security rules and procedures.

> **Procedure 3.2.1.4.**   Review overall organization policies to determine whether there is guidance covering the importance of data security as well as actions to be taken in the event of security violations.

Objective 3.2.2.   Appropriate software tools should be installed to assure that only authorized persons can access or update critical files or data.

> **Procedure 3.2.2.1.**   Review the tools available for minicomputer system security and determine that appropriate options have been installed to establish security levels and monitor improper access attempts.

> **Procedure 3.2.2.2.**   If a minicomputer menu based security system is used, evaluate whether the menus are structured to provide access only on a "need to know" basis.

> **Procedure 3.2.2.3.**   Determine that the information security software installed covers all major application areas including user retrieval languages and on-line processors.

Figure 3.2. Control objectives and audit procedures for reviewing minicomputer data and program access controls

Objective 3.2.3.   Improper software attempts should be monitored for subsequent action, and controls should exist to limit such access attempts.

Procedure 3.2.3.1.   Select a sample of improper access attempts from system reports and follow up on the actions taken by data processing personnel for these reported violations.

Procedure 3.2.3.2.   Review procedures for assigning terminal access codes, menu screens, and personal passwords to determine whether they are changed on a periodic basis.

Procedure 3.2.3.3.   Select several individuals who have either terminated employment or changed job responsibilities and determine that computer system access rights have been changed.

Objective 3.2.4.   Controls should exist over input documents as well as over the processing of sensitive applications.

Procedure 3.2.4.1.   Review procedures for determining that all batch input documents are received as scheduled and only by appropriate functions, including:

The use of published schedules for submission of input
Input logs to record the submission of input data
The use of batch header tickets with authorizing initials or signatures

Procedure 3.2.4.2.   Select several key or sensitive applications and consider the types of controls that have been installed over processing, including:

Appropriate marking and control logs for receipt of input documents
Controlled distribution of system output reports
Adequate disposal of all carbon sheets, waste paper, and other materials that may contain sensitive data

Procedure 3.2.4.3.   Review output report distribution procedures to determine that only authorized persons receive output reports.

Objective 3.2.5.   Controls should exist to limit access to the computer system through either on site use or telecommunication transmissions.

Figure 3.2. Control objectives and audit procedures for reviewing mini-computer data and program access controls *(continued)*

Procedure 3.2.5.1. Observe physical security arrangements as well as traffic patterns near the minicomputer system and assess the adequacy of physical security controls given the nature of the system.

Procedure 3.2.5.2. Interview appropriate personnel to understand and document the telecommunications network including any dial-up lines.

Procedure 3.2.5.3. Determine the adequacy of controls over any dial-up telephone lines, including the use of call-back devices to control improper incoming calls.

Procedure 3.2.5.4. If any dial-up lines are used by maintenance or software vendors, determine that access protocols are changed on a regular basis.

Objective 3.2.6. Controls should exist over tape files, diskettes, and other media to prevent unauthorized use or removal from data centers.

Procedure 3.2.6.1. Review procedures for storing and controlling any tape files, diskettes, or other removable media.

Procedure 3.2.6.2. Determine that labels, volume and serial numbers, and other identifiers consistent with the minicomputer operating system are used for all files.

Objective 3.2.7. Separate program libraries should exist for test and production purposes, and access to production libraries should be limited to authorized persons.

Procedure 3.2.7.1. Discuss program library procedures with responsible data processing personnel to determine that controls are adequate.

Procedure 3.2.7.2. Develop an understanding of the software or operating system package being used to control program libraries and assess its adequacy.

Objective 3.2.8. Changes to production program libraries should be approved by authorized persons.

Procedure 3.2.8.1. Review procedures for updating programs to the production library and determine that there is a proper level of review and approval.

Figure 3.2. Control objectives and audit procedures for reviewing mini-computer data and program access controls *(continued)*

Procedure 3.2.8.2. Review a sample of run documentation or the operating systems job control language to determine that production applications are not being run from program test libraries.

Procedure 3.2.8.3. Review production and test library listings and compare them to identify any possible production applications being run from test libraries.

Procedure 3.2.8.4. Review program maintenance records to identify newly modified programs and verify that updates to production libraries have been properly authorized.

Objective 3.2.9. Emergency procedures should exist for correcting production application problems during non-prime shifts and for subsequently updating production libraries with the changes.

Procedure 3.2.9.1. Review procedures allowing for emergency maintenance fixes of production processes where test library modules may be used temporarily.

Procedure 3.2.9.2. From production records, select several applications where nighttime emergency maintenance was performed and determine that the applications again are running on production libraries.

Procedure 3.2.9.3. Inquire into the use of SUPER-ZAP type programs which allow changes to object modules without corresponding source code module changes.

Objective 3.2.10. Key files and program should be backed up and stored in an off-site location on a periodic basis.

Procedure 3.2.10.1. Review procedures for backing up production files, program libraries, and operating system products and assess the overall adequacy.

Procedure 3.2.10.2. Determine that one or more generations of files are stored in an off-site location.

Procedure 3.2.10.3. Visit the off-site file storage location and, using production documentation, verify that selected key files are stored in the location.

Figure 3.2. Control objectives and audit procedures for reviewing mini-computer data and program access controls *(continued)*

The above steps will not provide complete assurance that all program changes are authorized. However, if the auditor periodically reviews logged changes and questions any discrepancies, programmers will take more care to document and log any production program changes.

Objectives and Audit Procedures for Data and Program Access Controls

The auditor should devote considerable attention to access controls over data files and programs when reviewing small business minicomputer systems controls. Figure 3.2 contains control objectives and audit procedures for this portion of the review. If substantial control weaknesses are found here, it will be difficult to reduce control risk assessments for any specific applications processed on the system. If anyone can easily modify a production program, it is difficult to place any level of reliance on the controls built into the application.

A strong compensating control in many small business computer environments is the use of purchased software, which cannot be changed easily. Applications are delivered to user data processing functions in object or runtime versions, where changes that can alter controls but still allow applications to operate are difficult to make.

Systems Development Controls

Much of this chapter has discussed control issues presented by the small size and informality of organizations using business system minicomputers. In many instances, the auditor must look for compensating controls when it is not cost-effective to suggest installation of controls normally found in larger computer systems. In the area of controls over the systems development process, however, controls similar to those in a larger system should be in place, with a somewhat scaled-down set of procedures.

Chapter 2 describes the Systems Development Methodology (SDM) procedures which should be installed on a larger computer system. The SDM creates a control environment through which data processing applications are requested, developed, tested, and implemented. A well-functioning SDM provides an effective control over the entire process of implementing new applications, whether developed in house or from purchased software. The auditor's reliance on this SDM is discussed in Chapter 7.

The auditor often finds that a formal SDM does not exist in a small business minicomputer data processing department. Systems or program modification requests are often initiated through casual conversation with users and are not documented. In many instances, the data processing function sets its own priorities for new or revised applications. This informal process often creates systems, either developed in house or purchased, which do not necessarily meet user requirements, may have weak controls due to a lack of testing, and may not follow overall organization priorities.

The auditor should gain an understanding of the procedures used for developing new or revised systems within the organization. Figure 3.3 provides objectives and methods for reviewing these procedures. As discussed in Chapter 2 for larger systems, there is no "correct" SDM. However, the methodology in use should provide adequate systems development procedures, and, more importantly, some form of SDM should be followed.

There is no correct SDM for a small business computer system, and there is also no typical one. Minicomputer systems development processes range from attempts to emulate SDMs found in larger computer systems to no SDM at all. This is often a good area for the auditor to make both control and efficiency recommendations.

Small business minicomputer departments often attempt to implement formal SDMs when required by a parent corporation. The parent may have a formal SDM within its mainframe environment and require that all subsidiary data processing departments follow it as well. If the SDM installed at the parent organization is one of the older, paper-intensive sets of procedures, it can bog down the small minicomputer department.

While the auditor may want to report lack of compliance with the central or parent SDM, the auditor should also evaluate the SDM for its efficiency and practicality. For example, some SDMs offered by outside vendors "require" numerous unique forms to be completed before a new information system can be designed. Many forms are difficult for the smaller data processing organization to use. When encountering such an SDM at a small data processing organization, the auditor might offer some efficiency recommendations to reduce the number and complexity of forms to a workable level. Chapter 7 provides guidance in using and evaluating these SDMs.

While the auditor sometimes finds one of these unwieldy SDMs, it is more common that the auditor finds little or no formal SDM in the small business minicomputer data processing organization. This is an area where an auditor can make effective recommendations for better controlling and improving the systems development process. These recommendations might include:

1. *Install a system work request form.* This document initiates requests for new systems or modifications to existing applications. It should contain space to document needs, appropriate levels of management approval, and data processing disposition of the request. An example is

shown in Figure 3.4. The strict use of this form will eliminate casual requests and create a better-controlled development environment.

2. *Establish a user steering committee.* Many smaller, less sophisticated data processing organizations still report to the Controller, who sets all priorities for new projects. When this happens, user groups other than accounting often are unsatisfied. A good audit recommendation may be to establish a steering committee composed of key management members to set overall priorities for new systems efforts.

3. *Follow design, programming, and documentation standards.* While there is no single correct way for a small data processing organization to create new systems, the auditor should recommend a consistent approach for new systems design, programming standards, and documenting systems. When applications consist of purchased software packages, there should be internal quality control procedures to ensure that those packages satisfy requirements and come with adequate documentation.

4. *Recommend internal audit pre-implementation reviews.* Chapter 7 discusses the process of reviewing new systems under development. Although these auditor reviews normally are performed for major, large applications, they may be appropriate for reviewing controls in some major minicomputer applications. The auditor might suggest during a general controls review that communication links be established so that the audit function is informed when such new systems are initiated.

5. *Establish user testing and acceptance procedures.* Just as it is important to have formal procedures for initiating, developing, or acquiring new applications, there should be SDM procedures to ensure that users test new applications and determine that they meet requirements.

Each of the above SDM procedures can easily be installed in a small business data processing environment. Together, they will form an effective SDM for the smaller, more informal organization. The auditor then follows this SDM when reviewing purchased package implementations or new systems under development, as explained in Chapter 7.

Operations Controls

Chapter 2 indicated that many traditional data processing operations control concerns are not always as significant in modern, larger data centers. For example, auditors once recommended that supervisory personnel review console logs as a computer operations control. In the modern, larger data process-

Control Objectives and Audit Procedures
for Reviewing Minicomputer Systems Development Controls

Objective 3.3.1. If the data processing organization does any in-house systems development work, there should be some Systems Development Methodology (SDM) procedure which defines the steps necessary to initiate, develop, and implement new data processing applications.

Procedure 3.3.1.1. Interview key data processing personnel to gain an understanding of the SDM used by the organization and assess the commitment to that SDM.

Procedure 3.3.1.2. Review documentation covering the SDM and determine that it adequately describes procedures for initiating and developing new applications.

Procedure 3.3.1.3. Select a new system under development and determine whether the organization's SDM procedure is being followed.

Objective 3.3.2. If the organization purchases virtually all of its software applications, there should be a limited set of SDM procedures to help evaluate, select, customize, and implement the software.

Procedure 3.3.2.1. Interview key data processing personnel to gain an understanding of the procedures used to evaluate, select, customize, and implement purchased software packages; determine if these procedures are sufficient to be the basis of an SDM.

Procedure 3.3.2.2. Select a newly purchased software package and determine whether established procedures were followed.

Objective 3.3.3. There should be a data processing steering committee, comprised of members of user management, to set priorities for the development or purchase of new software applications.

Procedure 3.3.3.1. Review the membership of the steering committee and determine that all major user areas are represented.

Figure 3.3. Control objectives and audit procedures for reviewing minicomputer systems development controls

Procedure 3.3.3.2. Determine the frequency of steering committee meetings and assess whether it is adequate for the level of new systems projects.

Procedure 3.3.3.3. Review the project priorities established by the steering committee to determine whether they are consistent with the data processing long range plan.

Procedure 3.3.3.4. Review the project status reports or other status data submitted by data processing to determine whether they are adequate.

Objective 3.3.4. Internal Audit should be aware of steering committee activities so they can set their own priorities for any pre-implementation reviews.

Procedure 3.3.4.1. Assess Internal Audit's awareness of steering committee activities by comparing steering committee plans and decisions with Internal Audit planned system review activities.

Procedure 3.3.4.2. Determine whether Internal Audit needs to increase participation in steering committee activities to better understand new systems priorities.

Objective 3.3.5. The first step of the organization's SDM should include a project initiation phase where general objectives, make versus buy options, and expected benefits of a new systems project are defined.

Procedure 3.3.5.1. Review project initiation SDM documentation for several new systems and determine that:

> Users have prepared project requests defining requirements
> Users have approved initial project objectives
> Any make versus buy decision was documented adequately
> Preliminary plans and budgets were developed

Procedure 3.3.5.2. Interview several users responsible for preparing new project requests and determine their understanding of SDM initiation procedures.

Objective 3.3.6. If the organization develops its own systems, the

Figure 3.3. Control objectives and audit procedures for reviewing mini-computer systems development controls *(continued)*

SDM should give adequate attention to input, output, and processing controls as well as application security considerations.

Procedure 3.3.6.1. Review design phase documentation for several applications under development or recently implemented to determine that adequate SDM procedures are being followed.

Procedure 3.3.6.2. Determine that the data processing organization has developed general programming standards or guidelines and assess whether these standards are being followed.

Procedure 3.3.6.3. Interview members of the systems and programming staff to assess their commitment to new systems controls and to systems security.

Procedure 3.3.6.4. For the applications selected in Procedure 3.3.6.1., determine that data processing is giving adequate attention to the review and approval of design phase documentation.

Objective 3.3.7. The process for selecting any software package should consider alternative product offerings and should include detailed evaluations of potential software products.

Procedure 3.3.7.1. Review procedures for identifying and selecting software packages including:

Literature searches to identify potential software products
Reference checks with other users of the software to determine satisfaction
Technical reviews of software products to determine that requirements are met
Economic analyses of software products to determine that cost and budget objectives have been met

Procedure 3.3.7.2. Review procedures for evaluating and testing control and security features for any new software products under consideration.

Procedure 3.3.7.3. Select a representative new software package installed on the system and determine that proper selection and evaluation procedures were followed.

Figure 3.3. Control objectives and audit procedures for reviewing minicomputer systems development controls *(continued)*

Objective 3.3.8. New applications projects should be controlled through a project management system which reports planned milestones, actual accomplishments, and estimated times to complete for the various phases of the project.

Procedure 3.3.8.1. Review procedures used to manage systems projects and determine that they include estimates of planned time and resources as well as actual results.

Procedure 3.3.8.2. Select several systems projects, either in-house developed or purchased software packages, and determine that established project management procedures are being followed.

Objective 3.3.9. Prior to implementation, in-house developed or purchased packages should receive comprehensive testing with active participation by key users of the new application.

Procedure 3.3.9.1. Determine that formal test plans or other procedures have been established for new applications testing.

Procedure 3.3.9.2. Select several recently implemented applications to determine if formal test plans were developed and followed.

Procedure 3.3.9.3. Assess the involvement of users in new applications testing.

Objective 3.3.10. Significant new applications should receive a post-implementation review by either members of data processing or Internal Audit to determine if initial objectives have been achieved.

Procedure 3.3.10.1. Perform a limited post-implementation review of one or two recently implemented applications to determine whether original objectives were met and whether controls are adequate.

Procedure 3.3.10.2. Determine that the data processing function gives adequate priority to correcting newly implemented applications that initially do not meet original design objectives.

Figure 3.3. Control objectives and audit procedures for reviewing mini-computer systems development controls *(continued)*

REQUEST FOR INFORMATION SYSTEMS SERVICE

DATE	
LOG NUMBER	

REQUESTOR'S NAME	PHONE	DATE PREPARED	AUTHORIZED SIGNATURE(S)
USER CONTACT IF DIFFERENT FROM REQUESTOR			ACCOUNT
DEPARTMENT			

TYPE OF REQUEST ☐ NEW SYSTEM DEVELOPMENT   ☐ SYSTEM MODIFICATION   ☐ SPECIAL PROJECT

ATTACHMENTS ☐ YES ☐ NO	REQUESTED COMPLETION DATE

DESCRIPTION OF SERVICES REQUESTED

THIS SECTION TO BE COMPLETED BY I.S. INITIAL ESTIMATE/SCHEDULE INFORMATION

REQUEST TO BE HANDLED BY: (SPECIFY FUNCTIONAL AREA)	PROJECT ASSIGNED TO/DATE	DATE ACKNOWLEDGEMENT SENT TO USER	PROJECT/TASK NUMBER ASSIGNED
USER ACCOUNT/SYSTEM	ESTIMATED HOURS	ESTIMATED START DATE	
BILLING INFORMATION		ESTIMATED COMPLETION	

PROJECT COMPLETION INFORMATION

		DATE	COMMENTS
	PROJECT LEADER		
APPROVAL ACCEPTANCE SIGNATURE	SUPERVISOR		
	REQUESTOR		

Figure 3.4. Systems work request form

ing center, this is not typically an appropriate control recommendation because the log is maintained automatically. However, such control concerns may still be applicable in the smaller computer center.

Computer operators can present a significant control risk in the smaller operation. Operators often bypass system controls, such as a file-label check, or insert incorrect run parameters through console log entries. Smaller computer systems often do not have software and hardware tools to monitor operator activities closely. For example, smaller systems usually do not have strong program library software maintenance tools.

The auditor should have operational control objectives for smaller business systems similar to those discussed in Chapter 2 for larger systems. However, the auditor's procedures for these smaller systems will be somewhat different. Because of a lack of hardware and software control tools, the auditor should spend more time reviewing and understanding the computer operations procedural controls. Figure 3.5 outlines objectives and audit procedures for reviewing operational controls in the smaller minicomputer data processing system.

Figure 3.5 includes control objectives and procedures for both systems programming and technical support even though these were considered separately for larger systems, because the smaller computer system typically does not have the specialized support staff that would be found in the larger center. The auditor should expect to find little if any modification of the operating system or other activities more common at a larger computer system.

Much of the smaller computer hardware and operating systems software is designed so that a traditional systems programmer is not necessary. For example, the IBM AS/400 operating system is designed to "tune" itself more effectively than could be accomplished by many systems programmers. If the auditor finds a systems programmer on the staff of a smaller system, the auditor should refer to control procedures discussed in Chapter 2. If the small business computer system is sufficiently large, however, and there is no such function, the auditor may recommend such a function to improve system efficiency and performance.

A prime concern in the smaller computer center is often simply whether the data processing organization is keeping the operating system and related software products updated. Computer vendors regularly supply upgrades to the operating system, and many vendors of software packages do the same for their products. If such upgrades are not installed on a fairly regular basis, the system can quickly become out of date and unreliable.

Control Objectives and Audit Procedures
for Reviewing Minicomputer Systems Operations Controls

Note: These objectives and procedures are designed for general controls reviews of small business minicomputer systems where there may be a limited staff and where the equipment does not have strict environmental requirements. For operations reviews of larger, "mainframe" computer centers, refer to Figure 2.9.

Objective 3.5.1. Computer equipment should be located in a limited access and environmentally controlled facility.

Procedure 3.5.1.1. Discuss computer systems physical and environmental control procedures with data processing management to determine equipment requirements, current policies, and future plans.

Procedure 3.5.1.2. Tour computer room facilities and observe physical security strengths and weaknesses including:

The existence of locking mechanisms, if appropriate, to limit computer room access to authorized individuals

The placement of computer equipment to limit access by unauthorized individuals

The use of separate power transformers and air conditioning units, if appropriate for the type of equipment, to provide environmental protection

The general location of the computer room facilities within the overall building such that it is outside of heavy traffic patterns

The existence of fire detection equipment including zone controlled heat and smoke detectors

The existence of an overall building fire protection system as well as appropriate local fire extinguishers

Procedure 3.5.1.3. Considering the size and requirements of the computer system, review computer room temperature, humidity, and other environmental controls and assess their adequacy.

Procedure 3.5.1.4. Briefly review maintenance records to ascertain that physical and environmental controls are regularly inspected and maintained.

Figure 3.5. Control objectives and audit procedures for reviewing minicomputer systems operations controls

Objective 3.5.2. Production processing should be scheduled to promote efficient use of computer equipment consistent with the requirements of systems users.

Procedure 3.5.2.1. Through interviews with operations management, develop an overall understanding of computer processing schedules including on-line and batch production work as well as any end user computing.

Procedure 3.5.2.2. Review procedures for scheduling regular production jobs including the use of manual logs or any automated job scheduling tools.

Procedure 3.5.2.3. Match a limited number of scheduled production jobs against actual completion times to determine if production schedules are being followed.

Procedure 3.5.2.4. Determine whether the operating system allows for job classes or priority codes, and if so, determine that they are used to give proper priority to critical production jobs.

Procedure 3.5.2.5. Evaluate procedures for running "rush" or rerun jobs to determine whether they are consistent with overall production schedule requirements.

Objective 3.5.3. Operations instructions should exist to allow operators to correctly process normal production as well as to respond to errors.

Procedure 3.5.3.1. Review documentation standards for production applications to determine that they provide operators with information regarding:

Normal operations procedures including instructions for special forms, tape files, and report disposition

Application restart and recovery procedures

Responsible programmer/analyst and user contract names for resolving production application problems

Procedure 3.5.3.2. Review procedures for turning over new applications or revisions to computer operations to determine that established operations standards are being followed.

Figure 3.5. Control objectives and audit procedures for reviewing mini-computer systems operations controls *(continued)*

Procedure 3.5.3.3. Select a sample of production applications and determine that the operating documentation is current by comparing run listings with published documentation.

Objective 3.5.4. Computer operators should not be allowed to change programs independently or initiate production jobs without authorization.

Procedure 3.5.4.1. If there is a separate computer operations staff, interview operations personnel to determine that they do not have programming responsibilities.

Procedure 3.5.4.2. Determine that data processing policies exist which prohibit computer operations personnel from performing programming tasks or running unauthorized jobs.

Procedure 3.5.4.3. Given the class of small business computer system used, determine that production source libraries can not be accessed without detection by operations personnel.

Procedure 3.5.4.4. Determine that there are data processing procedures for reviewing periodically the contents of console log files or otherwise monitoring improper operator use of computer equipment.

Procedure 3.5.4.5. If the data processing function operates with a very limited staff, determine that critical actions, such as program changes, are reviewed and monitored by someone outside of data processing.

Objective 3.5.5. Procedures should exist to allow for emergency program modifications when error conditions prevent critical production applications from being processed.

Procedure 3.5.5.1. Review and assess the adequacy of procedures for changing production programs or procedure libraries when emergency situations require special handling.

Procedure 3.5.5.2. Determine that all emergency processing activities are documented properly and are subject to subsequent management review.

Procedure 3.5.5.3. Select several documented emergency program fixes and determine that the necessary changes were subsequently added to production processing libraries.

Figure 3.5. Control objectives and audit procedures for reviewing mini-computer systems operations controls *(continued)*

Objective 3.5.6.   Logs or records of computer systems activity should exist to monitor both regular and abnormal computer operations.

Procedure 3.5.6.1.   Determine that a log file procedure exists to record all computer systems activity including:

Jobs and programs run in normal production as well as reruns

Abnormal terminations of any jobs or programs

Operator commands and other data entered through system consoles

Computer initiated console queries and operator responses

Procedure 3.5.6.2.   Determine that computer activity logs are reviewed periodically, that exception situations are investigated, and that the results of those investigations are documented.

Objective 3.5.7.   When batch jobs are run, procedures should exist to determine that only authorized input data is submitted at scheduled times.

Procedure 3.5.7.1.   Determine that schedules exist for the submission of critical input data batches and that procedures exist to follow-up on missing data.

Procedure 3.5.7.2.   Determine that all input data batches are initialed or otherwise identified to verify that they have been submitted by authorized individuals.

Procedure 3.5.7.3.   Review a limited sample of batch applications to determine that batch balancing techniques with user supplied control totals are used.

Objective 3.5.8.   Controls should exist to determine that computer system outputs are distributed only to authorized users.

Procedure 3.5.8.1.   Determine whether users or data processing personnel are responsible for reviewing output controls and assess whether those control reviews are being performed.

Procedure 3.5.8.2.   Review procedures for monitoring system

Figure 3.5. Control objectives and audit procedures for reviewing minicomputer systems operations controls *(continued)*

output reports to determine that they appear complete, there are no printer errors, and correct forms are used.

Procedure 3.5.8.3.   If output reports are distributed to users, determine that they are distributed only to authorized persons.

Procedure 3.5.8.4.   If users pick up output reports, determine that reports are controlled so that only authorized users can pick up their own reports.

Procedure 3.5.8.5.   Determine that sensitive or confidential reports are placed in sealed envelopes prior to distribution.

Objective 3.5.9.   Controls should exist to ensure that only correct data files are used for production processing.

Procedure 3.5.9.1.   Determine that all files are internally labeled and that production applications process only current data files.

Procedure 3.5.9.2.   Review procedures for inventorying and controlling computer tape and disc files.

Procedure 3.5.9.3.   Determine that procedures exist for monitoring tape usage and for recertifying and cleaning tapes when appropriate.

Objective 3.5.10.   There should be a job accounting system to monitor the use of computer resources and to charge or report on this usage to benefiting users.

Procedure 3.5.10.1.   Review and document the computer job accounting system to determine how charges are collected, allocated, and reported.

Procedure 3.5.10.2.   Interview several selected systems users to determine whether they have an understanding of the job accounting system.

Figure 3.5. Control objectives and audit procedures for reviewing minicomputer systems operations controls *(continued)*

Small Business Minicomputer Controls

Prior sections have emphasized the informality found in a small business minicomputer data processing system. This informality often makes the auditor's task more difficult than when reviewing controls in the larger data center, because management might not understand the need for controls. The auditor should attempt to exercise a bit of creativity when making controls improvement recommendations for the smaller data center. The auditor should consider analogies to manual control concepts and determine if substitutes are in place. In addition, compensating control procedures should be suggested when more cost effective than traditional control procedures.

MICROCOMPUTER BUSINESS SYSTEMS CONTROLS

Auditors frequently encounter personal or microcomputer systems for significant business data processing functions. These systems may be used to support departmental computing functions or sometimes even provide the entire data processing capability for smaller business units. Improvements in the integrated circuit technology used to build microcomputers, as well as improvements in packaged software, have caused microcomputers to be a common feature in the typical business environment.

Controls over microcomputers are discussed in several other chapters. Chapter 11 discusses the use of microcomputers to support special reporting needs such as word processing or spreadsheets where there is little or no interaction between the microcomputer and any mainframe system. Personnel from one or more departments may use this microcomputer for various applications, while all "official" data files might be maintained elsewhere on a mainframe computer.

Chapter 4 discusses distributed processing where microcomputers are linked to central mainframe machines for file uploading and downloading or are linked to other microcomputers for file sharing. If there are significant files on the microcomputer, they may be refreshed by the mainframe. Some applications run on the microcomputer include various spreadsheet or other analysis programs.

This section discusses situations where microcomputers are used for separate business data processing applications, including general ledger, accounts receivable, payroll, and other normal accounting functions. Some very powerful vendor-supplied software packages are available to perform these functions, and key accounting records are maintained on microcomputer files.

Machines used in this manner may be located at separate, smaller organizations or at autonomous units of larger organizations.

When the auditor is asked to review the controls surrounding such a microcomputer business system, many objectives and procedures previously outlined in this chapter for minicomputer systems are applicable. There are, however, unique audit procedures associated with microcomputer systems because of their special characteristics. Also, there are areas where minicomputer control objectives and audit procedures are not applicable. For example, minicomputer SDM procedures often are not needed for microcomputer systems because software usually is purchased rather than designed and programmed in house. When, for example, a sophisticated accounting package can be purchased for $50, it makes little sense to design and program a unique system.

The following sections outline areas where audit objectives and procedures for a microcomputer system for business data processing will be different from those used by the auditor for a minicomputer business system.

Unique Characteristics of the Microcomputer Business System

Many microcomputer systems installed today have far greater capabilities than typical minicomputer systems of just a few years ago. When used to support a business data processing function, however, they have unique characteristics when compared to the typical minicomputer business data processing system. These include:

1. *Simpler configurations.* Minicomputers usually are installed with multiple user terminals or multiple peripheral devices, such as tape drives or printers. A microcomputer system usually has a much simpler configuration. Typically there is a single processor with several built-in disc drives and a single printer. Although capabilities are increasing, the typical microcomputer system will have limited multiprocessing capabilities.

2. *Considerable operator interaction.* In a minicomputer system, the operator often starts a sequence of jobs and then walks away from the machine. Microcomputer systems usually require much more operator interaction, with the operator responding to program prompts, inserting diskettes to save files, or reviewing controls printed on reports before proceeding to the next program step.

3. *Transportable computer media.* Minicomputers typically use magnetic tapes or eight-inch floppy discs as external file storage media. Micro-

computer systems use much smaller discs or cassette tapes. Such microcomputer files can be easily slipped into a pocket, creating a special security concern.

4. *Considerable user system knowledge*. Through the hands-on use of similar machines in school or home, many personnel have a basic knowledge of microcomputer operations. They can boot an operating system, search and copy files, or process programs, in contrast to a typical minicomputer system where most users have little knowledge of the system beyond limited operating system functions and menu-based applications.

The above characteristics provide the auditor with some unique control concerns when reviewing a microcomputer business system. When the microcomputer is used as a business data processing device, it has control characteristics quite similar to a minicomputer business system but with significant differences. Also, the microcomputer business system should not be considered the same as the user-friendly departmental machine found throughout business organizations.

Organization Controls

If a microcomputer is the only device used for an organization's data processing requirements, the auditor will generally not find a separate data processing department. Rather, one person, often with other job responsibilities as well, will be responsible for inputting data, running microcomputer applications, and otherwise maintaining the system. This person probably will be attached to a user department such as accounting.

Sometimes other persons are also involved with running the systems. Clerical personnel may input certain transactions to selected screen formats, while outside specialists may perform program or system upgrade functions. Those specialists may include outside consultants or even personnel from the dealer that sold the system. Often, the auditor will find that the system is a "mystery" to most persons in the organization.

The auditor should determine who has access to the microcomputer business system and how those persons use it. Particular attention should be given to any end users, such as hobbyists, who have the necessary skills to access protected files or programs. Their use of the business system machine should be restricted. If there is considerable spreadsheet processing or other related activity on the machine, the auditor should recommend that a separate machine be installed for these purposes.

An organization control important for this type of processing is systems documentation. This should include the computer operations procedures for running systems and backing up files. In addition, the auditor should determine that there is documentation for the "AUTOEXEC.BAT" batch programs which often run sets of application programs. While the documentation for microcomputer applications need not be extensive, it should be sufficient to allow someone with little knowledge of the specific applications to run them.

When reviewing a microcomputer system's general organization controls, the auditor should consider efficiency issues. For example, some divisions of a larger organization install an independent, microcomputer-based data processing system even though similar systems are available through a central computer facility. If the microcomputer does not appear to be handling jobs efficiently and other processing resources exist, the auditor might raise the issue in an audit report to management.

Figure 3.6 discusses auditor control objectives and audit procedures when reviewing general organizational controls for a microcomputer business system. This list is not all-inclusive, however, and the auditor should also refer to Figure 3.1 on minicomputer systems.

Data and Program Access Controls

Microcomputer operating systems traditionally have been designed to be relatively "user friendly," with numerous tools built into the system to allow the user to access such things as the contents of files. While the novice user may not be aware of these tools, it is not difficult to learn their usage. Thus, it is not difficult to access or modify data in an unprotected microcomputer application.

Fortunately, password-based controls are increasingly available on many microcomputer applications. In addition, there are numerous "DOS Shell" type security packages now available. These products literally build a program "shell" or security program around the operating system and all microcomputer files and require the user to enter a password to penetrate that shell. Such packages are not expensive, and the auditor should encourage their use. The password-based shell programs also will help prevent individuals within the organization who have similar computers at home from copying application programs or from introducing unauthorized software for the system.

The best way to limit access to programs and data in a microcomputer business system is simply to limit the use of the machine. This can be done through the use of locks during off hours and management controls during regular operating hours.

Control Objectives and Audit Procedures
for Reviewing Microcomputer General Controls

Note: These objectives and procedures for microcomputer general controls are designed for reviews of free-standing microcomputer systems used for small business data processing systems. For a micro - computer that is part of a distributed processing network, with files refreshed from other processors, general controls are outlined in Figure 4.19, Objectives and Audit Procedures for Distributed Processing General Controls. *For a microcomputer that is used primarily for ad hoc, end user applications, general controls are outlined in Figure 11.2,* Objectives and Audit Procedures for End User Microcomputer General Controls.

Objective 3.6.1. One individual or function should be responsible for the overall operation of the microcomputer system.

Procedure 3.6.1.1. From interviews or discussions with management, identify the individual or function responsible for microcomputer operations.

Procedure 3.6.1.2. If one management individual is responsible for most microcomputer functions, assess whether that individual has an appropriate level of influence on processing decisions with adequate outside review.

Procedure 3.6.1.3. Determine that the function responsible for microcomputer operations understands computer system integrity and control issues.

Procedure 3.6.1.4. Review policies and procedures developed for microcomputer operations and determine that they include:

Policies prohibiting the use of illegal copies of software
Policies against employee copying of system software
Procedures to periodically back up key files and remove them to secure locations

Objective 3.6.2. If multiple, free-standing microcomputer systems with similar functions exist in the organization, the hardware and software configuration of each should be consistent.

Figure 3.6. Control objectives and audit procedures for reviewing microcomputer general controls

Procedure 3.6.2.1. Determine whether hardware standards or software product standards exist for multiple, free-standing microcomputer systems within the organization.

Procedure 3.6.2.2. Determine whether the microcomputer system under review is in compliance with overall organization standards.

Procedure 3.6.2.3. If the microcomputer under review is not configured according to organization standards, determine whether there are mitigating reasons to justify the exception.

Objective 3.6.3. Persons operating the microcomputer system should have adequate training in the use of the system and key applications.

Procedure 3.6.3.1. Review procedures for training new microcomputer operators and for providing refresher training when required.

Procedure 3.6.3.2. Review records to determine whether key operators have received training on key applications.

Procedure 3.6.3.3. Interview selected microcomputer system operators to assess their understanding of the hardware and software.

Objective 3.6.4. Documentation for microcomputer applications and operating system processors should be available for microcomputer system operators.

Procedure 3.6.4.1. Determine that vendor supplied or locally developed documentation is available for microcomputer operations.

Procedure 3.6.4.2. Determine that technical documentation, which would allow one to write programs or change files, is not available for microcomputer operators.

Procedure 3.6.4.3. Briefly review application documentation and determine that it includes:

Explanations of all error codes or conditions

Procedures for normal processing and such less used situations as the year end close

Figure 3.6. Control objectives and audit procedures for reviewing microcomputer general controls *(continued)*

Detailed descriptions of application control points and procedures for verifying those controls

Objective 3.6.5. The microcomputer system should be located in an environmentally friendly location with appropriate power and air conditioning controls.

Procedure 3.6.5.1. Observe the location of microcomputer equipment and determine whether it is guarded against excessive heat, dust, or other disturbances.

Procedure 3.6.5.2. Determine whether electrical line protectors and standby, uninterruptible power supplies are necessary for the equipment environment.

Procedure 3.6.5.3. Observe storage procedures for microcomputer diskettes or other media and determine that the media is properly protected.

Objective 3.6.6. Microcomputer files should be backed up on a regular basis with copies moved to an off-site location.

Procedure 3.6.6.1. Review procedures for backing up files with persons responsible for microcomputer operations.

Procedure 3.6.6.2. Review records to determine that logs are being maintained for all system backup processing.

Procedure 3.6.6.3. Check the versions and dates on several selected backup diskettes to determine that all critical files are backed up.

Objective 3.6.7. Controls should be in place to prevent the introduction of unauthorized programs to microcomputer systems.

Procedure 3.6.7.1. Review procedures for introducing new software to the microcomputer system and assess management's understanding of the danger of computer viruses.

Procedure 3.6.7.2. List selected directories of microcomputer program libraries and verify that sampled programs are properly authorized.

Procedure 3.6.7.3. Review directories for the presence of unprotected text editors which can change data or programs without detection.

Figure 3.6. Control objectives and audit procedures for reviewing microcomputer general controls *(continued)*

Objective 3.6.8. Security procedures should be installed to prevent unauthorized use of the equipment as well as to protect equipment from theft or damage.

Procedure 3.6.8.1. Determine that locks or other devices are used to prevent unauthorized persons from using the machine during off shift hours and to prevent tampering with circuit boards or other equipment mounted in the machine chassis.

Procedure 3.6.8.2. Determine that a "DOS Shell" type security system is used to restrict system use to authorized users.

Procedure 3.6.8.3. Verify that passwords associated with the security software are changed on a periodic basis.

Objective 3.6.9. Telecommunication links with other computer systems should be limited to authorized applications.

Procedure 3.6.9.1. Examine the microcomputer equipment to determine whether a modem is part of the configuration, and if so, determine its purpose.

Procedure 3.6.9.2. Verify that the modem is logically turned off or physically disconnected when not in use to prevent unauthorized access attempts.

Objective 3.6.10. Arrangements should be in place to provide service to the microcomputer hardware and software as required.

Procedure 3.6.10.1. Review procedures for maintaining computer hardware equipment and assess whether these arrangements are appropriate given the criticality of applications processed.

Procedure 3.6.10.2. Interview personnel operating the equipment to determine if telephone "Help Desk" services are used.

Procedure 3.6.10.3. Determine that documentation records are maintained for all system problems including a description of the problem resolution.

Figure 3.6. Control objectives and audit procedures for reviewing microcomputer general controls *(continued)*

Figure 3.7 contains auditor objectives and procedures for reviewing controls over access to data and programs in a microcomputer business system. The focus of these procedures is the microcomputer business system; for end user computing, the auditor should refer to the end user controls and procedures in Chapter 11.

Operations Controls

Operations controls are important for the microcomputer business system. These controls are centered around the following control procedures:

Procedures to back up programs and data files

Procedures to allow continuous data processing

Procedures to ensure the correct processing of applications

Program and Data Backup Procedures

Because microcomputers generally tend to be so reliable and "user friendly," many users do not believe their systems can fail. However, hard discs can crash and become unusable, or floppy diskettes can be damaged and become unreadable. If files are not backed up on a periodic basis, considerable time could be spent in recovering from media failures.

Diskettes can be backed up through simple file copy procedures, while hard discs can more easily be backed up through cassette-like streamer tape devices or through the use of Bernoulli Box-type removable disc drives. The auditor should recommend that such backup devices be installed and should perform procedures to determine that they are being used.

Controls to Ensure Continuous Data Processing

Controls that allow continuous data processing are the use of electric surge protectors and uninterruptible power supplies. Minor fluctuations in electrical power or power failures can cause considerable damage to a microcomputer system. Auditors have recommended such devices for years on larger, mainframe systems, but management has often rejected the recommendations because of their high cost.

Microcomputer electric surge protectors can be purchased for less than $100 while uninterruptible or standby power systems can be purchased for around $1,000. When a microcomputer is used as the main data processing machine for an entity, an auditor recommendation to install both is appropriate.

Control Objectives and Audit Procedures for Reviewing Microcomputer Data and Program Access Controls

Note: These objectives and procedures are designed for reviews of free-standing microcomputer systems used for business data processing functions.

Objective 3.7.1. Responsibility for overall microcomputer security should be assigned to one function that has responsibility to ensure that good practices are followed.

> **Procedure 3.7.1.1.** Interview the person or function responsible for microcomputer security and access controls to determine the level of understanding of the importance of the function.

> **Procedure 3.7.1.2.** Review any published procedures covering microcomputer access controls and, through interviews, determine that the procedures are being followed.

> **Procedure 3.7.1.3.** Determine which persons in the organization have access to the microcomputer system and assess whether they appear to need access.

> **Procedure 3.7.1.4.** Observe office procedures to assess whether unauthorized persons operate the microcomputer system.

> **Procedure 3.7.1.5.** Review evening or other off-shift procedures to determine that the microcomputer system and program files are secured against unauthorized access.

Objective 3.7.2. Password and other controls should be employed to restrict data file access to authorized persons.

> **Procedure 3.7.2.1.** Review key applications processed on the microcomputer system and determine whether the software contains its own password controls and whether those passwords are being used.

> **Procedure 3.7.2.2.** Assess whether it would be appropriate to install "DOS Shell" type security system to restrict system access.

> **Procedure 3.7.2.3.** Determine that passwords are changed on

Figure 3.7. Control objectives and audit procedures for reviewing microcomputer data and program access controls

a regular basis and are encrypted or otherwise protected on system files to restrict access. Also determine that passwords are separately assigned for read/write and read only functions.

Procedure 3.7.2.4. Determine that all applications are accessed through system menus rather than through operating system prompts.

Procedure 3.7.2.5. If key data files are maintained on diskettes or other removable media, determine that they are locked up or otherwise restricted to limit unauthorized access.

Objective 3.7.3. Microcomputer systems should be backed up on a regular basis with earlier backup versions stored in a location away from the computer system.

Procedure 3.7.3.1. Review procedures for backing up files and programs and determine if procedures are adequate.

Procedure 3.7.3.2. Trace several backup files to the off-site location to determine that proper back-up versions are in place at that location.

Objective 3.7.4. Access to microcomputer programs should be restricted to prevent unauthorized use or modification.

Procedure 3.7.4.1. If locally developed applications are used, such as database or spreadsheet macro-command applications, determine that the source code is "hidden" or otherwise protected to prevent unauthorized access.

Procedure 3.7.4.2. If package software is used for applications, determine whether there are procedures for preventing improper software modification attempts.

Procedure 3.7.4.3. Determine what procedures are in place to prevent the introduction of an unauthorized operating system, through a diskette drive, which could bypass established system security procedures.

Procedure 3.7.4.4. Review procedures to prevent unauthorized programs or program viruses from being introduced to the microcomputer system.

Figure 3.7. Control objectives and audit procedures for reviewing micro-computer data and program access controls *(continued)*

Objective 3.7.5. Procedures should be in place to prevent the unauthorized copying of programs and data.

 Procedure 3.7.5.1. Assess whether a policy exists against unauthorized software copying and whether personnel using the microcomputer understand that policy.

 Procedure 3.7.5.2. If the microcomputer is equipped with a modem, review controls to prevent unauthorized copying or other manipulation through modem lines.

 Procedure 3.7.5.3. Observe work areas around the microcomputer system to assess whether improper copy versions are stored in work areas.

Figure 3.7. Control objectives and audit procedures for reviewing microcomputer data and program access controls *(continued)*

Processing Controls

Procedures to assure the correctness of data processing include both system and application controls. Techniques used for larger systems, such as logs to monitor error messages and system failures, are appropriate for a microcomputer system. The auditor should look for procedures to ensure that applications run to completion and are run with the correct data, and that application controls are properly operating. Objectives and procedures to assure the correctness of processing for microcomputer applications are outlined in Figure 3.8. A more detailed discussion of these controls can be found in Chapter 5.

PROCESS OR NON-BUSINESS SYSTEMS CONTROLS

In many larger organizations, mini and microcomputers are found in many areas beyond the data processing operations, being used for purposes other than typical business information systems needs. They may be located in engineering laboratories, manufacturing control operations, marketing departments, and other areas. The machines may be used for process control, automated design work, statistical analysis processing, and other applications. Some of these machines are totally dedicated to specific control applications while others may be used for a variety of tasks. This proliferation of data processing machines has come about in many organizations because of their

**Control Objectives and Audit Procedures
for Reviewing Microcomputer Applications Controls**

Note: These objectives and audit procedures cover the overall use of microcomputer applications in significant business data processing areas. Chapter 5, Auditing Data Processing Applications, *discusses more detailed procedures, including testing, for more comprehensive reviews of individual applications.*

Objective 3.8.1. The auditor should develop an overall understanding of key input, output, and processing controls in order to assess the degree of control risk associated with the application.

   Procedure 3.8.1.1. Interview key users of selected applications and review their processing documentation to gain a general understanding of application functions and controls.

   Procedure 3.8.1.2. Determine that the selected microcomputer applications have adequate error checking procedures, audit trails, and other controls to provide assurance that processing results are correct.

   Procedure 3.8.1.3. Using one of the selected key applications, review, with key users, procedures to reconcile input to output.

   Procedure 3.8.1.4. Briefly document the selected applications using block diagrams or functional narratives.

Objective 3.8.2. Microcomputer applications should be run on software packages or in-house programs which are authorized, tested, and approved by the organization.

   Procedure 3.8.2.1. Determine that procedures are in place to evaluate new microcomputer applications before placing them into production.

   Procedure 3.8.2.2. Select several applications resident on microcomputer library files and determine that software packages are authorized.

   Procedure 3.8.2.3. If diskette copies or hard disc resident libraries are used for applications programs, determine that

Figure 3.8. Control objectives and audit procedures for reviewing microcomputer applications controls

the original software diskettes exist and are stored in a secure location.

Objective 3.8.3. The application being run should be appropriate to a microcomputer environment given the other overall data processing resources in the organization.

Procedure 3.8.3.1. Consider the types of significant applications run on the microcomputer system and determine whether they are appropriate given the overall data processing resources within the organization.

Procedure 3.8.3.2. Evaluate microcomputer resources, such as disc capacity or memory, in light of the requirements of several key applications and determine whether the microcomputer has sufficient resources to process applications efficiently.

Procedure 3.8.3.3. Review any sensitive applications run on the microcomputer system, such as payroll or purchase order generation, and determine whether system security controls are appropriate for those classes of applications.

Objective 3.8.4. Procedures should be in place to ensure that key microcomputer applications have run properly to the completion.

Procedure 3.8.4.1. Select one or more key microcomputer applications and review user logs or other procedures to provide assurance that applications have been run to completion.

Procedure 3.8.4.2. For the applications selected, review application program procedures to determine the types of controls to monitor successful completion.

Objective 3.8.5. There should be an adequate separation of responsibilities between prime users of the microcomputer application and system operators or programmers.

Procedure 3.8.5.1. Determine that there exists a separation of duties, given the size of the organization, between those who operate the applications, those who can change programs or tables, and those who rely on application outputs.

Procedure 3.8.5.2. If a small organization, determine the level

Figure 3.8. Control objectives and audit procedures for reviewing microcomputer applications controls *(continued)*

of involvement of any owner or key manager in microcomputer application operations.

Procedure 3.8.5.3. Determine whether controls or procedures exist to prevent unauthorized users from operating the system or introducing key application transactions.

Objective 3.8.6. There should be sufficient application documentation to provide users of the application with an understanding of operations and controls.

Procedure 3.8.6.1. Review documentation for selected key microcomputer applications to determine that:

The documentation covers both normal operations and special processing, such as year end closing procedures

There are sufficient descriptions of application error codes and suggested responses to those errors

The documentation is current and covers all major application functions

Procedure 3.8.6.2. If the selected application documentation consists of on-line "help" screens, assess whether those screens are adequate.

Procedure 3.8.6.3. Determine that procedures exist for maintaining master or backup copies of key application documentation.

Objective 3.8.7. If the microcomputer application is being run on package software, only current versions of the software under signed product licenses should be used.

Procedure 3.8.7.1. Review files to determine that there are signed licenses for key application software products.

Procedure 3.8.7.2. Through interviews, determine that key users understand the terms of software licenses including any restrictions on copying the software.

Procedure 3.8.7.3. Select several key microcomputer application packages and determine that current versions are being used, or if not, that there is a sufficient reason for the use of older versions.

Figure 3.8. Control objectives and audit procedures for reviewing microcomputer applications controls *(continued)*

Objective 3.8.8. If the microcomputer application is run with in-house developed software, there should be adequate technical documentation covering the application.

Procedure 3.8.8.1. Review procedures for documenting in-house developed software and assess the adequacy.

Procedure 3.8.8.2. Select several applications and determine that they comply with in-house documentation procedures.

Procedure 3.8.8.3. If applications have been written using spreadsheet or database packages, determine that "spreadsheet audit" packages or similar tools have been used to document them.

Objective 3.8.9. Policies and procedures should exist to prevent the introduction and use of unauthorized software copies.

Procedure 3.8.9.1. Determine that there are organization policies covering the improper copying of application software.

Procedure 3.8.9.2. Observe microcomputer operations and briefly survey program files to assess whether illegal copies are being used.

Procedure 3.8.9.3. Interview key microcomputer users and managers to determine that they are aware of the risks, such as computer viruses, of introducing unauthorized programs to the microcomputer system.

Objective 3.8.10. The microcomputer should be configured to provide reliable, continuous service.

Procedure 3.8.10.1. Determine that cassette streamer tape units or other devices are used to provide for ease of generating backup copies.

Procedure 3.8.10.2. Given the location where the microcomputer is located, determine whether environmental controls, such as dust covers, are adequate.

Procedure 3.8.10.3. Determine that electrical surge protectors or stand-by power systems are used to provide for application integrity.

Figure 3.8. Control objectives and audit procedures for reviewing microcomputer applications controls *(continued)*

relatively low cost, the familiarity of many professionals with data processing techniques, and the inability of traditional data processing departments to support specialized needs.

Although these computer systems are not used for traditional business data processing needs, such as maintaining accounts receivable records, they often support applications critical to the organization. For example, an engineering computer may support new-product design work. Systems backup and integrity concerns may be as great in this type of environment as in the typical business data processing center.

The auditor's role in these specialized data processing operations will vary with management's direction and the auditor's review objectives. While the external auditor will have little involvement with reviews of such data centers, the internal auditor can often play an important role here. Some of the areas where the auditor can make a contribution in reviewing such computer systems include:

Backup and recovery procedures. Auditors have stressed the need for computer system backup and recovery procedures in business data processing systems for years. Yet, specialized technical professionals, for example, those managing a marketing research computer center, often have not considered such issues. The auditor can help those users by recommending improved backup and recovery procedures.

Application system controls. Many specialized computer systems use unique applications purchased from technical vendors or developed in house. Application controls may not be strong. With even a limited knowledge of the purposes of such an application, an auditor often can make basic recommendations for system controls improvements.

Appropriateness of certain applications. Some specialized data centers become "shadow" business data processing shops in addition to their normal, specialized functions. For example, an engineering laboratory computer center may decide it can write a better job cost or estimating system and expend considerable efforts to develop one. This may raise issues ranging from the quality of regular data processing systems to the proper utilization of engineering personnel. The auditor should apprise management of such issues. If the computer center is used for extensive end user computing activities, the auditor should consider the control procedures discussed in Chapter 11.

Control of computer assets. Specialized data centers may acquire unnecessary data processing equipment in the name of research, with management unaware of any problem. Through a review of utilization logs or other records, the auditor can raise underutilization or control issues to management.

Before attempting any review of such a specialized computer center, the auditor should obtain a rough familiarity with the functions of that operation. For example, if the auditor plans to review a dedicated computer aided design and manufacturing (CAD/CAM) computer center, the auditor should attempt to understand roughly the terminology, general workings, and objectives of CAD/CAM. In this way, the auditor will not have to confront a barrier of technical jargon when first visiting the computer center and its personnel.

Reviews of specialized computer centers are not recommended for the less experienced auditor. They require translating control concepts from normal business data processing situations into specialized controls. This can be accomplished best after the auditor is experienced in reviewing business data processing computer centers.

Figure 3.9 contains objectives and procedures for reviewing specialized mini and microcomputer process centers. These objectives and procedures focus on the general controls areas discussed, where auditors should be able to make significant management contributions. The creative auditor should explore other areas in the course of the review if appropriate. (See Chapter 11 for more information.)

In future years, the auditor will encounter more of these specialized computer centers, many of which will use micro or super-minicomputers. The creative auditor can make an increasingly valuable contribution to management by performing operational reviews over these computer centers on a periodic basis.

Control Objectives and Audit Procedures for Reviewing Specialized, Non-Business Systems Controls

Note: These objectives and procedures are designed for general controls reviews of specialized mini and microcomputer systems used for non-financial applications such as marketing research, engineering, or manufacturing planning. When the computer system is used also for financially significant applications, the other tables of objectives and procedures in this chapter and Chapter 2 should be used.

Objective 3.9.1. The specialized process computer system should be used for only non-financial operations applications.

Procedure 3.9.1.1. Interview responsible members of management to determine the purposes and functions of the computer system.

Procedure 3.9.1.2. Review computer systems configuration charts, selected application documents, and other documentation to develop a general understanding of the purposes of the system.

Procedure 3.9.1.3. Review capital approval documentation or management approved project plans to determine that the computer system has been approved by upper management.

Procedure 3.9.1.4. Observe data processing activities and review lists of applications to determine that the computer system is used only for its designated functions.

Procedure 3.9.1.5. Assess whether personnel use the process computer for their own end-user applications, and if so, consider the information center controls objectives and audit procedures presented in Chapter 11.

Objective 3.9.2. All specialized computer applications, despite their special purposes, should have sufficient input, output, and processing controls.

Procedure 3.9.2.1. Develop an understanding of the functions, purposes, and control structures of key computer system applications.

Procedure 3.9.2.2. Review procedures for determining that

Figure 3.9. Control objectives and audit procedures for reviewing specialized, non-business systems controls

applications have run to completion correctly using proper input data and files.

Procedure 3.9.2.3. Interview key application users to determine their understanding of application controls.

Procedure 3.9.2.4. If key application users also process the applications run on the system, determine whether logging procedures or other controls exist to prevent improper modification of files or processing results.

Procedure 3.9.2.5. Develop an understanding of the types of error conditions that may result from specialized processing activities, and determine that adequate steps are taken to correct those errors.

Procedure 3.9.2.6. Review any long range plans to determine whether computer system will continue to be used for the same general functions.

Objective 3.9.3. Specialized computer programs should be maintained on controlled libraries and any programming activities should follow good information systems practices.

Procedure 3.9.3.1. Determine that secured program libraries are being used that prevent programs from being modified without proper authorization.

Procedure 3.9.3.2. If most applications are built around purchased software products, determine that current versions of those vendor products are being used, proper license agreements have been signed, and that vendor documentation appears to be adequate.

Procedure 3.9.3.3. If applications are developed using in-house programming resources, determine that standards exist for authorizing programming efforts as well as for documenting and testing applications.

Procedure 3.9.3.4. Select several recent programming tasks and determine that department programming standards were followed, or if standards do not exist, that good general programming practices were followed.

Procedure 3.9.3.5. Determine that controls are in place to limit program library access to authorized individuals.

Figure 3.9. Control objectives and audit procedures for reviewing specialized, non-business systems controls *(continued)*

Objective 3.9.4.   The computer center should have procedures for backing up files and for restoring operations in the event of an extended interruption in services.

Procedure 3.9.4.1.   Determine that one person or function is responsible for backing up key data files and program libraries.

Procedure 3.9.4.2.   Review procedures for backing up files, including a visit to any off site location, and determine whether procedures are appropriate given the functions of the specialized computer system.

Procedure 3.9.4.3.   Review contingency planning procedures and determine whether they appear adequate.

Objective 3.9.5.   The specialized computer system should be operated in accordance with management's overall direction and should make good use of organization resources.

Procedure 3.9.5.1.   Determine that performance reports are prepared covering the computer system and communicated to appropriate levels of management.

Procedure 3.9.5.2.   Review computer system performance reports to determine that the system is being utilized properly.

Procedure 3.9.5.3.   Assess whether the specialized microcomputer system should operate as currently structured or differently as part of the overall information systems organization.

Figure 3.9. Control objectives and audit procedures for reviewing specialized, non-business systems controls *(continued)*

Controls in the Distributed Network

INTRODUCTION

Data processing literature currently is filled with such words as "connectivity" and "network architectures." There is a growing realization that computer systems must have the ability to *connect* and communicate with each other in a *network*. The networked system is quite different from the traditional on-line, mainframe-based system where a central processor has a series of attached terminals with all files and programs located at the central processor. In the modern computer network various processors are linked together, and data processing responsibilities are distributed among several machines.

This chapter first discusses various architectures or strategies used for distributed processing and for telecommunications. The chapter then discusses the types of controls the auditor should expect to find in several of the more common data processing networks in modern organizations. Some of these network types exist within the workplace and are used to support activities such as office automation.

While distributed processing implies a closed network where various computer processors are linked directly, electronic data interchange (EDI) is a variation on that concept. In an EDI network, outside vendors or customers dial into a computer system to communicate data, such as product orders. This chapter also defines and discusses controls for organizations involved in EDI transactions. EDI is becoming increasingly important for many organizations, and the auditor needs a good general understanding of its control implications.

The areas of data communications, telecommunications, and distributed processing are changing rapidly. This chapter will give the auditor an introduction to this field and present the relevant audit control concepts. Auditors needing in-depth knowledge in the field should consult a detailed book on telecommunications and enroll in various short courses offered by local universities.

DISTRIBUTED PROCESSING NETWORKS

The concept of distributed processing originated in the early 1970s. At that time, large mainframe computers handled all processing tasks ranging from input data validation through central file updating. The typical terminal supplying data to the mainframe was "dumb" in that it could only transmit and receive messages. In the early 1970s, terminals were developed possessing "local intelligence" with the ability to perform local processing tasks such as data validation. The idea of distributed processing was born.

This section introduces some common variations of distributed processing networks for multi-area data processing systems. Local area networks serving offices and work groups are discussed as well as EDI networks. To evaluate controls in a distributed environment, the auditor needs a basic understanding of the type and architecture of the distributed processing network being reviewed.

Distributed Processing Architectures

The introduction to this chapter explained how distributed processing systems originated with a central processor allowing some processing to take place at local, intelligent terminals. These local terminals handled limited functions such as data validation or collection. The central processor would then poll these local terminals to collect the validated data or to download other data for local analysis and report generation. This type of configuration is called a centralized, or star, distributed processing network. Figure 4.1 shows such a configuration.

Star distributed processing networks typically are used for retail or shop floor data collection systems. The local terminals or nodes in the network

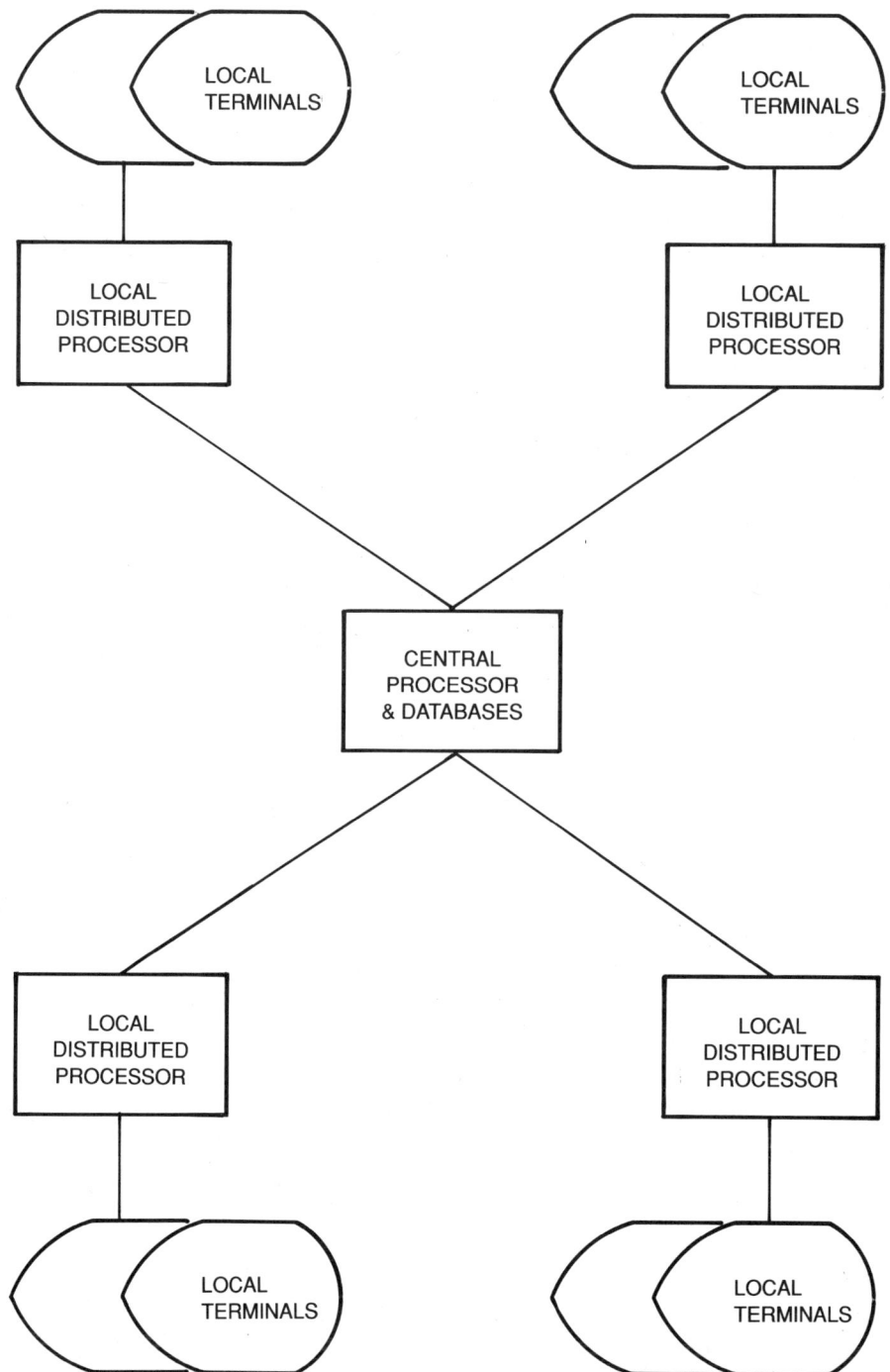

Figure 4.1. Centralized, or star, distributed processing configuration

process local data and then transmit it in summary fashion to the central processor. Some data may be retained at the remote nodes for local needs.

While star distributed processing removes some data validation responsibilities from the central processor, that central machine is still responsible for the overall distributed network. The local nodes often have limited processing responsibility and cannot communicate with one another except through the central machine.

As organizations expanded and added multiple data centers, the star network has often proved ineffective for true distributed processing. Various data centers often need to transfer data between one another although no one data center acts as the hub for all others. The result is a multicentered distributed processing configuration as in Figure 4.2.

In a multicentered configuration, each main processor is connected to other processors or terminals as required. However, the mainframe processors do not communicate with one another on a peer basis. When one mainframe transmits data to another mainframe application, the transmitting mainframe computer acts as a dumb terminal. In fact, the mainframes are not talking directly with one another. Both computers communicate through their respective network or terminal controllers.

The multicentered distributed processing network was necessary because, until recently, many mainframe computers could not communicate with one another on a peer to peer basis. One computer system had to be "in charge" and then treated the other computers as little more than powerful terminals. With peer to peer communications, each computer system treats each other computer system on the network as equal.

International standards organizations have recently adopted a peer to peer communications protocol called LU6.2, which many computer manufacturers have implemented. As a result, various large mainframe computers as well as mini and microcomputers can talk with one another as peers. IBM calls their protocol System Application Architecture (SAA). This type of peer to peer network is shown in Figure 4.3.

Peer to peer distributed networks offer the most flexibility in distributed computing. However, they also present potentially serious systems design and systems control problems. Because no single processor is "in charge," anarchy is possible. It may be some time before true peer to peer distributed networks are implemented with programs and databases shared from processor to processor with no major conflicts as to which is in charge.

There are many variations to the three network configurations discussed above. For example, two mainframe processors can be connected using peer to peer communications, and each could have a star network attached to it.

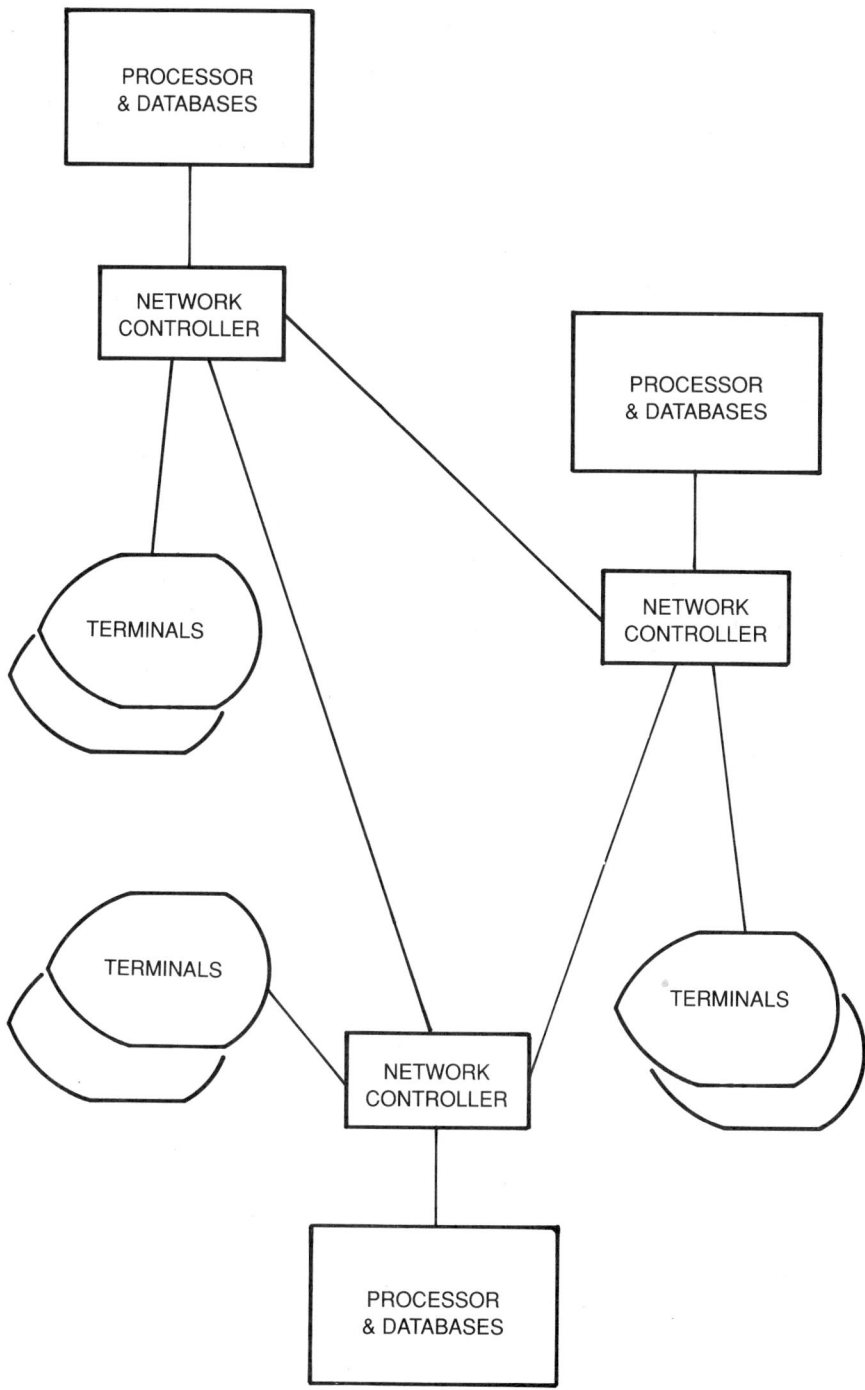

Figure 4.2. Multicentered distributed processing configuration

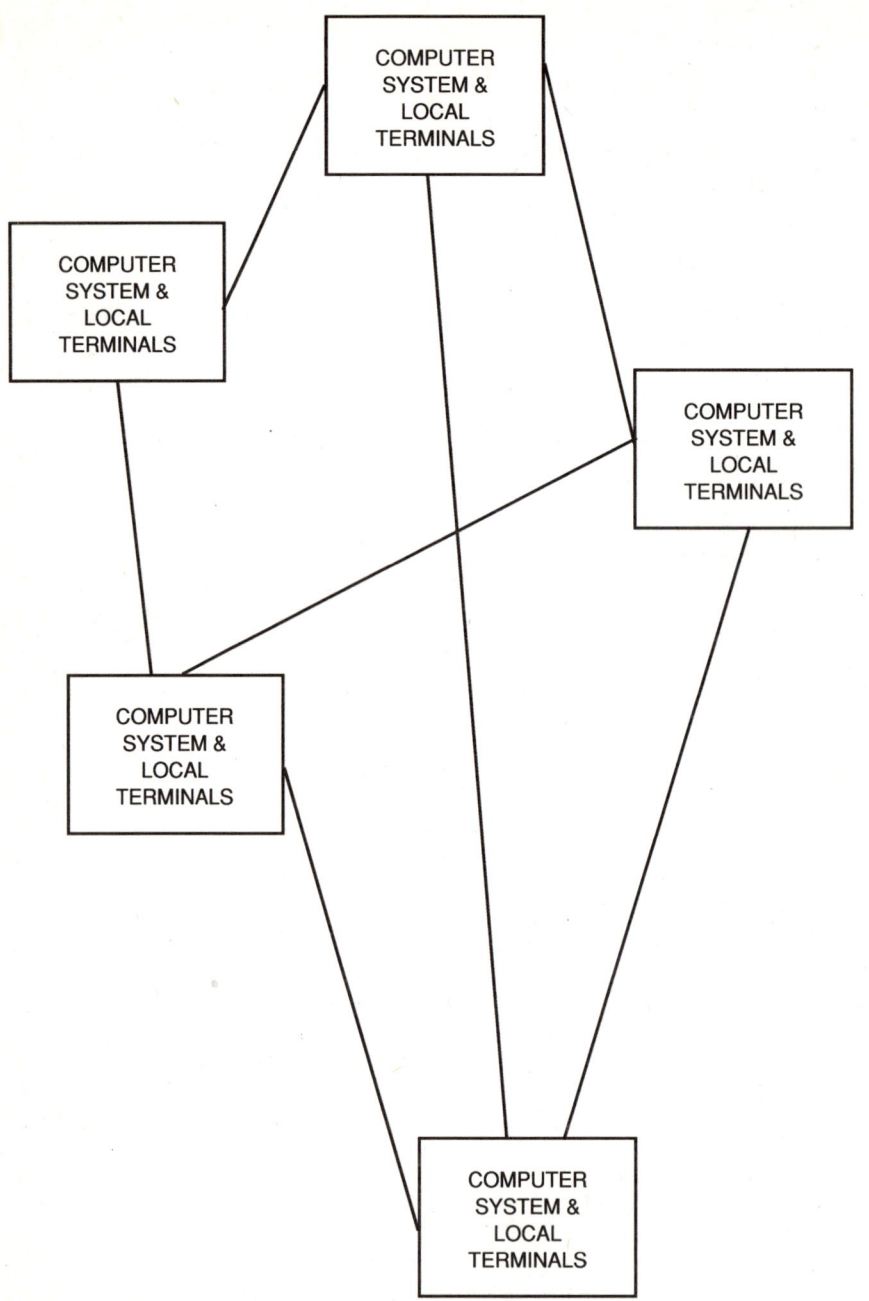

PEER TO PEER DISTRIBUTED SYSTEM—EACH COMPUTER SYSTEM CAN DISTRIBUTE
AND SHARE WORK WITH ANOTHER CONNECTED SYSTEM AS A PEER.

Figure 4.3. Peer to peer distributed processing configuration

The auditor will see more use of distributed processing systems in the future. Some of the attractive advantages of such systems are:

Local processing autonomy. Depending upon overall system and network design, local units can run applications to establish real time control and satisfy local needs while contributing data to meet needs of other organization applications.

More efficient use of smaller computers. In a growing organization, a centralized data processing system requires larger and larger central computer systems. The needs of local units in a distributed network can be satisfied by mini or microcomputers at less cost.

Increased access to data and processor resources. Because telecommunication lines make processors in a distributed network readily accessible, both data and computer resources can be shared. For example, a local processor can download a computationally intensive application to a larger mainframe on the network for processing rather than consuming local resources.

There are also potential control problems and concerns with distributed networks. Many of these concerns deal with the overall auditability, security, and control over applications implemented on such a network. These are explained below.

Local Area and Workgroup Processing Networks

The modern office typically uses a variety of electronic devices, such as word processors and microcomputers, to perform day to day functions. Just as large computers link into national or worldwide networks, many organizations link office automation equipment into local area networks (LANs).

These networks usually are called LANs when they are used to link all office devices. Sometimes, they are called workgroup networks when they are restricted to a small office or department. LANs represent a localized form of distributed processing and have the following characteristics:

Short transmission distances. Most LANs cannot operate over distances greater than one mile. As a practical matter, usually they are limited to a single building or office complex.

Capability of sharing office automation resources. LANs can connect microcomputers, office automation equipment, and specialized equipment such as high quality printers or data storage devices. Thus, all word process-ing devices on a LAN could print their outputs on a single, high quality laser printer located on the LAN.

Multivendor connectivity. LANs can connect a variety of office automa-tion or other devices including word processors, microcomputers, large scale computer systems, and even gateways to other telecommunications networks.

Typically, auditors have given little attention to office automation systems. The perception has been that critical applications (such as the general ledger or the accounts receivable system) resided on the large mainframe computer system while local office automation or word processing systems contained no critical applications. With office devices connected on a LAN and potentially connected to other computer networks, however, a LAN often presents poten-tial security and control issues. It may be possible for an unauthorized user to access a microcomputer connected to a LAN which can then tie into a major mainframe computer system also attached to that same LAN. In addition, other microcomputer-based devices can be connected to a LAN to create the effective power of a much larger computer system with more significant control concerns.

This section describes some fundamental concepts behind local networks and their audit, control, and security considerations. In performing general controls reviews of the data processing function in an organization, the auditor should consider controls and procedures surrounding the local area network and the office automation systems.

LANs introduce new concepts to the auditor accustomed to standard data processing. When the auditor discusses a LAN with its designers, the follow-ing concepts may be encountered:

Network architecture or topology. These terms refer to the ways data processing equipment can be connected into a network. Various architec-tures, topologies, or strategies can be used to connect office devices on a LAN. The most common are the ring or bus networks as shown in Figure 4.4. Each can be expanded, through branches on the bus or through multiple intersecting rings.

Signaling strategy. LAN managers frequently use the terms *baseband* and *broadband* when discussing LANs. These refer to the way data is communicated across the LAN network. Baseband transmission might be considered similar to direct computer-to-computer high speed channel

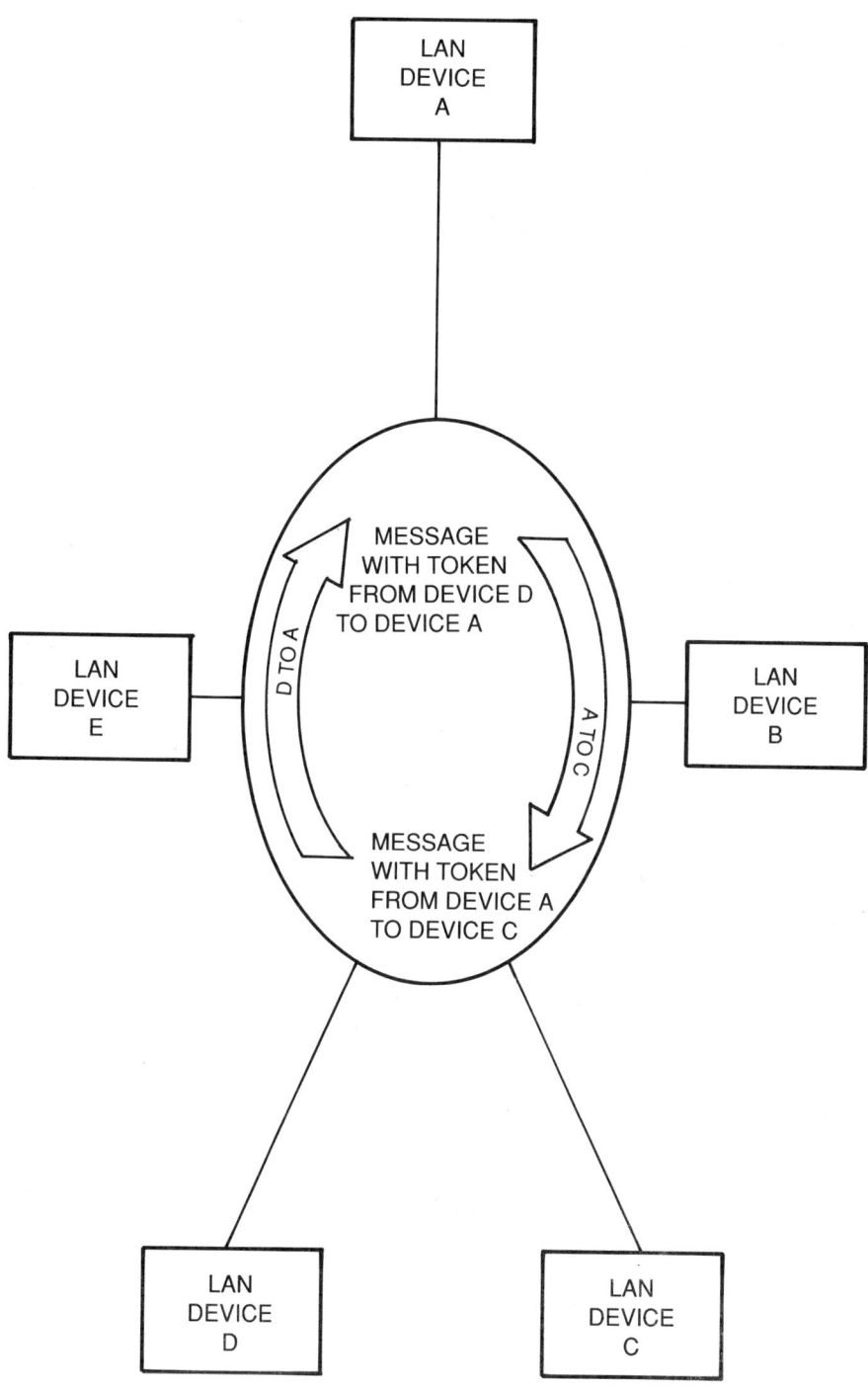

Figure 4.4. Token ring architecture

communications. Data can be transmitted at high speeds of approximately 10 megabytes per second (Mb/s) but over limited distances of less than a half mile. Broadband communication is slower, at 1 Mb/s, but multiple applications can be transmitted across the same communications channel, permitting the LAN to combine voice, data, and even video images. Data devices require a modem to transmit in broadband.

Wiring strategy. LANs can operate over coaxial (coax), twisted pair, or optical fiber wiring. Coax refers to the shielded coaxial cable used for cable television and other high speed communications. While it has the advantage of allowing broadband communications, it requires special installation work to be placed in the office. Twisted pair wiring refers to the normal telephone lines already installed in many buildings. It is inexpensive and relatively easy to install. The newest type of wiring now used for some sophisticated LAN networks is optical. Rather than a physical wire, a thin filament of fiberglass carries laser generated optical signals. This works for high data rates but only for point to point networks. An optical fiber line must start at one device and end at another; it cannot have branches along the route. At the present time, optical fiber LAN systems are expensive to install and found only in specialized applications.

CSMA/CD LAN access method. A LAN consists of a variety of different devices connected to a single wire and capable of sending messages to one another. A communications access method is needed to direct messages to the correct destination without conflict with other messages. The two access methods used are called "Carrier-Sense Multiple-Access with Collision Detection" (CSMA/CD) and "token ring." Using the CSMA/CD method, each device on the LAN attempts to send its message down the LAN at any time. If one message collides with another, it backs off from the attempted transmission and tries again after a few microseconds to see if traffic has cleared. One can think of the CSMA/CD access method as being similar to an expressway. If a given auto on an entrance ramp detects ongoing traffic, it waits until traffic has cleared. Once there is a break, the auto merges into the flow of expressway traffic. The commonly used Ethernet is an example of CSMA/CD transmission.

Token ring LAN access method. In a token ring network, each message on the LAN is passed from node to node following addresses that go with the transmission. When it arrives at a given node, it may be held awaiting another message about to be transmitted from that node. A token ring system is like a single railway line with a series of stations with side tracks. Trains wait at each station until controllers decide which train to send down the line next.

Auditors frequently encounter LANs in the modern organization. They may be located in the office or in specialized work areas such as an engineering laboratory. Data processing organizations that use CASE, or computer aided systems engineering, discussed in Chapter 7, often have programmer work-bench terminals connected in a LAN. In addition, smaller organizations may distribute data processing activities by tying a series of microcomputers together in a LAN rather than concentrating them onto a larger computer.

Electronic Data Interchange Systems

Many manufacturers use a system called "Just in Time" (JIT) for ordering parts and supplies for manufacturing processes. In a JIT system, the manufacturer tries to minimize parts inventories by requiring vendors to deliver parts and supplies only as they are needed, or "just in time," to add to the manufacturing process. The manufacturer often uses electronic data transmission networks to transmit these needs to suppliers. This process is called "Electronic Data Interchange" (EDI).

EDI systems are becoming more frequent in many industries. EDI is used to transmit orders, advise customers of shipping data, do billings, transfer funds, or exchange data. Hospitals use EDI to order supplies; insurance companies use the technique to pass policies to co-insurers; and trucking companies use it for tracking deliveries. While EDI systems cannot be described strictly as distributed network systems, they present some of the same control issues found in many internal distributed systems.

If the auditor's organization does not participate in an EDI network, the auditor often becomes involved when management considers the use of EDI and asks for help or advice. A key customer may inform the organization that to continue selling to that customer orders must be accepted via EDI and shipping data must be transmitted back in the same manner. A management group aware of the vulnerabilities associated with dial-up telecommunication networks may ask the auditor for advice on potential risks regarding this EDI approach.

The auditor probably encounters even greater uses of EDI as we move progressively toward a paperless society. Organizations find it more cost effective to use telecommunications and data processing systems to communicate rather than sending paper to and from one another. As sophistication of these EDI systems grows, they increasingly resemble classic distributed systems.

EDI systems can be classified as either "one-to-many," "clearinghouse," or "incremental trail" systems. Each will be described separately.

One-to-Many EDI Systems

The one-to-many EDI system centers around one organization and its main computer system. The other participants are suppliers and customers or dealers of that organization. This type of EDI system is used by automobile manufacturers to communicate with major suppliers. A similar system is used by several microcomputer manufacturers to communicate with their dealer networks. A one-to-many EDI system is shown in Figure 4.5.

In a one-to-many EDI procurement or replenishment system, the following steps typically take place:

1. At the central organization, replenishment orders are assembled on computer files rather than being printed.
2. At some mutually agreed time, the central organization's computer dials up or accesses leased communications lines to transmit orders to each of its suppliers, after a password or electronic "handshake" exchange to assure that the proper supplier has been identified.
3. The supplier processes the order as if it were entered as batch transactions through the supplier's local order entry system; and the supplier, depending upon the system, transmits an order acknowledgment back to the transmitting central organization.
4. When the supplier is ready to ship, the supplier's invoice data is captured on a file and transmitted back to the central organization.
5. The central organization may transmit payment advices back to the supplier while also transmitting payment through a bank clearing system.

The auditor may encounter a one-to-many EDI system at a central organization that communicates with its suppliers or dealers, or at one of the latter communicating with a central source. In either event, there can be potential security, audit, and control concerns with this type of EDI system. These concerns usually involve controls over file transmissions, system and file format incompatibilities, and lack of formal paper audit trails.

Different industry standards currently exist for EDI file format protocols. The American National Standards Institute (ANSI) established a standard for data interchange which is followed by many industrial groups. Other industry groups or even large EDI individual organizations have established their own standards.

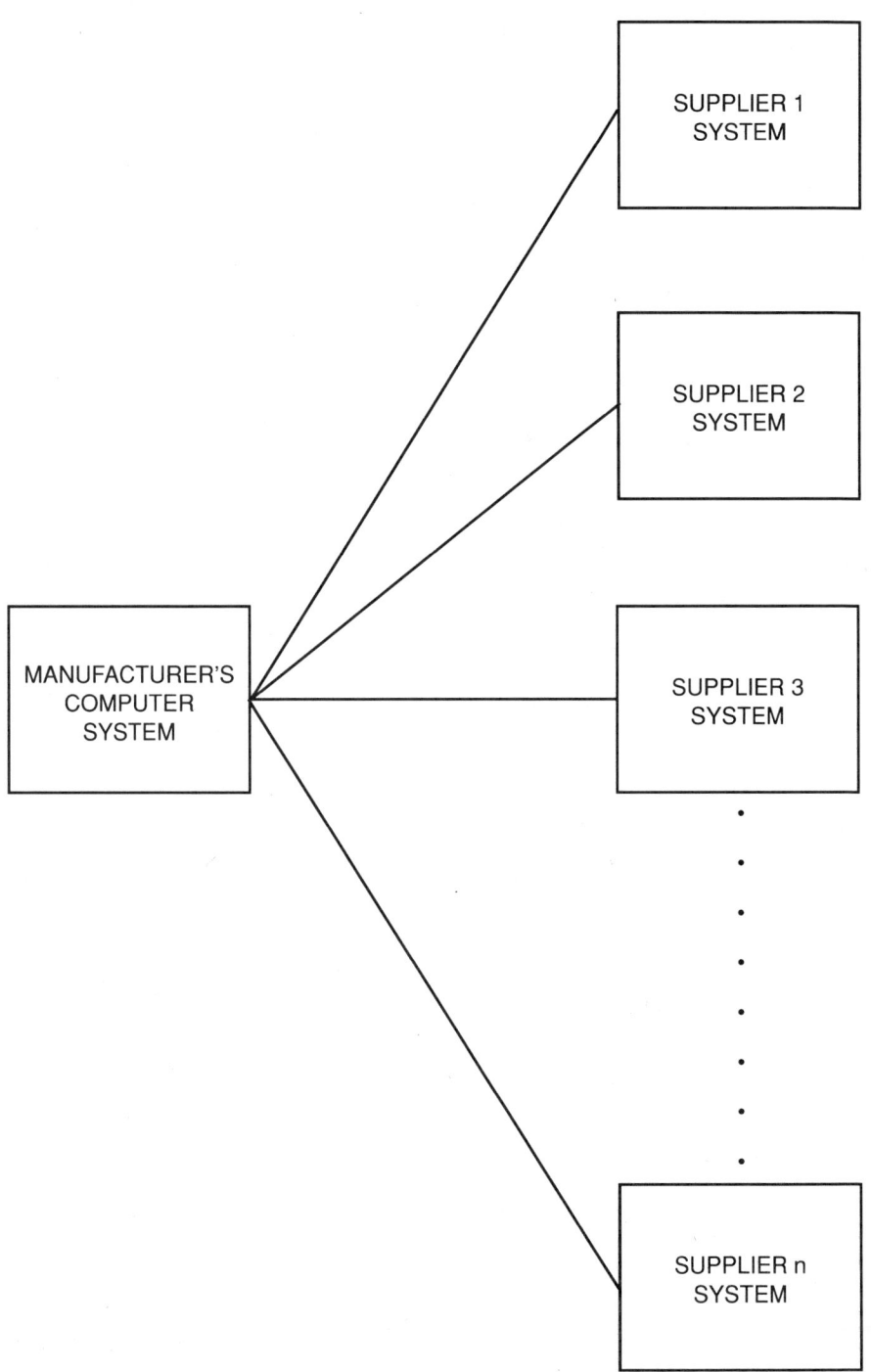

Figure 4.5. A one-to-many EDI network

Clearinghouse EDI Systems

Some EDI networks have been formed by industry groups rather than a single organization. A trade group for manufacturers, parts suppliers, and dealers in a given industry may band together to use EDI. Since each member of such a group may not be dominant, the group may use a central computer site to act as a clearinghouse for all EDI activities. Often, a major computer timesharing service is used to supply that clearinghouse function.

A clearinghouse EDI network is shown in Figure 4.6. In many respects, it operates like an old-fashioned, small town post office. Delivered letters are placed into a box for each recipient who visits the post office to pick up messages. Recipients have a set of "personal keys" or passwords to their boxes for security purposes.

Clearinghouse EDI systems require that all participants follow a common set of standards, a good system for information security, and a system for reporting usage. The latter is important for allowing the clearinghouse to distribute costs among its members.

When the auditor encounters a clearinghouse EDI system, the auditor should ask to see a third party or service auditor's report covering that clearinghouse. This type of report is typically prepared by an independent public accountant for users. The report will discuss the controls environment at the EDI clearinghouse.

Incremental Paper Trail EDI Systems

In the two EDI system types discussed above, a document was exchanged between the two parties to a transaction. Many systems, however, require multiple levels of documentation where many parties contribute small elements of information to the process. An example is a system to support international shipments. Freight forwarders, shippers, port authorities, and customs officials all need to add or extract certain data from the documentation package.

An incremental paper trail EDI system can be viewed as a sophisticated electronic mail system. However, rather than simply transmitting and embellishing verbal messages, such a system passes electronic data from one party to the next. There are only a few such systems in existence today. However, the auditor will see many more in the future.

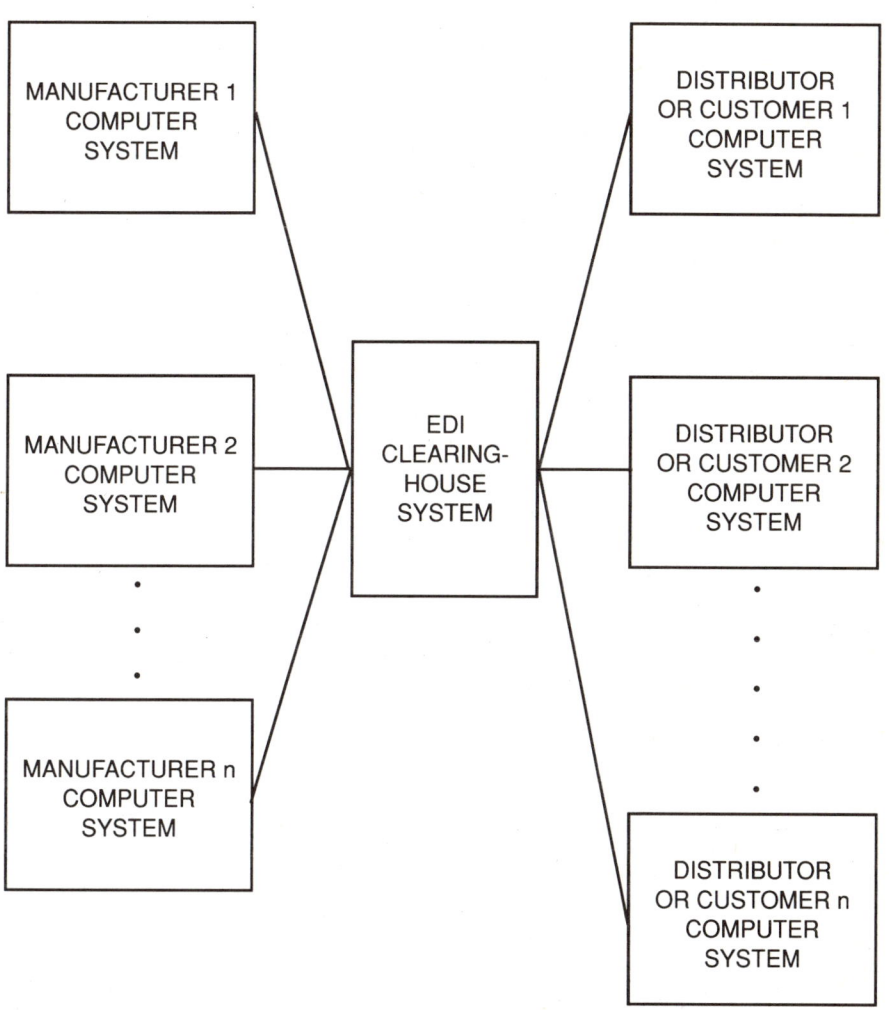

Figure 4.6. A clearinghouse EDI network

TELECOMMUNICATIONS NETWORK CONTROLS

The previous section introduced distributed processing systems, local area networks, and electronic data interchange systems. Each of these links computer systems through telecommunication networks. Security and control over telecommunication networks are essential to ensure the overall integrity of the distributed system.

This section introduces some basic telecommunications concepts as well as data processing networking principles. Also, the section discusses control procedures and objectives for auditing telecommunications systems. Telecommunications is a specialized area. In many respects, technology is changing even faster for telecommunications than for computer systems. Because of new and evolving technologies, situations arise where the auditor may want expert help in performing a detailed review of controls within a complex telecommunications network. However, this section should provide sufficient information for the auditor to perform a preliminary review of a telecommunications network supporting a distributed processing system.

The Evolution of Telecommunications Networks

The first telecommunications networks for data processing were developed to allow better transmission of batch data to central computer sites. In the early days of business data processing, organizations often had remote sites keypunching transactions, loading them on tape files, and then mailing or shipping the tapes to central computer sites.

By the mid-1960s, however, devices became available to transmit data over regular telephone lines. These first telecommunication links were off line, but data could be brought to the central computer more rapidly. These devices were slow and had relatively high error rates.

By the late 1960s, many data centers began transmitting data directly to their central computers. These remote devices acted as card readers and transmitted transactions through modems over high speed leased telephone lines to central computers. This type of processing configuration is shown in Figure 4.7. It is called *decentralized* processing. The actual computing remains at the central computer, but some preparatory work takes place at decentralized sites.

Decentralized processing, or remote job entry (RJE) processing, introduced few new control concerns for the auditor. Networks were generally fairly simple, the communications software incorporated facilities to check for

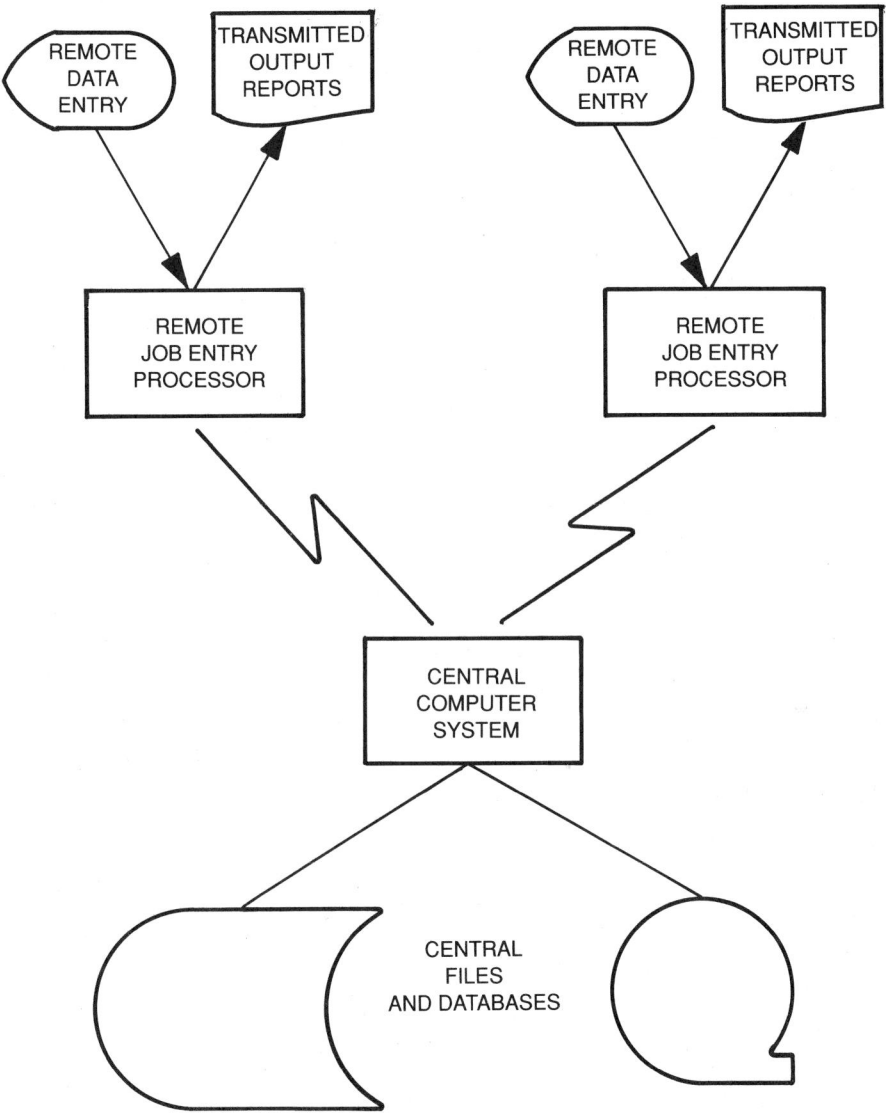

Figure 4.7. A decentralized processing network

transmission errors, and decentralized processing was used for batch processing applications that could be easily verified with conventional procedures. Many data centers continue to use some decentralized processing applications today despite more sophisticated telecommunications networks.

With the growth of data processing capacity in larger organizations and the growth of on-line systems in most organizations, telecommunications networks have grown more complex. Central computers are connected in a peer to peer distributed network as shown in Figure 4.3. A variety of telecommunications approaches are used to make that connection. They may use the same leased lines (through modems) as described for distributed processing, or communications may take place via packet switching networks, microwave channels, satellites, or other methods.

An example of this advanced telecommunications network is shown in Figure 4.8. While the typical data communications network does not have all these elements together, the auditor generally encounters more than one in a given computer center. Some, such as dial-up lines and connections from LANs, have greater control implications than others, such as leased lines.

One of the common telecommunications links is called "packet switching." This term refers to public networks, such as Telenet or Tymnet, established for public data communications. These networks cover virtually all major cities in the world and process data communications messages as the post office system processes letters. When using packet switching, the sender's and recipient's logical addresses are included with the transmitted data message. The message is then routed the way the packet switching system determines to be most efficient, similar to the way a letter is routed by the postal system.

Network Architectures

Figure 4.9 shows a complex network with many different types of users and systems. When each of these links requires a different software controller, the communications control network becomes quite complex. For example, IBM over time has developed a variety of telecommunications access methods for different types of interconnections. These are referenced by the following acronyms:

TCAM   Telecommunications Access Method
BTAM   Basic Telecommunications Access Method

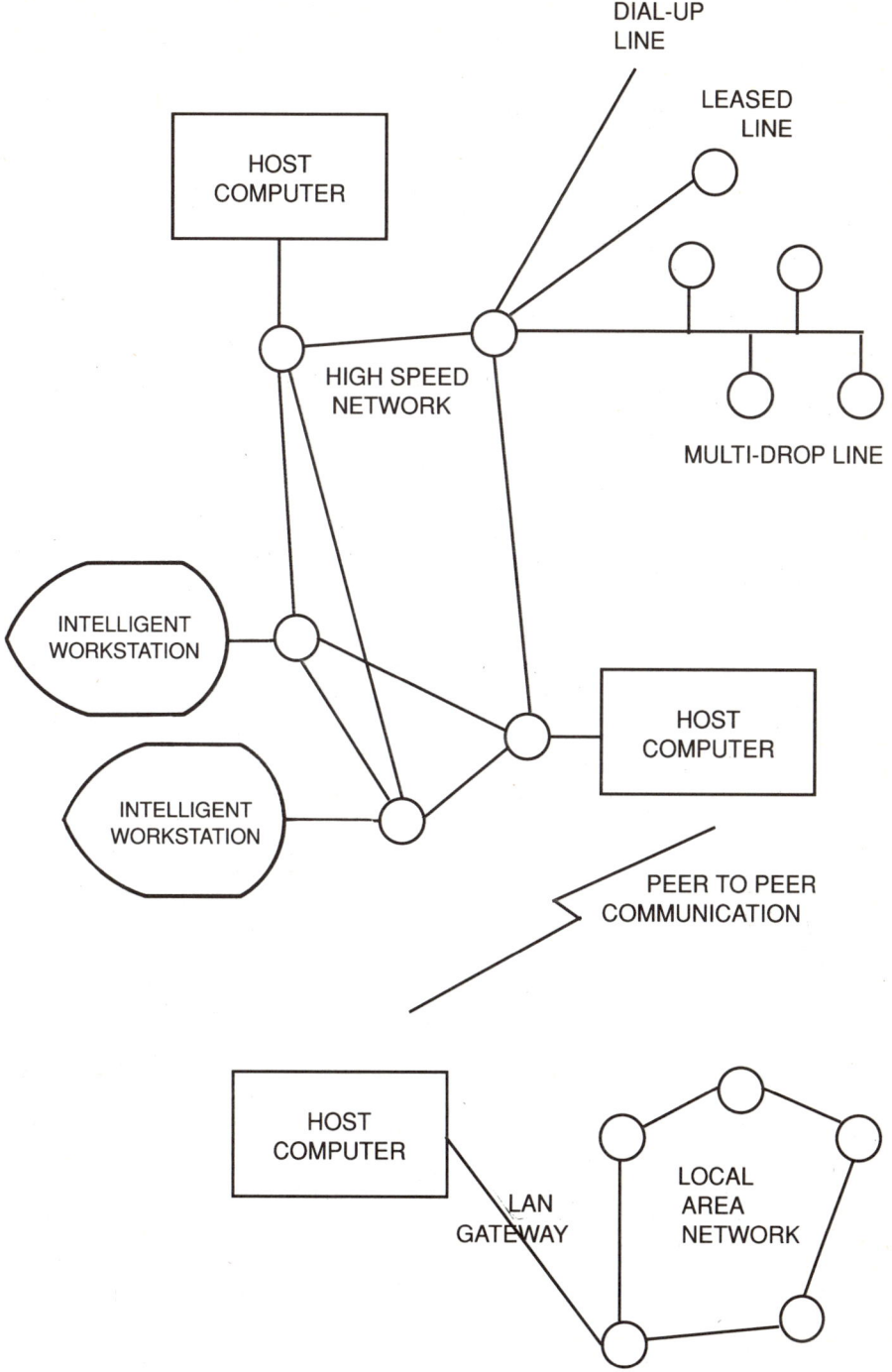

DIAL-UP
LINE

LEASED
LINE

HOST
COMPUTER

HIGH SPEED
NETWORK

MULTI-DROP LINE

INTELLIGENT
WORKSTATION

INTELLIGENT
WORKSTATION

HOST
COMPUTER

PEER TO PEER
COMMUNICATION

HOST
COMPUTER

LAN
GATEWAY

LOCAL
AREA
NETWORK

Figure 4.8. An advanced telecommunications network

RTAM   Remote Telecommunications Access Method

VTAM   Virtual Telecommunications Access Method

Because these access methods may control different network lines, application programs may be required to go through several methods to connect to a particular system. This creates a situation as described in Figure 4.9. To improve efficiency, many larger data processing organizations with complex telecommunications networks have installed "network architecture" software. This software has different product names, depending upon the vendor. It is called System Network Architecture (SNA) for IBM equipment, DECNET for DEC, and so forth. A network architecture environment is shown in Figure 4.10.

Network architecture software removes the need for specialized linkage packages such as BTAM. The architecture handles all teleprocessing and introduces efficiencies for medium to large system users. One can visualize network architecture as a series of layers or protocols which must establish agreement with one another and the other point before data communication can begin.

The International Standards Organization (ISO) has developed what is called a seven-layer model for a network architecture. Although other architectures, such as IBM's SNA, differ slightly, the ISO model provides a representative example of network architectures. One can think of this architecture as similar to a national postal system. The various levels are as follows:

Level 1—Physical layer. This level refers to the physical, electrical, functional, and procedural characteristics necessary to connect physical circuits between communications equipment. Following the post office analogy, the physical layer defines the size, color, and placement of mailboxes.

Level 2—Link control layer. This is the next higher level of communication, where blocks of data are transmitted over the physical layer. This layer provides for reliable interchange of data through such things as flow and error controls. In the post office model, the letter which is placed in the mailbox must comply with certain size and cost rules.

Level 3—Network control layer. Where alternate routes may be taken between two points on a network, this layer of software determines the most appropriate route considering traffic conditions and cost factors. Just as when we mail a letter, users at either end do not need to worry about the routing from one end to the other.

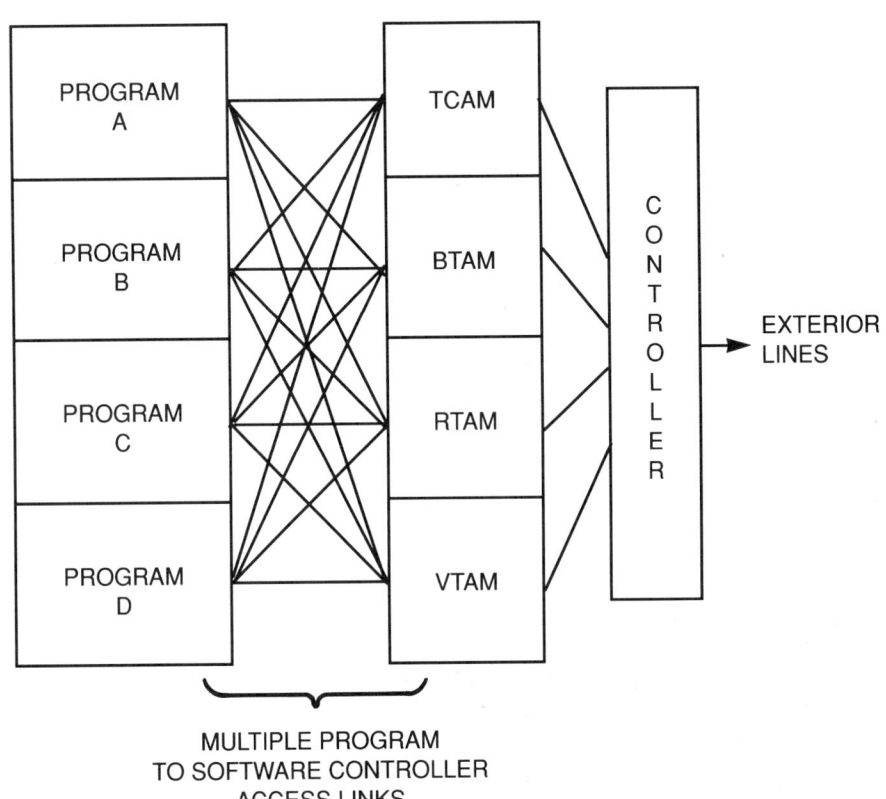

Figure 4.9. Connections to multiple access method

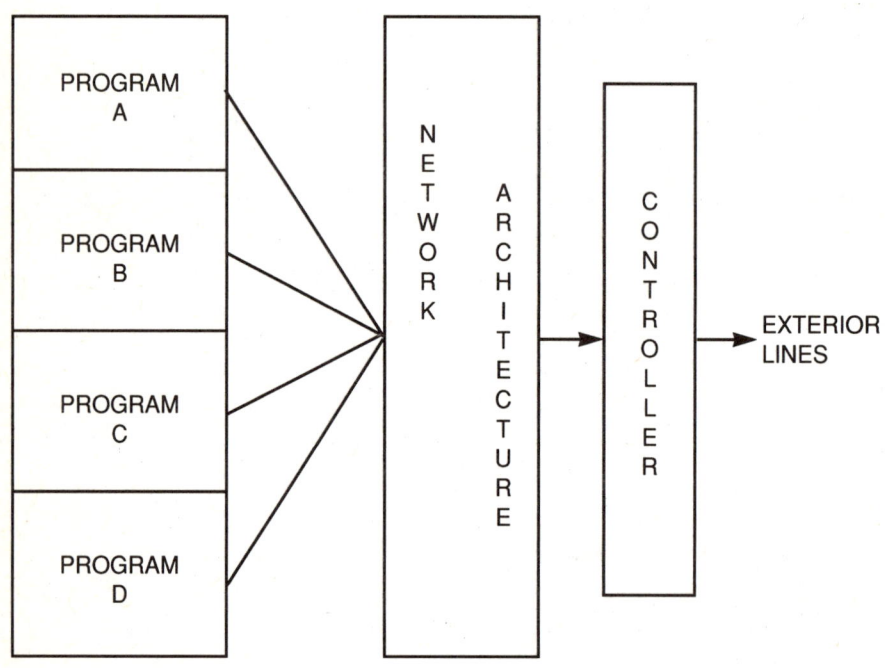

Figure 4.10. Telecommunications network architecture

Level 4—Transport integrity control layer. This level imposes end-to-end integrity controls to ensure that data is not lost or corrupted during a transmission. In the post office sense, there are rules for types of routings and types of carriers.

Level 5—Session layer. This layer defines log-on procedures between two users. The layer supports interaction between cooperating partners or processes. In the postal example, personal addresses and ZIP codes are required.

Level 6—Presentation layer. This layer is essentially a high level programming language where screen display aids, tab formats, and other aids to make data more meaningful are added. Encryption and decryption, if used, take place here. This would correlate to color and size of the envelope.

Level 7—Process layer. This layer carries out the application- and management-related aspects of the communication. Examples would be the manual operation of an automatic teller machine, accessing a remote file, or a microprocessor controller linked to a computerized control system. In the post office analogy, this is comparable to the text of the letter.

The ISO model, while somewhat conceptual, helps the auditor understand the various components in the network architecture that must interface with each other to establish telecommunication connections and controls. Figure 4.11 illustrates how these network layers communicate with each other. Each of the layers must be consistent with others at a node, and each must communicate with the corresponding level or layer at the other communication node. Only after these links have been established can data communications take place.

The security of a network and its data can be affected by errors or failures in any layer of the network structure. Some may be caused by equipment or software failure while others may be caused by telecommunications network management system failures. The auditor should be particularly concerned about the latter.

Some potential network errors and failures that can occur, classified by ISO levels, are listed below. Some errors, such as security violations, can occur at any of several levels, depending upon the type of violation.

Physical layer
   Electrical circuit failure
   Modem problems
   Wire tapping or electronic eavesdropping

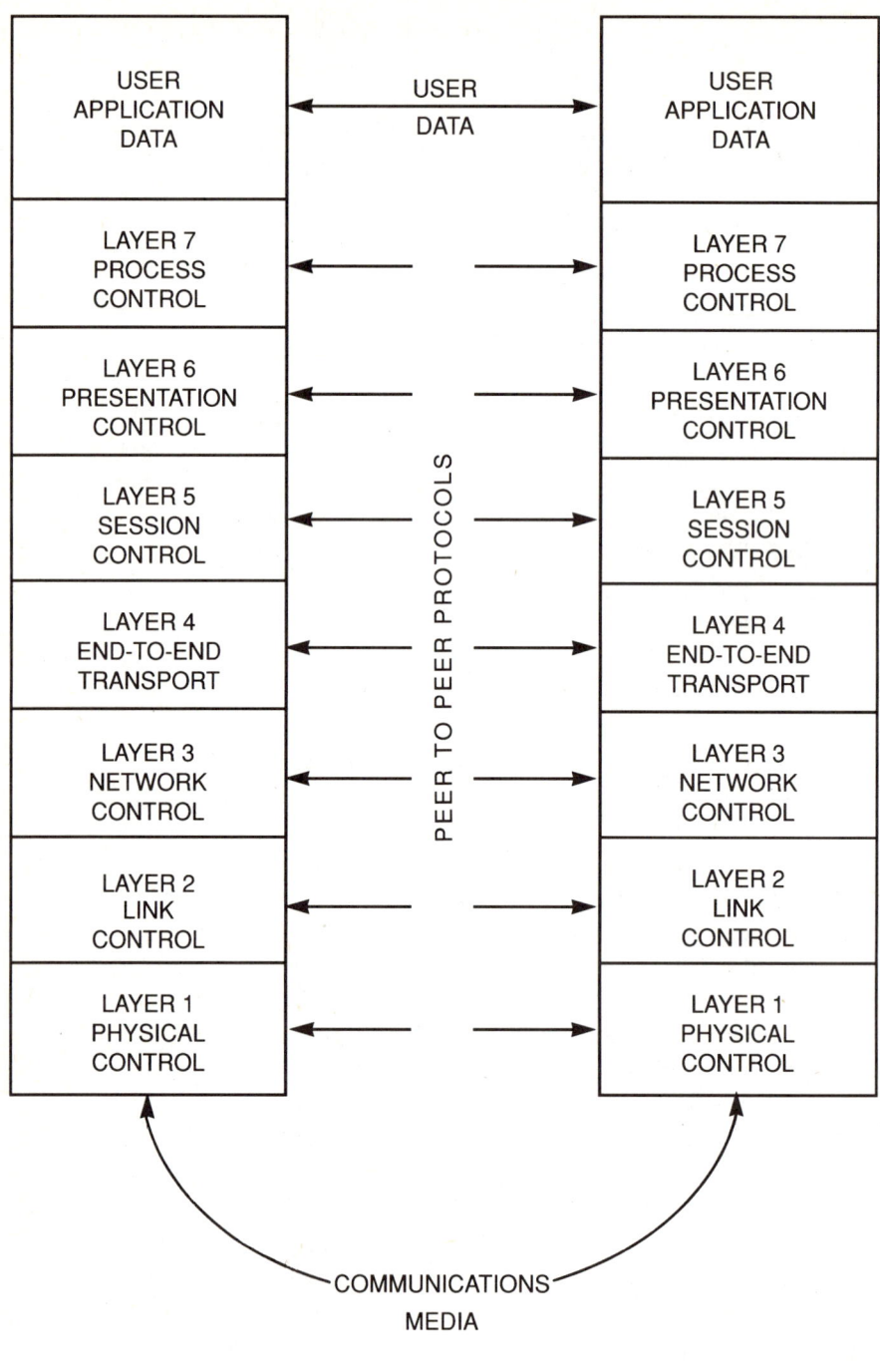

Figure 4.11. OSI model network architecture

Link control layer
  Data bit transmission errors
  Faulty frame or message headers

Network control layer
  Traffic congestion
  Invalid or unauthorized call setups or disconnections
  Failures at network nodes

Transport integrity control layer
  Garbled, incomplete, or lost messages
  Duplicate delivery of messages
  Network security violations
  Network deadlocks

Session layer
  Invalid or unauthorized session requests
  Security violations
  Unauthorized requests or unavailable data

Presentation layer
  Failure to establish appropriate session resources
  Failure to perform encryption or decryption
  Security violations
  Unauthorized requests or unavailable data

It is important for the auditor to have a general understanding of network architectures. Many data processing organizations are installing them, and they introduce a new language of terms and acronyms. Just as auditors once had difficulty talking with data processing professionals in the early days of computers, trained auditors today may face similar difficulties in talking with telecommunications experts if the auditors lack a basic understanding of the terminology.

Auditing Telecommunications Controls

As part of any review of controls in a larger data processing organization, the auditor should include a review of telecommunications controls. This is often a critical control area because some standard, good operating practices existing for other areas of data processing organizations may not exist for the telecommunications function. For example, data processing management generally has a good understanding of management controls required for computer room operations, but often does not realize that these same controls are essential for the management and operation of the telecommunications function as well.

All too often, telecommunications functions within a data processing organization are not given sufficient and proper levels of management attention. Because some equipment resides in the computer room, computer operations may be given responsibility for the network. Because telecommunications management is a rather technical area, sometimes the systems programming function is given partial responsibility. While this arrangement works if there is only minimal telecommunications activity, it usually should be structured as a separate function within the data processing organization. It will be a worthwhile audit and control recommendation to management to establish a separate telecommunications control function within data processing.

The first step for the auditor in reviewing telecommunications controls is to gain an understanding of the network. This can sometimes be difficult because of poor documentation. Programmers, for example, are used to being asked for documentation and do document their work, albeit sometimes grudgingly. Telecommunications managers seldom are visited by auditors and may not be accustomed to audit questions.

Because of the frequent lack of auditor review of telecommunications functions and because of its technical nature, telecommunications management often is skeptical of the auditor's skills and qualifications. Their comments, in effect, ask the auditor, "What do you know about teleprocessing?" This chapter, it is hoped, will give the auditor an introduction to the subject.

The auditor may want to consider an audit of the telecommunications functions as a three-step process:

1. Auditing the telecommunications management function
2. Auditing telecommunications network physical controls
3. Auditing telecommunications network logical controls

The sections following will discuss each of these audit areas and present tables of control objectives and audit procedures. The auditor may want to consider an initial review of the telecommunications function as a separate audit or one combined with distributed processing controls. In subsequent periods, it can be included in the overall review of data processing general controls.

Auditing the Telecommunications Management Function

In the earlier, simpler days, managing data communications and teleprocessing was not that difficult. Data processing management called in the telephone company to install equipment and solve any line or communications problems. With deregulation and the formation of many telephone equipment and carrier operating companies, managing this function is no longer an easy task. A typical data processing organization will use a variety of different vendors to serve telecommunications needs, and when there are troubles, internal management help often is needed.

Auditors should look for a strong telecommunications management function within data processing. This is an important element of the overall control environment. If the network is sufficiently large, consideration should be given to a separate function on the same level as systems programming or database administration. If there is less emphasis (or in a smaller organization), a designated individual should be given responsibility for managing the data processing telecommunications function.

A typical telecommunications management function should be responsible for the following broad areas:

Overall management of the network, including maintaining inventories of equipment and setting standards for allowing new equipment or nodes onto the network.

An ongoing program for monitoring network usage, logging problems, and reporting them, as appropriate.

Periodic reviews of the telecommunications network to better balance loads, achieve cost savings, and install optimal equipment.

Active participation in the systems development (SDM) process, to determine that telecommunications needs and standards are considered for new production applications.

The above responsibilities do not include voice communications. However, some organizations combine voice and data communications under one management function. While many voice telecommunications functions are similar to the management control functions listed above, voice telecommunications controls are not included in the scope of this book. This separation will probably change in future years as voice and data communications become more closely integrated.

Figure 4.12 contains objectives and procedures for auditing the management of the data processing telecommunications function. These audit procedures are written for a large data processing organization. In the smaller organization (or the organization with a more limited network) the auditor will generally find the objectives applicable, but the procedures must be adjusted to the organization's relative size.

Control Objectives and Audit Procedures
for Reviewing Telecommunications Management Controls

Objective 4.12.1. There should be a strong telecommunications management function with the authority to establish standards and procedures.

Procedure 4.12.1.1. Through discussions and review of organization charts, determine that telecommunications management reports at a sufficient level to have the authority to manage and control the function.

Procedure 4.12.1.2. If data and voice telecommunications are separate, assess through interviews the level of coordination between the two functions.

Procedure 4.12.1.3. Determine that job descriptions, defining duties, skill requirements, and reporting responsibilities, exist for all personnel in the telecommunications organization.

Procedure 4.12.1.4. Determine that telecommunications departmental standards exist for such areas as:

Approved equipment types, such as terminals, which can be introduced to networks

Authorization procedures for introducing new equipment to networks

Schedules and procedures for authorizing the introduction of lines and terminals outside normal operating hours

Procedures for the use of any dial-up data lines

Procedure 4.12.1.5. Review any contracts with outside telecommunications network suppliers and other vendors to determine that responsibilities and liabilities have been defined.

Procedure 4.12.1.6 Review the extent of long range plans for telecommunications, including:

Integration of voice and data communications facilities

Planning for higher speed communication devices such as fiber optics, microwave, or the use of satellites

Planning for the integration of telecommunications cables

Figure 4.12. Control objectives and audit procedures for reviewing telecommunications management controls

and equipment in any new building under construction by the organization

Procedure 4.12.1.7.   Review the organization's overall information systems disaster recovery plan (see Chapter 10) and determine that adequate consideration has been given to planning for telecommunications backup and recovery.

Objective 4.12.2.   Adequate records should exist to document and control telecommunications equipment inventories and to manage equipment changes.

Procedure 4.12.2.1.   Determine that inventory lists exist covering all data communications equipment including modems, controllers, terminals, lines, and related devices.

Procedure 4.12.2.2.   Determine that network diagrams are maintained to document both physical and logical connections between telecommunications and other data processing equipment.

Procedure 4.12.2.3.   Select several items of telecommunications equipment, both inside and outside the data processing area, and trace them to inventory records and to network diagrams to determine that records appear to be correct.

Procedure 4.12.2.4.   Review change procedures for such matters as adding a new terminal or changing a port assignment, and determine that they appear adequate and consistent with other data processing operations change procedures.

Procedure 4.12.2.5.   Determine that a formal testing procedure exists covering the introduction of any new equipment or changes to the telecommunications network.

Procedure 4.12.2.6.   Select several new items of equipment or changes to the network over a recent period and determine that formal testing procedures were followed.

Objective 4.12.3.   Procedures should exist for monitoring telecommunications network usage, for making adjustments to improve performance, and for recording and resolving any problems.

Procedure 4.12.3.1.   Determine that performance standards

Figure 4.12. Control objectives and audit procedures for reviewing telecommunications management controls *(continued)*

have been established covering such areas as terminal response times and error checking.

Procedure 4.12.3.2.   Determine that activity is monitored within the on-line system and that adjustments to improve performance are made as appropriate.

Procedure 4.12.3.3.   Review procedures for identifying, documenting, and taking corrective actions for any telecommunications failures; assess whether these procedures appear adequate.

Objective 4.12.4.   Procedures should exist for tracking telecommunications costs and charging them back to appropriate persons or organizations.

Procedure 4.12.4.1.   Develop an understanding of the data communications accounting and chargeback system and assess its adequacy.

Procedure 4.12.4.2.   Select several items of telecommunications equipment, such as modems or terminals located in user areas, and determine that they are recorded properly and the usage is charged back.

Procedure 4.12.4.3.   Determine that billings from outside telecommunications carriers and other vendors are reviewed on a regular basis and that disputed items are resolved promptly.

Objective 4.12.5.   The telecommunications management function should take an active role in the design and development of new, on-line applications to ensure that telecommunications standards are followed.

Procedure 4.12.5.1.   Determine that telecommunications management is involved in pre-implementation planning for new information systems applications which may have a telecommunications impact.

Procedure 4.12.5.2.   Determine that telecommunications capacity planning considerations are taken into account in the design and implementation of new applications.

Figure 4.12. Control objectives and audit procedures for reviewing telecommunications management controls *(continued)*

Auditing Telecommunications Network Physical Controls

Telecommunications physical controls are often a weak link in the overall system of data processing general controls and telecommunications controls. All too often, an otherwise effective disaster recovery plan gives only cursory attention to the telecommunications network. Similarly, telecommunications management may install good controls over access to the network but ignore physical security controls for telecommunications devices.

The auditor often finds that all lines for a network are centralized in a single location. They are exposed to a potential physical disaster along with the rest of the computer equipment in that situation. In addition, the recovery of lost leased lines often takes a very long time. Although in earlier days the dial-up lines were offered as an alternative for recovery, they are insufficient to handle telecommunications traffic in today's data processing center.

Data processing telecommunications management often does not give proper attention to the physical security over lines. In addition to the potential for vandalism, they can be tapped easily. Even though strong access control software has been installed, it is possible to buy inexpensive components at an electronic supply store that will allow a person to wiretap a data communications line. Doing this enables "electronic piggybacking," a form of wiretapping discussed in Chapter 9.

Figure 4.13 contains objectives and procedures for reviewing telecommunications physical security controls. Just as data processing management placed little emphasis on physical security in the early days of business data processing, the auditor may find little emphasis on telecommunications physical security today. However, as auditors begin to point out the need for such controls and as management realizes the importance of the networks to the overall organization, auditors can expect increased emphasis on telecommunications physical security in the future.

Auditing Telecommunications Network Logical Controls

All network users should be approved by management to perform only a limited set of authorized transactions. The network architecture should detect unauthorized users and block them from access to the network. In addition, the network architecture should monitor the integrity of the data transmitted through error detecting codes.

Auditors will often find, in reviewing a larger telecommunications network, that it has grown by accident rather than design. Terminals once installed are

**Control Objectives and Audit Procedures
for Reviewing Telecommunications Physical Security Controls**

Objective 4.13.1. Communications equipment should be maintained in a secure area to prevent improper access.

   Procedure 4.13.1.1. Determine that communications equipment is kept in a secure locked room with access limited to authorized individuals.

   Procedure 4.13.1.2. If some communications equipment, such as a communications controller, is kept in the computer operations area, determine that physical security within that area also is adequate.

   Procedure 4.13.1.3. Obtain a list of persons authorized to enter the communications equipment room and determine that only persons with the responsibility and knowledge to use that equipment are authorized to enter the facility.

Objective 4.13.2. Communications lines should be properly shielded and protected to prevent improper tampering.

   Procedure 4.13.2.1. Observe the areas near the communications equipment room and determine that all lines are out of sight.

   Procedure 4.13.2.2. Determine that communication lines within the equipment room and elsewhere are labeled with a code maintained by telecommunications management, rather than with a physical description.

   Procedure 4.13.2.3. Review any procedures requiring the shielding of cables and terminals to prevent electrical emanations which could be intercepted and read by an unauthorized person.

   Procedure 4.13.2.4. Determine that the data communications network is checked on a periodic basis for active or passive wiretaps.

Objective 4.13.3. Communications test equipment, used to monitor the network, should be secured to limit improper access.

   Procedure 4.13.3.1. Review the purposes and functions of

Figure 4.13. Control objectives and audit procedures for reviewing telecommunications physical security controls

communications test equipment used for resolving data communications problems.

Procedure 4.13.3.2. Determine that adequate controls exist over the telecommunications test equipment used to monitor lines and fix problems including:

Procedures restricting use of the equipment to authorized persons;

Logging facilities built into the equipment to monitor activities;

Proper approval and documentation of entries of data to a communications line to correct a problem.

Objective 4.13.4. Information systems disaster recovery planning should give adequate attention to the recovery of data communications facilities.

Procedure 4.13.4.1. Review the overall disaster recovery plan for information services (see Chapter 10) to determine whether adequate attention has been given to restoring and recovering data communications facilities.

Procedure 4.13.4.2. Determine if contingency plans exist for a "telecommunications only" disaster, such as the failure of a carrier switching station, and assess the adequacy of those plans.

Procedure 4.13.4.3. Review the telecommunications facilities at any contingency plan "hot site" or "cold site" to determine if adequate consideration has been given to physical security at those locations.

Objective 4.13.5. Controls should exist over data communications dial-up lines to prevent improper access to the system or network.

Procedure 4.13.5.1. Determine that "call back" or "see through" security devices have been installed over all dial-up lines to the computer system to prevent unauthorized access attempts.

Procedure 4.13.5.2. Determine that any dial-up numbers are on a three digit exchange number different than the organiza-

Figure 4.13. Control objectives and audit procedures for reviewing telecommunications physical security controls *(continued)*

tion's main telephone exchange to prevent accidental line detection.

Procedure 4.13.5.3. Review procedures for authorizing users to access the dial-up system and determine that all users have been approved.

Procedure 4.13.5.4. Determine that dial-up telephone numbers are changed on a periodic basis.

Figure 4.13. Control objectives and audit procedures for reviewing telecommunications physical security controls *(continued)*

seldom reviewed for usage or access rights. This results in a telecommunications network exposed to excessive cost, potential unauthorized access, and erroneous record keeping.

While the auditor with a good understanding of basic control techniques easily can review and make effective recommendations for administrative controls, logical integrity controls may require additional technical expertise. The auditor must develop a general understanding of the network architecture installed within the data processing function. That understanding should focus on features that are or can become available. The auditor can then determine if they have been implemented.

The types of logical integrity controls that the auditor should consider follow. Some, such as error checking, should always be present, while others, such as encryption, can be on an as-needed basis.

1. *Transmission error checking.* Error checking codes should be used to detect bit errors in transmissions due to line noise or power fluctuations. These codes should include echo checks so that a message transmitted to a remote site is echoed back for verification.

2. *Automatic rerouting.* Some architectures can switch different data blocks of a single message along alternate routes, so that a wiretapper can get only random fragments of a message.

3. *Audit trail logging.* User network activity should be recorded through a logging facility to assist in the reconstruction of data files and transactions for network nodes. These logs also help detect illegal access attempts.

4. *Store and forward capability.* The network controllers should be able to store messages when lines are busy and then forward them when the receiving station is free. This capability helps avoid loss of messages or transmissions.

5. *Encryption*. Encryption is encoding transactions at transmission and decoding them upon receipt. It is particularly important for applications involving sensitive data, such as a wire transfer system.

Figure 4.14 contains objectives and procedures for auditing telecommunications logical controls. To perform this step effectively, the auditor should have a general knowledge of the network architecture in place as discussed earlier.

CONTROLS OVER DISTRIBUTED PROCESSING APPLICATIONS

Prior sections of this chapter discussed telecommunications controls as well as variations of network architectures. The typical distributed processing system offers other control considerations for the auditor. This section discusses the unique aspects of a distributed processing system that will impact the auditor's review of application controls.

Auditors normally review application controls after the general controls review in a data center. The applications review often involves gathering evidence by tracing selected input documents through the application to determine that processing is correct. This general approach presents the auditor with a greater challenge when reviewing a distributed processing application.

When an auditor is reviewing a distributed processing application at a given data center, the auditor may find no paper trail or a very limited one for the application at the center under review. Often, this paper trail may exist only at the remote, distributed sites in the network. Logistics and geography often make review of the controls at the remote, distributed processing sites impossible. However, the auditor must gain some level of assurance that the remote processing is consistent with the controls environment at the auditor's location.

If the distributed processing network is limited to several large computer systems on a peer to peer network, this may not be a great problem. The auditor or another colleague in the audit function may recently have reviewed general controls at that remote computer system. Although the specific distributed application may not have been covered in that review, the auditor will have available documentation about such controls as access to data and access to program libraries.

The problem becomes more difficult in a larger distributed processing network. For example, it is possible to build a star network distributed processing system consisting of many microcomputers performing remote

Control Objectives and Audit Procedures
for Reviewing Telecommunications Logical Controls

Objective 4.14.1. Passwords and other procedures should exist to limit and detect any unauthorized access attempts to utilize the telecommunications network.

Procedure 4.14.1.1. Determine that passwords and unique user IDs are required to sign on to all communications software.

Procedure 4.14.1.2. Review procedures for changing passwords and determine that codes are changed regularly.

Procedure 4.14.1.3. Determine that users are locked out of the system after a designated number of unsuccessful access attempts.

Procedure 4.14.1.4. Observe a terminal sign-on session and determine that:

Users are given no systems help prior to successful sign-on

Passwords are not displayed on user screens when entered during the sign-on

During the sign-on, users are informed of when they last signed-on to help identify potential unauthorized usage

Procedure 4.14.1.5. Determine that any hardware or software standard access protocols, such as those delivered with original installations, have been changed or deleted.

Objective 4.14.2. Error checking facilities should exist to detect transmission errors and establish retransmissions if appropriate.

Procedure 4.14.2.1. Determine that a software log is maintained for all transactions including errors and retransmissions.

Procedure 4.14.2.2. Discuss error checking procedures and facilities, such as parity tests, with telecommunications personnel and assess whether such procedures appear adequate and whether any have been disabled.

Procedure 4.14.2.3. Determine that all transmitted messages contain identifying data regarding source, date, and time.

Figure 4.14. Control objectives and audit procedures for reviewing telecommunications logical controls

Procedure 4.14.2.4. Determine that the communications software provides a facility for reporting all duplicate, distorted, or delayed messages and assess whether these exception items are being reviewed for corrective action.

Objective 4.14.3. If sensitive data is being transmitted, the system should use automatic rerouting facilities to limit any wire tap access, as well as controls to ensure that transmissions go only to authorized users.

Procedure 4.14.3.1. If sensitive data is being transmitted, determine whether the communications architecture allows automatic rerouting of messages so that single messages follow alternate routes and a wire tap perpetrator would find only random fragments.

Procedure 4.14.3.2. Determine that adequate controls exist to return sensitive data to authorized terminal and remote print facilities only.

Objective 4.14.4. All network activity should be recorded on a logging facility to assist reconstructing files and detecting unauthorized accesses.

Procedure 4.14.4.1. Determine that logging procedures exist to capture and help reconstruct all transaction activities.

Procedure 4.14.4.2. Determine that console log files are reviewed, through automated tools if possible, on a daily basis to monitor for improper access attempts.

Procedure 4.14.4.3. Review procedures for using log files to reconstruct improper access attempts and assess their adequacy.

Objective 4.14.5. Data encryption techniques should be used where there is a strong risk of improper access to sensitive transmissions.

Procedure 4.14.5.1. Determine if there has been a proper risk analysis (see Chapter 8) performed over data processing applications to identify areas where encryption controls might be appropriate.

Figure 4.14. Control objectives and audit procedures for reviewing telecommunications logical controls *(continued)*

Procedure 4.14.5.2. If encryption is being used, review control procedures over the generation and issuance of encryption keys and assess their adequacy.

Procedure 4.14.5.3. Determine that encryption keys are changed on a regular basis.

Procedure 4.14.5.4. Review procedures for transporting encryption keys from the generating site to the encryption devices and assess adequacy.

Procedure 4.14.5.5. Interview a selected user of an application which uses encrypted data and assess whether that user has a proper understanding of encryption procedures and of the sensitivity of the application data.

Figure 4.14. Control objectives and audit procedures for reviewing telecommunications logical controls *(continued)*

processing and transmitting results to a central computer site. Earlier systems designers minimized the control problem by having the central computer site simply download programs to the remote sites, which might have only retail, point of sale machines. The problem is more complex when those remote systems are, for example, microcomputers based on the 80386 or other powerful processors, which can have the power of much larger computer systems. Without some type of integrity controls, it is difficult to determine whether improper changes or procedures have taken place at these remote sites.

These comments may raise questions as to whether such distributed systems are auditable at all. The answer is generally yes, but special procedures must be performed. The auditor's first objective in such a review is to gather evidence about the distributed system and understand its controls. In addition to the auditor's review of controls over the telecommunications network, the auditor should understand the following areas:

Distribution of data. There should be a central point in the distributed network to control and coordinate the policies and procedures for the distribution of data within the system.

General controls at distributed sites. There should be procedures to ensure that distributed sites are following good general control procedures.

Distributed system security and auditability. Proper security controls should be built into distributed systems to ensure security and privacy of data. In addition, the auditor must be able to assess risk and report to management that a given distributed application is auditable.

Before starting a review of the distributed processing network, however, the auditor should determine that the system represents a true distributed processing system. "Distributed" is a popular and commonly misused data processing term. The system or application may be just a decentralized application where remote sites simply gather data for processing at the central computer system. The general controls environment for such systems would be similar to the environments described in Chapters 2 and 3.

Understanding the Distribution of Data

Data can also be distributed in several unique ways. The auditor should understand how data is distributed and controlled. Distribution in a distributed processing system usually takes one of the following forms:

Distributed processing only, with no data distributed. Most distributed processing systems distribute processing functions to multiple sites with the data controlled at a central administration site. Remote sites communicate with this central site for database retrievals and updates, and perform local processing only. In other instances, the central database is replicated on a periodic basis, and a copy is transmitted to the remote sites. These configurations are illustrated in Figure 4.15.

Split data systems. Some distributed systems split data among different locations. For example, a large organization may have each of its regional locations process against the same database version but with each location maintaining its own data. This configuration is shown in Figure 4.16. The unique feature of this type of a system is that data is shared among the various system nodes.

Separate data, or cooperative processing, systems. Some distributed systems have separate and different databases at each node with data shared among them. This environment, shown in Figure 4.17, might exist in a large organization where a manufacturing division maintains a separate database for its functions while other divisions maintain their own databases with certain processing functions and applications distributed among them. The auditor can expect an increasing number of applications to be based on this type of system.

While there is strong justification for each of the above types of distributed data systems, the auditor always should look for a central administration function over the distributed processing network. This role can be filled by the data administration function or the systems analyst responsible for the distributed processing system.

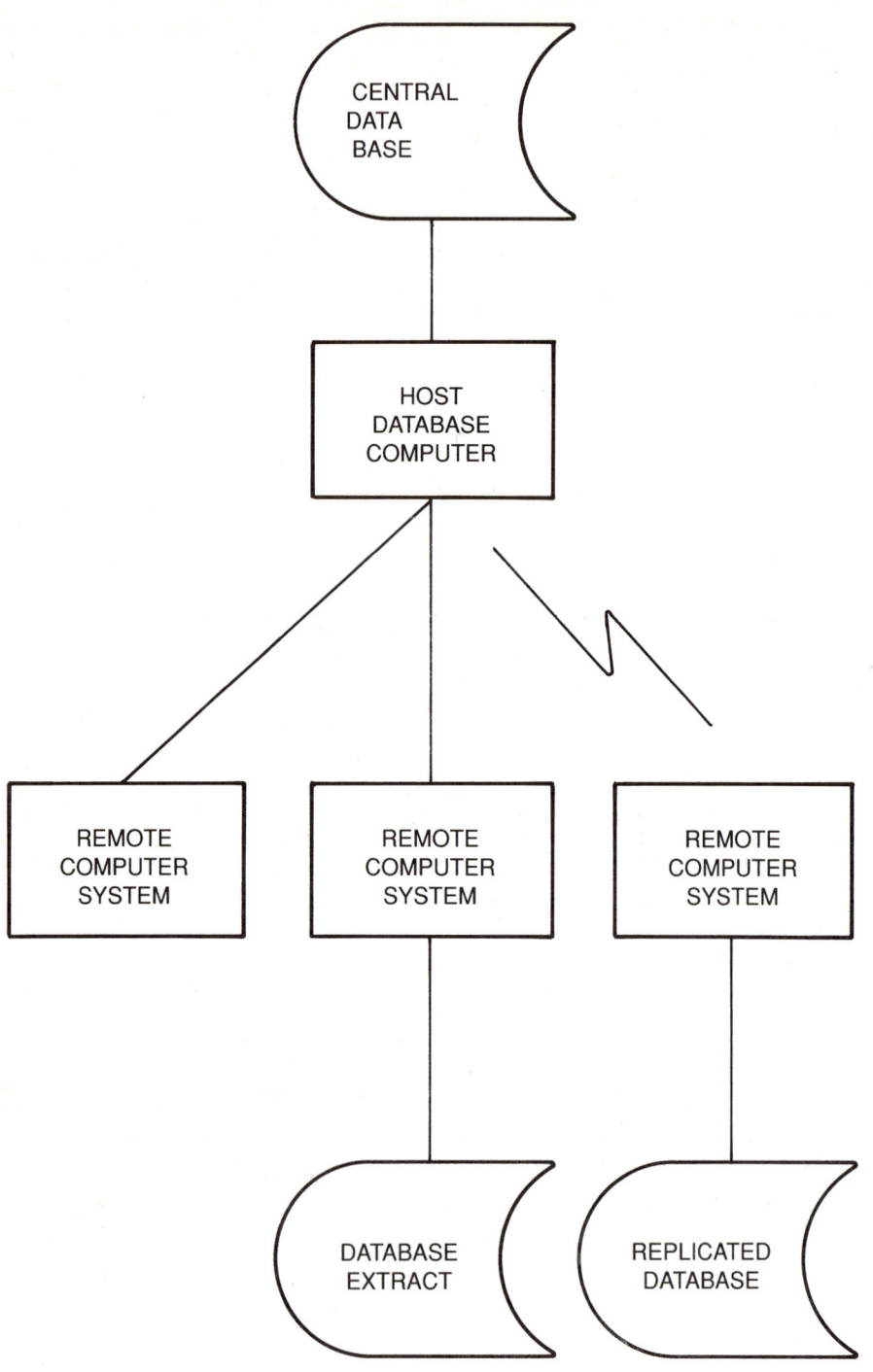

Figure 4.15. A multiple site distributed processing system

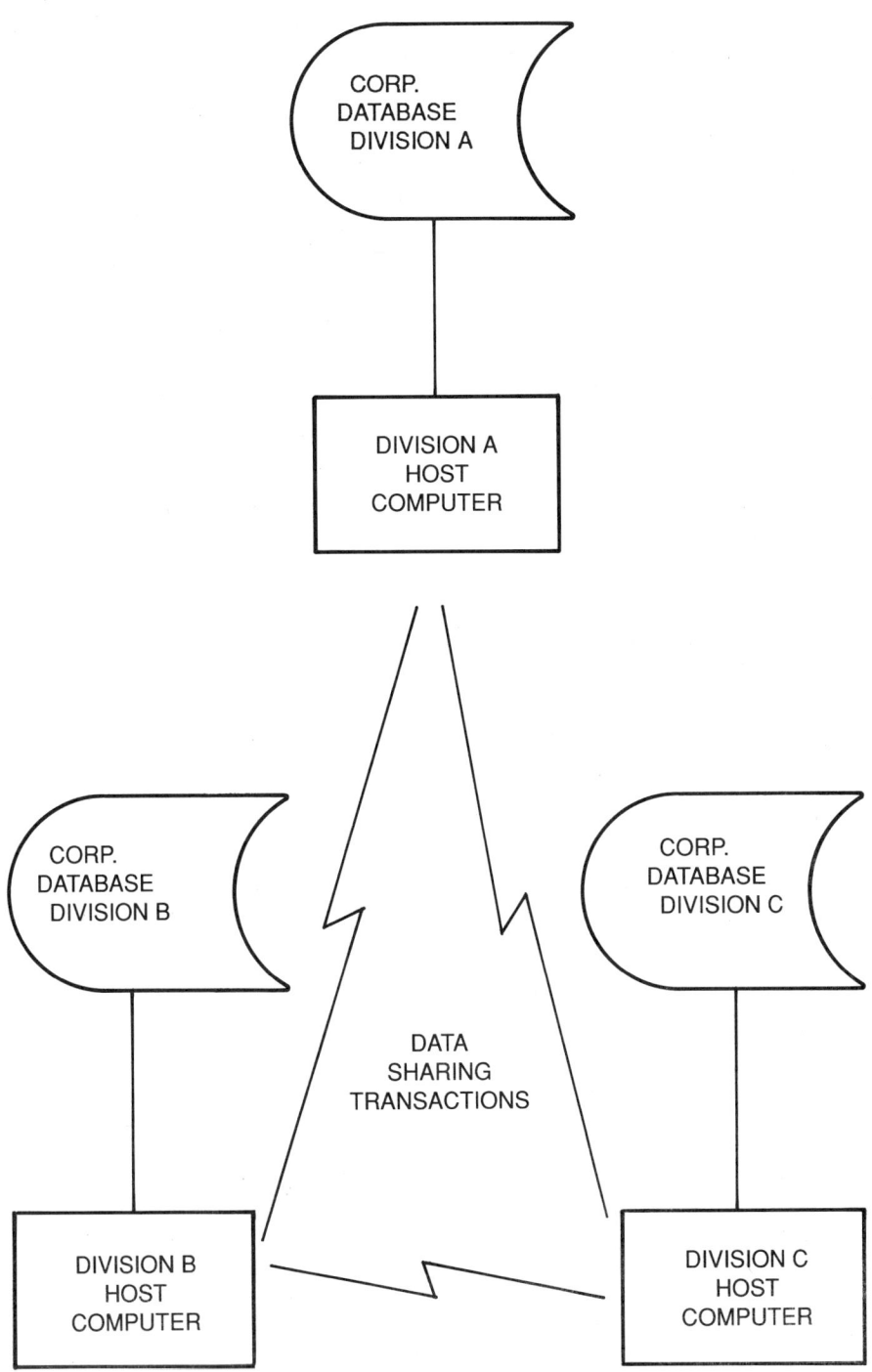

Figure 4.16. Split database distributed processing system

181

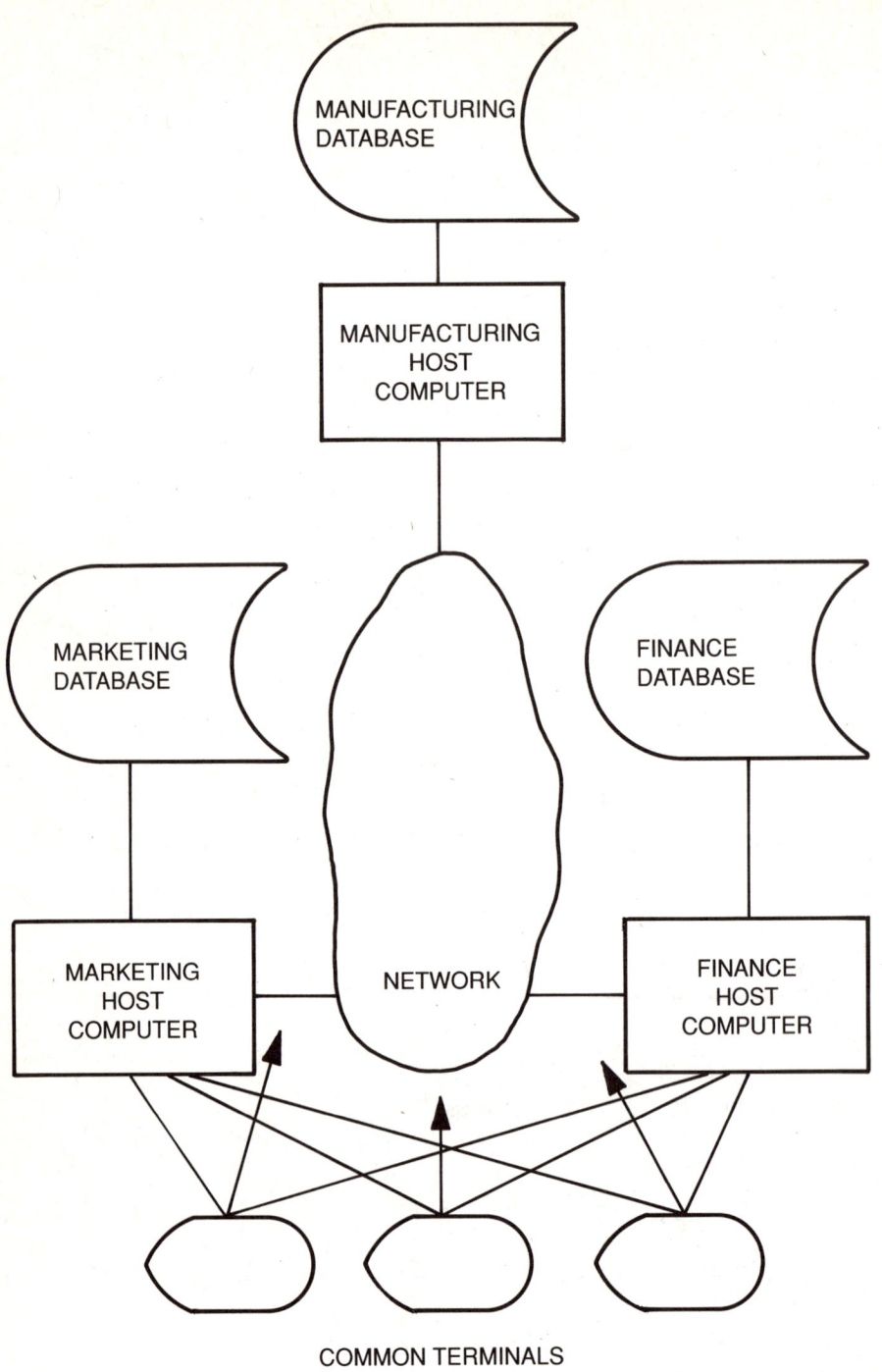

Figure 4.17. A separate database distributed processing system

The strongest control environment for a distributed processing system is a central database with distributed processing as in Figure 4.15. However, this environment can cause control problems. If two distributed processing locations try to update the same record or element of a database at the same time, "update interference" can occur. If the first one to read the record does not lock out the second, both will update the record independently, resulting in an incorrect record, as follows:

1. Location A reads record X and begins to update it.
2. Location B reads record X and begins to update it.
3. Location A writes out its updated record X.
4. Location B writes out its updated record X.

The final record should reflect both A's and B's actions, but in this example it will reflect B's activity only. This problem is solved by many database software products by locking out Location B until A's activity is complete. However, this can cause the problem of "deadly embrace" which occurs in the following situation:

1. Location A reads and locks record Y.
2. Location B reads and locks record Z.
3. Location A tries to read Z to complete update and finds it locked.
4. Location B tries to read Y to complete update but finds it locked.
5. A "deadly embrace" occurs and the system hangs in limbo!

Good systems design and a proper database management system will prevent many of these problems. However, they must be a concern when the database is centralized and processing is distributed.

Replicated or separated databases for distributed processing can result in many other control problems. When multiple copies of data exist, they can become quite different due to different update cycles or local machine failures. Unless the files are synchronized periodically, they risk becoming further and further apart and common program routines may no longer work.

The auditor will not solve these design problems in an applications review. The auditor can point out, however, that strong central administrative controls are necessary to maintain data integrity in a distributed processing system. The auditor can also point out system design and software problems which can lead to data integrity problems in a distributed processing system.

Figure 4.18 contains objectives and procedures for reviewing controls over

**Control Objectives and Audit Procedures
for Reviewing Distribution of Data Controls**

Objective 4.18.1. The distributed processing system should have a central data administration function to establish overall standards for the distribution of data across data processing applications.

Procedure 4.18.1.1. Develop an overall understanding of the distribution of data and the distributed processing environment including whether there are centralized, split, or distributed files and document the understanding.

Procedure 4.18.1.2. Review central control procedures over distributed processing applications and any distributed data files to determine that:

When multiple applications use a single database, one application or function is responsible for controlling the overall integrity of the database;

When databases are split between two applications or processing functions, there are procedures to monitor overall database integrity;

When portions of a database are downloaded to smaller computer processors or when data is uploaded from those processors, overall database integrity will be maintained.

Procedure 4.18.1.3. Determine that the central data administration function monitors data integrity and usage of each distributed database element.

Objective 4.18.2. A strong data administration and data dictionary function should be in place to control the distribution of data across application files.

Procedure 4.18.2.1. Determine that data dictionary tools are used to document and help monitor the responsibilities for various data items or elements among the distributed or cooperative processing systems.

Procedure 4.18.2.2. Review procedures for authorizing requested changes to databases from distributed applications

Figure 4.18. Control objectives and audit procedures for reviewing distribution of data controls

in order to take into account the requirements of all users of the distributed applications.

Procedure 4.18.2.3. Review procedures for testing and implementing changes to distributed applications and databases to determine whether test systems are used and whether all key aspects of the applications are tested.

Procedure 4.18.2.4. When a centrally developed set of application programs or program updates is distributed to multiple processing sites, determine that procedures exist to ensure that all program updates have been installed.

Objective 4.18.3. Audit trails should exist for all activities from distributed processing applications against their own and other shared distributed databases.

Procedure 4.18.3.1. Based on a survey of various hardware and operating system types, determine that facilities exist to maintain transaction log files for all elements of distributed data systems.

Procedure 4.18.3.2. Survey a sample of distributed application users to determine that transaction log files are being retained for a proper period of time.

Procedure 4.18.3.3. Determine that procedures exist at the central data administration facility to access audit trail logs from distributed locations in order to solve file problems.

Procedure 4.18.3.4. Review the integrity controls and procedures for recovery from hardware, software, or telecommunications line failures and assess their effectiveness.

Objective 4.18.4. Software controls should exist to prevent update interference on central databases in distributed systems.

Procedure 4.18.4.1. Review database management system controls to determine that there are lockout controls to prevent update interference.

Procedure 4.18.4.2. Review procedures for monitoring data integrity with selected users of the distributed processing application and assess their adequacy.

Figure 4.18. Control objectives and audit procedures for reviewing distribution of data controls *(continued)*

Objective 4.18.5.   Adequate consideration should be given to costs and benefits in the design of any distributed processing environment.

   Procedure 4.18.5.1.   Determine that adequate consideration is given to central system costs, telecommunications costs, and distributed systems costs when assessing the total costs of any distributed application.

   Procedure 4.18.5.2.   Review the documented cost savings for a given distributed application and determine if the savings are based on hard cost savings, possible future cost avoidances, or intangible future benefits.

   Procedure 4.18.5.3.   Select a documented cost versus benefit analysis from an implemented application for review, if available, and determine whether proposed benefits have been achieved.

Figure 4.18. Control objectives and audit procedures for reviewing distribution of data controls *(continued)*

the distribution of data. Many are stated in a general manner, and the auditor needs to understand the distributed processing application being reviewed (discussed in Chapter 5).

General Controls at a Distributed Processing Site

In many respects, general controls at a distributed processing site will not differ from the general controls discussed in Chapter 2 for large computer centers or Chapter 3 for mini and microcomputer sites. Each of these chapters discussed the need for good control practices within computer operations and the system development function. In a distributed processing environment, an additional control element is introduced in that the distributed site must also comply with overall control procedures necessary for proper functioning of the distributed network.

The auditor always should focus on the central administration of the distributed processing network when reviewing general controls at the central site or at one of the nodes on the distributed network. That central administration should issue policies and procedures for control and maintenance of the distributed applications. In addition, the central administration group should maintain control over common programs and updates issued to the distributed sites.

When the auditor is performing a general controls review at a computer center that is also a node on a distributed network, the auditor should look for compliance with overall distributed processing procedures. Since the auditor will only be visiting that or any node on an occasional basis, the auditor should look for the active participation of the distributed processing control function in monitoring the integrity of the distributed processing nodes.

Whether databases are distributed or centralized, particular attention should be given to backup and recovery procedures. Because of the complexity involved when many remote locations process and update common files, it is essential that there be ongoing file integrity. If the central database fails, without proper procedures it is likely that one or more of the distributed nodes will lose transactions. If there are replicated databases, system recovery and resynchronization will be difficult unless there are good recovery procedures at each of the nodes.

Figure 4.19 contains objectives and procedures for reviewing general controls over a distributed processing network. These procedures must be considered only as a supplement to the other general control procedures (discussed in Chapters 2 and 3).

Distributed Processing Security and Auditability

Both security and auditability issues cause concerns for the auditor when reviewing a distributed network or a distributed application. Because many users may be tied into an overall telecommunications network and have access to processing occurring on various nodes, there may be a greater security risk than in a conventional, centralized system. A related issue and auditor concern is the level of personal privacy rights maintained over data on the distributed network.

Auditability of distributed systems also causes a special concern. As discussed earlier in this chapter, auditors often look for paper based audit trails when reviewing an application. In a distributed system, that paper trail may be difficult or impractical to reconstruct due to the geographic and logistic construction of the distributed network. The auditor may need to investigate other methods of evidence gathering to assess control risk in a distributed environment.

Chapter 6 discusses many security issues that the auditor will encounter in any data processing system, whether distributed or not. While the objectives and procedures discussed there generally apply to centralized systems, the auditor may emphasize the security aspects of a review for distributed systems. Depending upon the type of distributed system, the auditor may increase the security emphasis in the following areas:

**Control Objectives and Audit Procedures
for Reviewing Distributed Processing General Controls**

Objective 4.19.1. Adequate procedures and documentation should exist and should be understood by all users of the distributed system.

Procedure 4.19.1.1. Review procedures and published documentation which covers both the distributed hardware processors and any distributed applications to determine that these appear adequate and up to date.

Procedure 4.19.1.2. Review procedures for sending documentation updates to remote distributed processing locations and determine whether adequate controls are in place to ensure that changes have been implemented.

Procedure 4.19.1.3. Visit a selected distributed processing site to determine whether users understand the operation of the distributed processing system and whether documentation is up to date and filed properly.

Procedure 4.19.1.4. Determine whether a "help desk" or similar procedure exists to answer distributed processing remote site questions and to solve user problems.

Objective 4.19.2. There should be adequate backup and recovery procedures throughout the distributed system to assure ongoing system integrity.

Procedure 4.19.2.1. Review published procedures, including schedules for backing up data from any separate distributed processing nodes and determine whether they appear to be consistent.

Procedure 4.19.2.2. Examine records covering any recent distributed system recovery involving two or more nodes and assess whether published recovery procedures appeared to be effective.

Procedure 4.19.2.3. Visit a selected distributed processing remote node site and review the status of file backups, compliance with published backup procedures, and off site storage of backup files.

Figure 4.19. Control objectives and audit procedures for reviewing distributed processing general controls

Procedure 4.19.2.4. Examine the information systems organization's overall disaster recovery plan (see Chapter 10) and determine if appropriate consideration has been given to the recovery of multi-node distributed processing systems.

Objective 4.19.3. Access to distributed program libraries and data files should be controlled to maintain the integrity of related application programs run on the distributed processing system.

Procedure 4.19.3.1. Determine that only object code versions of programs are made available to distributed processing sites to prevent local modifications.

Procedure 4.19.3.2. Review program library versions or files at several remote, distributed processing locations to determine that each has the correct version of key distributed application programs.

Procedure 4.19.3.3. Determine that password schemes or other information security procedures exist at all distributed locations to control access to data.

Procedure 4.19.3.4. If the distributed system includes the use of a fourth generation language or similar application, determine that procedures are in place to prevent unauthorized updating of data files with that processor.

Objective 4.19.4. Effective data security policies should be implemented at all distributed processing locations.

Procedure 4.19.4.1. Review physical security controls at selected distributed locations, and if security procedures are restrained by equipment limitations at any site, determine that compensating restrictions have been placed over operator actions.

Procedure 4.19.4.2. If the distributed application requires users to access other nodes in the distributed network, determine that adequate security procedures exist at each node.

Procedure 4.19.4.3. Interview selected users at a distributed processing site to assess their understanding of data security concerns regarding the distributed processing system.

Figure 4.19. Control objectives and audit procedures for reviewing distributed processing general controls *(continued)*

Objective 4.19.5. There should be overall change procedures for installing new hardware, data files, or program components to the distributed system.

Procedure 4.19.5.1. Review procedures for introducing changes to the distributed processing system and assess whether adequate attention has been given to maintaining overall system integrity.

Procedure 4.19.5.2. Determine that a test system has been established to simulate the distributed processing system and monitor the effect of changes to that system.

Procedure 4.19.5.3. Review procedures for resolving problems which may have been introduced to the distributed system due to an error in a system modification and assess the adequacy of those procedures.

Procedure 4.19.5.4. Review change documentation records at the central control site and assess their correctness through a review of the actual equipment and programs at a selected distributed site.

Figure 4.19. Control objectives and audit procedures for reviewing distributed processing general controls *(continued)*

Network user identification. Because there may be many and varied users on the distributed processing network, it is essential that users be identifiable by the system before processing begins. Many computer systems have strong information security controls over on-line users but far weaker controls over decentralized processing. With distributed processing, a decentralized user has some of the same rights as a conventional on-line user. Information security controls should be used for all incoming distributed processing transmissions.

Network activity monitoring. Because of the many users or processing nodes capable of accessing a distributed network, good tools to monitor all transaction activity are essential. They allow problems to be solved as they occur and provide a means to improve the overall performance of the network.

Distributed network recovery procedures. As mentioned previously, the distributed network should be protected physically from loss or destruction and should have good procedures for recovery of data and equipment in the event of an unavoidable loss.

Problems associated with distributed network security can multiply as the distributed network becomes more complex. For example, a microcomputer can be attached to a distributed processing node. That microcomputer may have access rights to the node for local processing purposes. However, it is often easy for that microcomputer to gain access, through authentication rights, to the central distributed processing site. The same microcomputer also could be equipped with its own modem so that others could dial into the microcomputer and, through it, penetrate the rest of the network. While these problems are not unique to a distributed network, they are exacerbated by the complexity of the network.

The auditor may find "see-through" security particularly appropriate for distributed networks. With this type of security, a user possesses a device containing an encryption algorithm and key unique to that individual. The host computer system has the same algorithm and grants authorization back to users through a comparison of unique passwords generated for each session. This concept of "see-through" security is discussed in more detail in Chapter 9.

An important element in the security program for a distributed network is an active procedure to monitor and report potential violations. If users of such a network realize that someone is monitoring their actions, they will be less likely to violate the system. For example, in a distributed banking system environment, the branch security office should receive a report from the central system reporting all access attempts from that branch.

Distributed system complexity often makes the problem of auditability at least as great as security or some of the other control problems discussed in this chapter. Paper audit trails may not exist, and systems may be sufficiently complex that normal audit procedures are difficult to perform. Normal procedures for evidence gathering, such as use of computer audit software, may be impractical in a distributed environment.

Before declaring a distributed system to be unauditable, however, the auditor should work with the distributed processing network administration function as well as application developers to determine what tools might be available for determining activity on the distributed network. These can be helpful in testing the controls in the system.

Figure 4.20 contains objectives and procedures for reviewing the auditability of a distributed system as well as planning a detailed applications audit of such a system. Because of the diverse forms that such systems take, the auditor should understand the application's network and the control characteristics. These objectives allow the auditor to develop detailed procedures for testing controls over the distributed application.

CONTROLS OVER OFFICE AUTOMATION AND LOCAL AREA NETWORKS

Auditors once gave little attention to office automation systems and equipment. Office automation equipment was thought of as little more than word processing devices which had little criticality in contrast to other data processing systems. However, the world of office automation is changing. This chapter describes local area networks (LANs) and how various devices can be connected to one another and to central computers to process messages, receive data from central computers, and act as distributed processing nodes. In addition, many offices are replacing traditional, dedicated word processing devices with microcomputers which can perform word processing and other data processing tasks.

Many auditability and security concerns that existed in the traditional data processing environment also exist in the office automation LAN. In some instances, they may be greater for the auditor today due to the following challenges:

Criticality of data. In a traditional data processing environment, it is relatively easy to determine which files or data are critical. However, office automation files may have data at various levels of criticality. A word

Control Objectives and Audit Procedures
for Reviewing Distributed Processing Applications Controls

NOTE: These objectives and procedures are designed to guide the auditor in performing a review of a distributed processing application. The Audit Procedures are general and must be tailored to the specific application being reviewed. Refer to Chapter 5, Auditing Data Processing Applications, *for more guidance on tailoring such general procedures to an application audit.*

Objective 4.20.1. Documentation should be available at a central location which describes the data and programs at all distributed processing locations or nodes.

Procedure 4.20.1.1. Determine that there is a central control point for the distributed processing application with responsibility for maintaining and circulating application documentation.

Procedure 4.20.1.2. Review procedures for maintaining and distributing documentation to processing nodes; assess the adequacy of those procedures.

Procedure 4.20.1.3. Determine that control procedures exist for the distribution of tabled data and other processing parameters across all nodes of the distributed processing application.

Procedure 4.20.1.4. Visit a remote processing node to determine that the documentation and any tabled program data is current and is understood by users at that node.

Procedure 4.20.1.5. Assess whether the distributed application contains any highly sensitive programs or files and consider the adequacy of procedures to protect them from tampering.

Objective 4.20.2. There should be consistent training as well as documented procedures so that application users at all processing nodes have a good understanding of application functions and can use the applications consistently.

Procedure 4.20.2.1. Review procedures for training users of the distributed application at each of the processing nodes to determine whether training is consistent.

Figure 4.20. Control objectives and audit procedures for reviewing distributed processing applications controls

Procedure 4.20.2.2. Determine that procedures for backing up the application and for restoring files are consistent.

Procedure 4.20.2.3. Review restore procedures for a multi-node application file to determine whether the procedures are in place and are understood by remote users.

Procedure 4.20.2.4. Review procedures for responding to questions or problems from remote node users through the use of such mechanisms as a central "help desk"; assess the adequacy of those procedures.

Objective 4.20.3. The flow of input transactions and data across all nodes of the distributed system should be traceable.

Procedure 4.20.3.1. Determine that all transactions throughout the distributed application are identified, at a minimum, with the name and location of the originator and the date and time of the transaction.

Procedure 4.20.3.2. Review controls over data and transaction file transfers across processing nodes to determine that all data is being transferred correctly.

Procedure 4.20.3.3. Determine that data entry integrity controls exist and that each entry point within the distributed processing application is responsible for the accuracy of its own set of input transactions.

Objective 4.20.4. The results of similar processing actions across distributed processing nodes should be consistent.

Procedure 4.20.4.1. Through a review of documentation and discussions with application users, assess whether the distributed processing application is processing similar data in a consistent manner throughout the network.

Procedure 4.20.4.2. Using central records, review program modules and versions located at the distributed processing nodes to determine that the same versions of key programs are used at all nodes.

Procedure 4.20.4.3. Review procedures for distributing application program revisions to processing nodes as well as pro-

Figure 4.20. Control objectives and audit procedures for reviewing distributed processing applications controls *(continued)*

cedures for affirming that revised programs have been installed.

Procedure 4.20.4.4. Construct a "test deck" of sample transactions (see Chapter 6) for testing processing at multiple processing nodes; arrange to process these audit test transactions and determine whether they produce consistent results for both normal and error status transactions.

Objective 4.20.5. Outputs from the distributed processing application should be made available only to those with a need for those outputs.

Procedure 4.20.5.1. Review application access and output report controls to determine that only authorized individuals are allowed to access application data.

Procedure 4.20.5.2. If users retrieve data or receive outputs from other nodes within the distributed application, determine that controls exist to log those outputs and to release them only to authorized persons.

Procedure 4.20.5.3. Determine that there is an output from the distributed application which provides a control report to monitor file and data activities at each of the processing nodes.

Figure 4.20. Control objectives and audit procedures for reviewing distributed processing applications controls *(continued)*

processing file may contain a very critical, confidential draft for a company reorganization as well as the less important announcement of an annual holiday party.

Limited system security controls. System access controls often are rudimentary, to encourage use of the office automation equipment. In addition, the typical office environment sometimes makes it necessary for employees to share log-on IDs.

Easy access from outside sources. By design, it often is easy to access an office automation LAN from remote locations. For example, if a microcomputer on the network is equipped with a modem, it can access or be accessed from outside by other similarly equipped microcomputers. It is then possible to download data to a less secure environment.

Informal security administration. There is frequently no function or person responsible for monitoring security violations and training users in office automation security. In addition, the systems backup and recovery procedures can be informal.

Limited physical security controls. Office automation equipment generally exists in an open environment with few physical security controls. Even key locks often are ignored.

In addition to the above, office automation users often have little appreciation for controls that already exist in the equipment and can be built into LANs. A large portion of the user population are clerical, administrative, and management personnel who have minimal interest in security features in their efforts to "get the job done." In addition, management may not be receptive to auditors reviewing controls and procedures over their work habits and those of their secretaries.

Despite the potential problems, auditors should consider scheduling periodic reviews of the office automation function and the use of office LANs. With the increasing growth of such systems and ease of system connectivity, this is an appropriate area for review in the modern data processing center. If the auditor can point out potential risks and vulnerabilities associated with this equipment, the auditor can capture management's attention and interest.

Figure 4.21 contains objectives and procedures for reviewing the office automation function and the office automation LAN in particular. By the nature of the data stored on such networks, such a review tends to be limited to assessment of written procedures. That is, it is not practical to gather evidence by footing word processing files!

Control Objectives and Audit Procedures for Reviewing Office Automation Controls

Objective 4.21.1. Critical data on office automation equipment or within the office local area network (LAN) should be protected from improper access attempts.

Procedure 4.21.1.1. Review the office automation equipment in use to determine the types of security and password controls available and assess both the use and effectiveness of those controls.

Procedure 4.21.1.2. Determine that guidance materials or procedures are in place and understood covering the security classifications of key office documents.

Procedure 4.21.1.3. Review a catalog list of office automation files which have not been flagged as being password protected and select a sample from that list to determine whether any should have been protected.

Procedure 4.21.1.4. Determine that office automation equipment passwords and user IDs are changed on a regular basis.

Procedure 4.21.1.5. Review procedures for backing up office automation equipment files and determine that backup diskettes are stored in a secure, off-site location.

Objective 4.21.2. Physical security controls over the LAN and the office automation equipment should exist and should be consistent with the criticality of the data processed and the equipment.

Procedure 4.21.2.1. Tour the office automation equipment area during an off-shift period and determine that:
> All machines are turned off and locked when not in use
> Draft reports, diskettes, and other materials are stored away from work areas and are secured
> Critical items of equipment, such as LAN file controllers, are located physically separate in secure facilities

Procedure 4.21.2.2. Determine that all office automation equipment is properly identified with asset tags, and trace a sample of that equipment to physical asset inventory records.

Figure 4.21. Control objectives and audit procedures for reviewing office automation controls

Objective 4.21.3. There should be a central control point within the LAN system to control files and to monitor the overall efficiency of the LAN and its processes.

Procedure 4.21.3.1. Interview the individuals responsible for administration of the LAN to determine that procedures are in place to monitor overall performance and to adjust file access privileges.

Procedure 4.21.3.2. Review procedures for archiving older word processing file documents and determine that these appear appropriate given any special document retention requirements for the office.

Procedure 4.21.3.3. Based on document retention policies within the office administration function, select from catalog records several older documents and determine that they can be retrieved easily.

Objective 4.21.4. Adequate controls should exist covering the import or export of data through any LAN gateways to other computerized systems beyond the LAN, the use of office automation equipment for non-business applications, and the introduction of non-authorized software into the LAN.

Procedure 4.21.4.1. Review the types and extent of gateways from the LAN to mainframe or other computer systems and assess any potential information security vulnerabilities associated with those gateways.

Procedure 4.21.4.2. If the office automation LAN contains microcomputers, determine whether any modems are attached to these devices, and review the types of password or other security controls associated with these devices to prevent improper access to other devices on the LAN.

Procedure 4.21.4.3. Determine that policies exist prohibiting the introduction of personal microcomputer programs to any LAN processor as well as prohibiting the use of the devices for personal purposes.

Objective 4.21.5. Any general business data processing applications operating on the office automation LAN should be subjected to

Figure 4.21. Control objectives and audit procedures for reviewing office automation controls *(continued)*

the same controls as any similar end user data processing application.

Procedure 4.21.5.1. Through interviews, determine the types of applications run on office automation devices coupled to the LAN, and identify any that appear to be for general business data processing purposes.

Procedure 4.21.5.2. If any general business data processing applications are run on the office automation LAN, determine if they appear to be appropriate for that operating environment and consider performing detailed applications reviews (see Chapter 6) of the more critical applications.

Figure 4.21. Control objectives and audit procedures for reviewing office automation controls *(continued)*

The auditor often makes a significant impact on the organization through a review of office automation controls. The ease of use of such systems often conflicts with good internal control procedures, and often, users are unaware of the control exposures associated with this environment. As office automation equipment and microcomputers become closer, and as equipment connectivity increases through the use of LANs, these reviews will increase in importance.

EDI AND DIAL-UP LINE CONTROLS

Many data processing organizations have established tight physical security controls over their computer equipment and good information security controls over their programs and data, but continue to have exposures through their use of dial-up telephone lines into the computer system. These lines may be used for electronic data interchange (EDI) networks or for remote system users that occasionally need to access the computer system.

The use of dial-up lines and EDI networks can be expected to increase. For example, many organizations are equipping field salespeople and representatives with laptop computers which can dial into the main processing center to report daily sales activity. Managerial and professional personnel often have microcomputers at home to access central machines for work activities. Some computer hardware and software vendors also request remote dial-up lines so they can diagnose and fix problems. Finally, more industry groups can be expected to use EDI procedures to transact business. For a list of control objectives and audit procedures for a review of the EDI functions within an organization, see Figure 4.22.

Control Objectives and Audit Procedures
for Reviewing Electronic Data Interchange Controls

NOTE: These objectives and procedures are for use in reviews of controls when an organization uses Electronic Data Interchange (EDI) techniques to either receive data, such as orders from customers, or to send data, such as to place orders with vendors. In addition to reviewing general EDI controls, the auditor should also consider reviewing the actual application (see Chapter 6) which uses the EDI data.

Objective 4.22.1. EDI transmissions should be based upon overall or specific industry standards governing the types and formats of the data messages transmitted.

Procedure 4.22.1.1. Develop an understanding of the protocols and data formats used for any EDI activity including:

American National Standards Institute record and file formats for EDI transmissions ANSI x.12

Standards established by an industry group such as the Automotive Industry Automation Group (AIAG)

Standards established by an EDI clearinghouse service bureau

Procedure 4.22.1.2. If more than one protocol is used for an organization's EDI activities, review efforts by the information systems organization to establish only one of these as a standard.

Objective 4.22.2. EDI transmissions should be sent and received according to preestablished schedules with outside sources.

Procedure 4.22.2.1. Review published schedules for EDI activity and then examine the actual operations logs to determine whether those schedules are being met.

Procedure 4.22.2.2. Determine that all EDI transmissions or receipts are followed by an appropriate electronic acknowledgment.

Procedure 4.22.2.3. If dial-up telephone lines are used for EDI activities, consider the controls over those dial-up lines

Figure 4.22. Control objectives and audit procedures for reviewing electronic data interchange controls

including the use of "time locks" to limit activity to specific scheduled time slots or the use of call-back devices.

Procedure 4.22.2.4. If EDI activity goes through a general computer system rather than through a specialized processor, review the information security controls in place which would prevent an improper incoming transmission from damaging other data or programs.

Procedure 4.22.2.5. Determine that passwords are used for all EDI activities and that these passwords are changed on a regular basis.

Objective 4.22.3. Procedures should be in place to detect EDI transmission errors and to request retransmissions when appropriate.

Procedure 4.22.3.1. Determine that sequence number controls are used on EDI transmissions to help detect lost, duplicated, or mis-sequenced data.

Procedure 4.22.3.2. Review software procedures in place to screen for incomplete or corrupted input transactions and review procedures for requesting retransmissions.

Objective 4.22.4. Logs should be maintained within computer operations documenting all EDI activity.

Procedure 4.22.4.1. Determine that operator logs are maintained for all EDI activities and that any unusual occurrences are documented on a computer operations problem report.

Procedure 4.22.4.2. Review procedures for acknowledgement of EDI transmission receipts by outside parties and determine that any unusual activity, such as externally requested retransmissions, is documented.

Procedure 4.22.4.3. Review procedures for backing up and saving EDI transmission tapes and assess their adequacy.

Objective 4.22.5. Applications which are based upon the transmission or receipt of data using EDI should have controls to detect unusual data items and to flag them for subsequent manual review.

Figure 4.22. Control objectives and audit procedures for reviewing electronic data interchange controls *(continued)*

Procedure 4.22.5.1. Select one or more applications used for EDI activities and determine that these programs contain "unusual item" checks which would, for example, flag and report an extremely large order quantity.

Procedure 4.22.5.2. Select a sample of unusual item reports generated from the EDI support applications and determine that these items have received proper follow-up and review attention from users.

Figure 4.22. Control objectives and audit procedures for reviewing electronic data interchange controls *(continued)*

Dial-up lines that lack proper controls can represent a vulnerability to the computer center, and the auditor should review for controls in this area. It is relatively easy for a "hacker" to use a personal microcomputer to find modem dial tones from dial-in lines and then take steps to break into the system. In the case of some larger, more prominent organizations, telephone numbers of dial-up lines have been listed on electronic bulletin boards, including required log-on commands!

EDI connections do not have to go through dial-up lines. If there is sufficient traffic, the computer center may consider using a leased line to the EDI partner. In addition, other systems-related controls can be built into the EDI applications to help prevent any dial-up intruder getting beyond the front end of the EDI process.

The auditor should suggest control procedures for any EDI network or connection that is established by the organization. Many of these procedures relate to security over telecommunications lines. Others refer to good control procedures for the EDI network.

Dial-up lines in general require a series of other security controls including file authorization, user authentication, and a strong system of password monitoring and maintenance. Depending upon the criticality of the system and its applications, dial-up lines should not always go directly into the host computer but sometimes into a front end processor. The procedures for securing dial-up lines are discussed in more detail in Chapter 9.

The auditor can expect that EDI, telecommunications networks, and connectivity will increase in the future. These trends introduce new risks, controls concerns, and procedures to the audit function. A modern audit specialist can no longer get by with simply a knowledge of data processing controls. That specialist must also develop a good general knowledge of telecommunications controls and procedures as well. This chapter was written to introduce some of these issues.

Section II

AUDITING DATA PROCESSING APPLICATIONS

CHAPTER 5

Auditing Data Processing Applications

INTRODUCTION

A typical business data processing organization processes a large number of applications ranging from relatively simple payroll to highly complex manufacturing control systems. General control procedures, as discussed in Chapters 2 through 4, apply to the entire data processing operation. There also are specific controls associated with each application. It is important that auditors understand how to evaluate and test controls over specific data processing applications. Often, application controls can be more critical than general data processing controls.

However, application controls also are very dependent upon the overall system of general controls discussed in previous chapters. Without effective controls over program library updates in a data processing organization, for example, it is difficult to rely on controls in a special application such as an order entry system. The application may have been designed to properly screen orders for valid credit approvals, but the application's programs potentially could be changed without management authorization to accept orders without proper credit approval.

A typical data processing organization may have a large and diverse set of applications, supporting a wide variety of functions within the organization, starting with accounting but including areas such as manufacturing, marketing, and engineering. These applications may be implemented with a variety of data processing technologies ranging from modern on-line terminals and database files to older batch-processing systems. Some may be developed in-house, while newer ones generally will be purchased software applications. In-house-developed applications may be written in a programming language such as COBOL or in a fourth generation language. Applications documentation and controls may range from being complete to almost nonexistent.

Typically, management is interested in audit findings dealing with specific applications controls. For example, it is difficult to draw management's attention to a finding dealing with general controls over a data dictionary, while an audit finding which discusses an incorrect discount calculation in a database-oriented accounts payable application is sure to draw attention. However, because of the relative complexity of many applications and because controls often reside within the application and are exercised in user areas, the audit of specific data processing applications often can be a challenge to the auditor.

This chapter discusses how the auditor can review effectively internal accounting controls in data processing applications. It explains how to select applications for review based on their control risk, develop an understanding of specific application controls, and evaluate and test those controls. Chapter 6 provides guidance on the next step—gathering evidence through detailed testing to complete the auditor's risk assessment.

SELECTING APPLICATIONS FOR REVIEW

The typical audit of a data processing organization cannot review the controls for *all* data processing applications. There are too many for an auditor's limited time. In addition, there is a low risk associated with many of them. Therefore, after performing the general controls review the auditor typically selects a limited number of specific applications to be reviewed. Some factors which impact the selection of these specific applications include:

Management requests. The auditor often is asked by management to review controls in newly installed data processing applications. This may be because management is concerned about a new application because of reported problems or the strategic criticality of the application. However, management requests are not always made for the correct reasons. For example, sales analysis reports may be incorrect due to bad data submitted from a reporting division, but management may perceive the problem to be lack of applications controls. They will ignore the bad data and consider it a "computer problem" and request a review. The auditor may be unaware of the data problems and perform a normal review. However, when the auditor is aware of mitigating circumstances, they should be considered before starting the review.

External auditor considerations. External auditors often are interested in specific data processing applications controls. In their attestation function, they must assess control risk and develop an understanding of certain critical applications. Many external, financially oriented auditors may not have the necessary technical skills to evaluate these data processing applications controls adequately and will ask specialists from their own organizations or from internal audit departments to perform application reviews. While upper management may have general concerns over a given application, the external auditors probably will have more specific concerns and internal accounting controls objectives.

Post-implementation applications reviews. Chapter 7 discusses the process of selecting and reviewing applications controls prior to implementation as well as performing post-implementation reviews after the new implementation. For some critical applications, auditors also perform a detailed applications review some time after the actual system implementation. If an application is financially and operationally significant, the auditors even may schedule limited controls reviews on an ongoing basis.

Other audit applications selection criteria. Following are some other factors to consider when selecting one data processing application over another for a detailed, internal accounting controls review:

Does the application impact significant assets?

Does the application represent a significant risk exposure to the organization?

Is the application a strategic system for organizational decision making?

Does the application support a function that will be reviewed later as part of an internal audit review?

Have various changes been made to the application system which were not part of any pre-implementation audits?

Have there been significant personnel changes in the organizations using the application?

The auditor typically is faced with a large number of application candidates at any time for review. Care should be taken in documenting the justification for selecting one application over another.

Frequently, data processing applications controls reviews are included as part of a general controls review of a data processing function. The auditor develops a detailed understanding of general controls and then reviews the controls surrounding one or more specific applications as part of a total review. Often this approach is also taken by external auditors.

Internal auditors routinely perform reviews of specific applications which support an overall functional area. The internal audit function may schedule an operational and financial review of the purchasing department. This then can be an appropriate time to review the application controls for the automated purchasing system supporting that department. This is an effective approach as the computer audit specialist concentrates on more technical issues surrounding the application while the other auditors help review related applications controls. Chapter 13 discusses how to integrate the computer audit and financial audit functions more effectively.

UNDERSTANDING AUTOMATED APPLICATIONS

Once an application has been selected for review, the auditor needs a preliminary understanding of the purpose or objectives of the application, the methods of data processing used, and the relationship of the application to other automated applications and end-user functions. Also, it will be necessary for the auditor to understand any special technical aspects of the application to perform a more effective review.

Often, this understanding is gained through reviews of past audit workpapers, if available, interviews with data processing and user personnel, and reviews of applications documentation. While prior workpapers can be helpful and the interview process allows the auditor to ask relevant questions, a review of applications documentation usually is the first step.

Reviewing Applications Documentation

There have been many changes in recent years in the way data processing applications are documented. In the earlier days of computerized applications, documentation usually consisted of detailed program and system flow charts and record layouts. This provided the programmer with documentation but was of little help to applications users or the auditor. In addition, frequently, the flow charts and record layouts were hand prepared and quickly became out of date because designers were reluctant to erase or to redraw hand-prepared charts for small changes.

Over time, applications documentation evolved into a more text and functional chart format. Decision tables and logic charts often were used to describe functions of individual programs while extensive text was used to describe the overall system. While this documentation is more functional and less technical, it also quickly becomes out of date. Often, programmers and system designers do not take the time to incorporate changes into this systems documentation.

In today's computer systems, the documentation process is easier. Powerful design and documentation tools are available on microcomputers or other machines to aid in the technical documentation process. These procedures, often called "Computer-Aided Systems Engineering" or CASE methods, are discussed in Chapter 7. Figure 5.1 shows an example of a program system chart produced using CASE software tools. Many such tools are available as microcomputer software packages. Using these techniques, detailed program flow charts are combined automatically into system level summarized charts. Changes introduced on one chart are updated to all others. In addition, word processing tools are available to provide textual documentaton which also can be easily updated.

The auditor will find varying amounts of documentation, ranging from very little to extremely large volumes, depending on the relative age of the application to be reviewed. Published documentation covering some popular outside vendor-supplied insurance or banking application systems fills dozens of volumes of descriptive text. Users treat such documentation as encyclopedic reference materials.

Applications documentation should be used as a first step toward gaining an understanding of the system to be reviewed. If aspects of the documentation are missing or out of date, the auditor probably should note the defect as a finding at the conclusion of review. However, this lack of documentation should not prevent the auditor from performing the review. When reviewing an application, the auditor normally should look for the following documentation elements:

Systems development methodology initiating documents. These are the initial project requests, any cost/benefit justifications, and the general design requirements analysis. Although many initial assumptions may change during systems design and implementation, these documents give

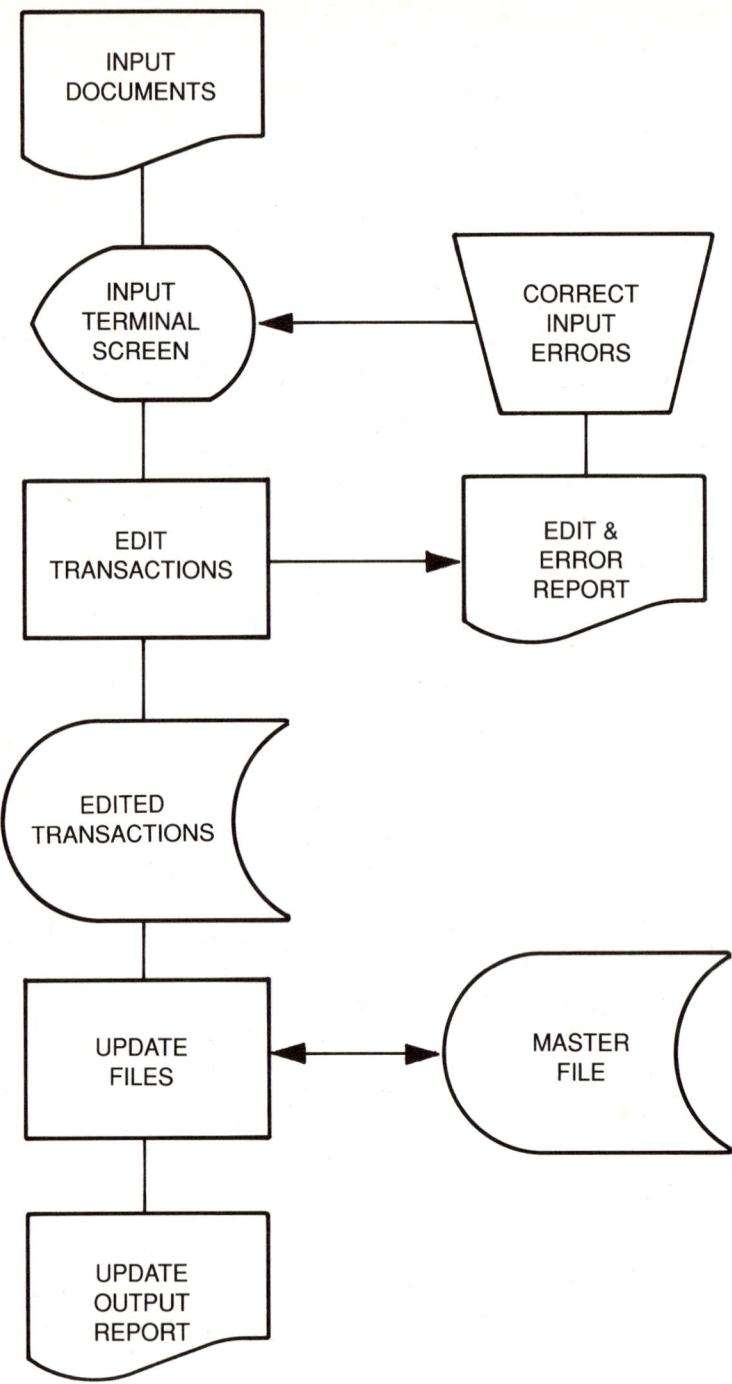

Figure 5.1 Example of a program system chart

the auditor an understanding of the preliminary applications design and control features.

Functional design specifications. This document should describe the application in detail. Each of the program elements, database specifications, and systems controls should be described. Any major changes made to the application since its original implementation should be reflected in this documentation. It allows the data processing department analyst or programmer to make changes or respond to user questions regarding the application.

Program change histories. There should be a log or documented record listing all program changes within the application. Some data processing departments keep this with the applications documentation while others maintain it in a central file cross referenced to the program source code. While this documentation is an essential element to control program changes, it also provides the auditor with some feeling for relative application stability. A large number of ongoing change requests for a given application often means that the application system is not achieving user objectives.

User documentation manuals. Along with the technical documentation, there should be documentation prepared for the application's end-user. With a modern, on-line system, much of this user documentation may be in the form of "HELP" or "READ ME" files. It should be comprehensive enough to answer most user questions.

The auditor reviews this documentation to gain a general understanding of the application and its controls and to develop questions for later interviews. Copies of key or representative sections should be taken for workpaper documentation. However, the auditor normally should not attempt to copy the entire documentation file for workpaper purposes, a step too often taken by auditors. It adds considerable bulk to workpaper files and does little to accomplish audit objectives!

Conducting an Application Walk-Through

After the auditor has reviewed prior workpapers, reviewed applications documentation, and interviewed users and data processing personnel to clarify any questions, the next step is a walk-through to verify the auditor's understanding of the application. The purpose is not to test application controls, which will be covered in a subsequent audit step.

An example will help explain the application walk-through process. Assume that the auditor has been asked to review the controls over an in-house developed on-line accounts payable application written in COBOL on a corporate mainframe computer. The organization is a manufacturing firm with sophisticated data processing applications. The accounts payable application was installed several years ago but was never reviewed during development. It was not considered to be a critical application when installed because labor costs rather than material costs were the major factor in the organization's cost structure. Now, the external auditors have reassessed application risk and have asked internal audit to review the application's internal controls.

The application receives input from the following sources:

Open commitments, from the purchasing system
Notifications of goods received, from the materials receiving system
Various payment transactions, from acounting terminals
Payables journal vouchers entered as batch data

Application data is recorded on a relational database file, and the application uses a table of values for calculating cash discounts.

The outputs to the application include the following:

Accounts payable checks
Transactions to the general ledger application
Transactions to cost accounting applications
Various control and accounting reports

The prime users controlling the system are in the general accounting department, although the purchasing department also inputs transactions to set up automatic vendor payments under preagreed terms.

This example will be used to demonstrate an application walk-through and will be referenced in other examples in this chapter. It describes a type of application common in a modern data processing environment. Steps to performing an application walk-through for this example are as follows:

1. *Briefly describe the application for workpapers.* Based on the auditor's review of the application's documentation, a brief description of the application should be prepared for the workpapers. This can be an update of the description prepared in earlier steps, or more detailed as appropriate. This documentation car. follow the general format used above to describe this

example, but with greater detail. The workpaper documentation should identify key subsystems—input screen formats, key data file names, and output report formats.

2. *Develop a block diagram description of the application.* This is a systems or functional level flow chart for the application. It should reflect the written description and should better illustrate the application flow concepts. Again, its purpose is to help the auditor gain a general understanding of the application being reviewed. Figure 5.2 is an example of such a system block diagram based on the example. This diagram should be used to confirm the auditor's understanding of the system with key data processing and user personnel.

3. *Select key transactions from the system.* One or several input transactions should be selected to walk or trace through the application. This selection would be based on discussions with users and members of the audit team. In the example accounts payable application, the auditor may select the automated transactions that the receiving system matches against records of open commitments to initiate payment, considering these to be the most significant application transactions.

4. *"Walk" a selected transaction through the system.* In the days of manual or less sophisticated data processing applications, a walk-through amounted to just what the words say. An auditor would take an input transaction form and walk it through each of the clerical stations or steps that normally would process it, to determine the processing procedures. In a modern computerized environment, this "walk" requires recording a transaction as it is entered into a terminal and following it through all subsequent steps. For example, in the accounts payable example the walk-through transaction was a receiving report entry indicating the receipt of a valid open purchase commitment. The auditor then reviews the open commitments module to determine if the transaction was recorded, then traces it to a properly computed accounts payable check and transactions for the general ledger system, including both the net amount and the cash discount taken. Routinely, applications testing is called "control risk assessment" or "compliance testing." The auditor is verifying that the application is operating in compliance with pre-established controls procedures. A test to verify that *all* accounts payable checks have been input to the general ledger, through a comparison of account balances or other methods, would be called "substantive testing" or "tests of financial statement balances."

5. *Modify system understanding as required.* The purpose of a system walk-through is to develop a basic understanding of system functions and

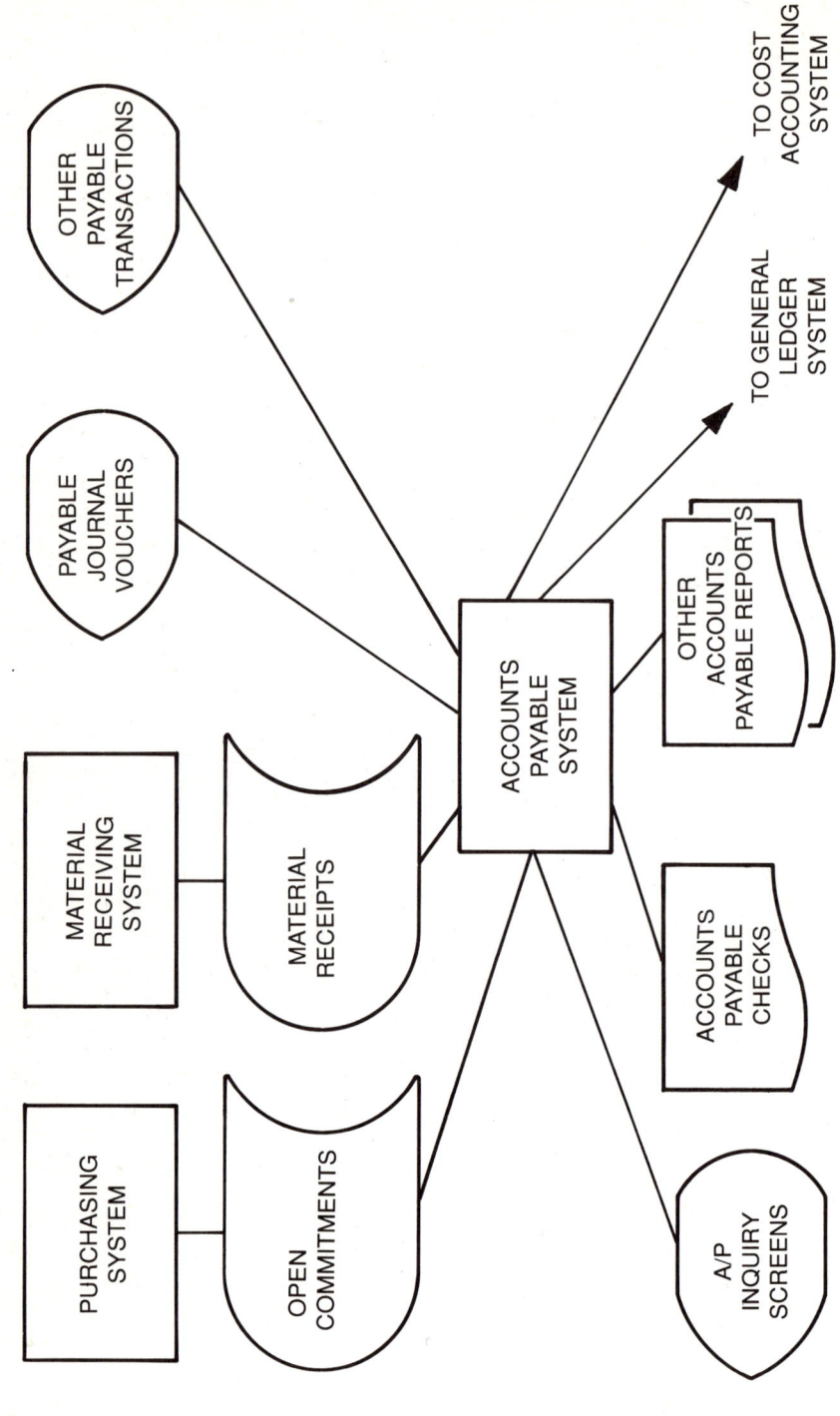

Figure 5.2. Block diagram of an accounts payable application

controls. It does not allow the auditor to test whether all transactions are working as described. However, if the auditor discovers that the walk-through transactions are not working as assumed, the preliminary audit documentation may need to be revised. Once revised, the auditor may repeat the above steps to reverify the auditor's understanding of system transaction flows.

These application walk-through steps are summarized in Figure 5.3. This process is an important first audit step in gaining an understanding of a computer-based application system, including its controls and relationships with other automated systems or functions. While it is not a substitute for detailed or substantive testing of an application, the walk-through identifies any potential control weaknesses and provides an understanding of the application to define and test control objectives for subsequent, detailed audit testing and evaluation procedures.

Developing Control Objectives and Audit Procedures

The next step for the auditor, after gaining a preliminary understanding of the data processing application, is to develop a set of detailed control objectives and audit procedures for completing the review of that application. This process is dependent upon the nature of the auditor's review as well as characteristics of the application. A computer audit specialist working with an external auditor would be concerned with an assessment of control risk and the ability of the application to process financial statements correctly. The audit procedures would test financial statement balances built up from the detailed transactions.

The internal auditor may have other objectives in reviewing data processing applications. Management may ask the auditor to review an application to determine whether it is making use of computer resources efficiently, or whether operational calculations associated with that application are performed correctly. Thus, before proceeding with the review, the auditor must determine the appropriate control objectives of this application review. The control objectives for the accounts payable example might include the following:

The system should have adequate internal accounting controls so that all transactions are processed correctly and only correct transactions result in output checks

The system should compute correctly cash discount terms to take only advantageous discounts

Application Walk-Through Steps

NOTE: These steps describe the procedures that an auditor might document while tracing receiving report transactions through the accounts payable system example described in this chapter.

1. Select a receiving report document from the receiving dock and record the purchase order number, vendor name and number, part numbers, quantities, and prices.

2. Observe the entry of the receiving report data into the accounts payable materials receiving screen. Observe that all data was entered correctly and that the system accepted the receipt.

3. Trace the receiving report data to a materials open commitments report or to the actual purchase order to verify that all receiving information is correct.

4. After the nightly accounts payables update, review the open items payables screen to determine that the selected document has been updated to accounts payable and that it appears to be correct.

5. Review records over the next several periods to ascertain that the vendor's invoice has been received and that the open item represented by the selected receiving report has been set up for payment. Verify that the payment authorization appears on proper transaction reports or screens.

6. Review tabled data for the selected vendor to determine that discounts and payment terms are being handled correctly.

7. Trace the item to the accounts payables check register to determine that the item has been paid. Also, trace the item through the various accounts payable output reports.

8. At the end of the accounting period, determine that the selected item, in either detail or summarized form, has been transferred to the general ledger system with all accounts properly relieved.

Figure 5.3. Application walk-through steps

The system should make efficient use of computer resources

The system should contain controls to assure that no improper or illegal payments are made

Of course, the particular control objectives chosen are influenced greatly by the overall objective for the review, as defined by management. For example, if management asked for a review of the accounts payable system because of a concern over illegal payments, the control objectives might shift from an emphasis on accuracy to an emphasis on authorization.

Before actually starting the application review, the auditor should give attention to documenting the objectives of the review. These objectives should be discussed with management or other persons requesting the application review, to verify whether the planned approach is on target and will satisfy the audit request. This discussion should take place even if the application review has been initiated by the audit department as part of a total review of a data processing function.

Figure 5.4 provides a general list of control objectives that are applicable to most data processing applications. Not all of the objectives are applicable in all review situations, however, and certain special review requests may require other, more detailed objectives. However, these provide the auditor with an initial set of control objectives for most application reviews.

PROCEDURES FOR AUDITING APPLICATION CONTROLS

It is easier to develop general objectives for reviewing a data processing application than to define the audit procedures for that review. Detailed procedures vary considerably depending upon whether the application uses purchased or in-house developed software, whether it is integrated with other applications, whether it uses modern technologies such as databases, and whether it is a significant accounting application.

In addition, the exact nature of the application can vary considerably. Although auditors were asked in the past to review only accounting-related applications, such as accounts receivable, accounts payable, or fixed asset systems, today they are asked to review applications such as material requirements planning, marketing analysis, or other non-financial decision support applications. In the future, they will be asked to review even more advanced applications, such as decision support or expert systems. This requires knowledge of the specific attributes of the application to be reviewed, as well as an understanding of new technologies.

Control Objectives for Reviewing an Automated Application's Controls

Note: These are generalized objectives for controls in an automated application. The auditor should tailor these to the specific application being reviewed and develop specific audit procedures to test whether the objectives are being accomplished.

Objective 5.4.1. There should be a clear transaction audit trail so that all transactions entered into the system can be traced to an output file or report.

Objective 5.4.2. Source documents or input screen formats should be designed to minimize errors and omissions.

Objective 5.4.3. Input transactions should be screened and validated for errors as close to the source of input as possible.

Objective 5.4.4. The application should make use of information systems department security software or other mechanisms to ensure that only authorized persons can access the application.

Objective 5.4.5. Where applicable, batch control procedures should be used for application inputs to assure that they are complete and correct.

Objective 5.4.6. Transaction error handling procedures should make use of suspense files and/or informative error reports to facilitate the timely and accurate resubmission of all corrected input data.

Objective 5.4.7. When the application receives inputs from other automated applications, run to run controls should be in place to ensure that inputs are complete and accurate.

Objective 5.4.8. The application should provide clear indications to computer operators or users whether it has run correctly or a processing error has been encountered.

Objective 5.4.9. Output reports and screens should be balanced back to control totals with the use of audit trails, if required, to facilitate tracing and reconciliation.

Figure 5.4. Control objectives for reviewing an automated application's controls

Objective 5.4.10. Output reports and screens should follow organization standards and should contain proper identification, processing dates, control totals, and other data as required.

Objective 5.4.11. Critical output documents should be produced only under proper supervision and should contain controls such as document numbers.

Objective 5.4.12. The application should satisfy the overall information and business requirements of its users.

Figure 5.4. Control objectives for reviewing an automated application's controls *(continued)*

The auditor should then follow five basic procedures in conducting an applications review:

Understand the objectives of the review

Review the application with key users

Describe the application for audit workpaper purposes

Identify application control points

Test key application controls

Each of these application review procedures will be discussed in following sections.

Understand the Objectives of the Review

Previously discussed were the various alternative objectives of a review of a data processing application. Before embarking upon a review of a specific data processing application, the auditor should understand the objectives of that review.

This understanding is particularly critical when the auditor begins a review of an application outside the normal realm of accounting applications. If management asked the auditor to review a new manufacturing materials requirements planning system, for example, that review could be intended to validate internal accounting controls, measure manufacturing materials parts flow efficiencies, validate system compliance with applicable regulations, or a combination of these. Rather than simply "looking at" the application, the auditor should define the objectives of the applications review to be performed and summarize them in a brief statement.

This statement must be discussed with audit management and the application user's management. Although this appears to be an obvious first step, it is often missed by auditors and management.

Of course, the objectives of an applications review may change during the course of the review. In a materials requirements planning systems review, the auditor might initially start with the objective of affirming the system's internal accounting controls. If potentially invalid transactions were encountered during the course of the review, the objective might change to fraud detection.

Review the Application with Key Users

The purpose of this procedure is to gain a general understanding of the application being reviewed. This is often accomplished by discussing the application with key user personnel or responsible systems personnel, or by reviewing application documentation. Normally, the auditor does all three.

Figure 5.5 lists some of the steps necessary to gain a general understanding of a data processing application. The amount of effort spent on each step depends upon the objectives of the review and the type of application being reviewed. A decision support application such as for capital budgeting probably will have a small group of key users with a thorough understanding of application procedures. If the application is a logistical support system, such as factory-floor data collection, it may be more difficult to identify the key system users.

The relative age of the application and whether it was purchased or developed in-house also cause the auditor's procedures to vary. An older, in-house-developed system may not have current documentation. However, because older technologies were used in the application design, a key user probably can walk the auditor through the system. Purchased applications often have a high level of vendor documentation to provide this understanding. If the application uses newer technologies, such as distributed databases, the auditor may rely primarily on data processing systems personnel to gain this general understanding.

Describe the Application for Audit Workpaper Purposes

After the auditor has discussed the application with key users and reviewed the documentation, the next step is to complete the audit documentation of the application. The auditor should have made workpaper notes throughout the previous steps, and now the auditor describes the understanding gained, including any notes for potential follow-up review work.

A. Understand the Purpose of Review Objectives

 1. Review the request for an application review or audit to determine the purpose of the review. Possible purposes are:

   An overall review of application control procedures to identify and document application controls

   A review to verify that application controls are operating properly, as described, through comprehensive compliance testing

   A review of computer processing or clerical efficiencies

   A review for compliance with either applicable regulations or organization policy

 2. Briefly review the application's structure and existing documentation to determine that audit objectives can be met. For example, the lack of source code for a purchased package application may limit use of computer assisted procedures to test the application.

 3. Consider potential factors which may limit the auditor's ability to achieve review objectives, such as time constraints or a lack of skills in certain technical areas, and assess whether it is necessary to modify preliminary review purposes.

 4. Document preliminary review purpose and review with requester or audit management to obtain agreement.

B. Review Application with Key Users and Systems Personnel

 1. Identify the key knowledgeable users of the application and request a general "walk through" of the application's functions and procedures.

 2. Based on the "walk through," assess the organization's decision dependence on the application. For example, determine whether the application is used for:

   Historical reporting to support financial statement preparation

   Decision support to help users make production decisions

   Logistical support to allow functions in the organization to work smoothly

 3. If the application requires extensive end user computing sup-

Figure 5.5. Steps to gaining a general understanding of an application

port, interview individuals responsible for that end user computing to determine levels of activity and control procedures.

4. If multiple departments or functions use the application, determine application control responsibilities and functions which span organizational lines.

5. Discuss the application with appropriate information systems personnel to determine their level of involvement and ongoing activities regarding the application.

6. Determine whether the application:
   Is an in-house developed or a purchased software package
   Uses a database or a traditional flat file structure
   Is a relatively new application
   Is written in a procedural language such as COBOL or a fourth generation language
   Has been programmed using a structured code or using earlier, more cumbersome program coding techniques
   Was developed as a centralized application or a distributed one, with processing taking place at remote sites

7. Review the status of overall application and program level documentation to gain a general understanding of supporting materials available for more detailed audit testing.

8. Review the change control history for the application including any current outstanding requests for modification.

C. Describe the Application for Workpapers

1. Document application data flows and interrelationships with other applications through a series of simple diagrams.

2. Extract copies of application documentation or reference them for workpaper documentation purposes.

3. Capture copies of key screens, reports, or related documents.

4. Review the documented general understanding of the application with information systems and key user personnel to determine that the general audit understanding is correct.

D. Identify System Control Points

1. Obtain a preliminary understanding of key application control

Figure 5.5. Steps to gaining a general understanding of an application *(continued)*

points and determine whether end users have the same understanding.

2. Identify specific application controls consistent with the overall
objectives of the review and with key user discussions.

3. Consider the criticality of identified application controls and
document the potential risk if any of those controls are not functioning.

E. Test Key System Controls

1. Based on overall objectives of the review, develop a preliminary
plan for testing key application controls such as:

A "walk through" test of one or a limited number of transactions to determine whether the application is functioning as
documented

A balancing or reconciliation test to determine whether run to
run controls and other application controls appear to be
correct

A compliance test of application functions where designated
functions or applications are selected for detailed testing

A more substantive test of transaction balances

2. Determine the timing requirements of the planned applications
controls tests. For example, it may be necessary to tie tests of
accounting controls to monthly cycles.

3. Depending upon objectives of the review and the test plan
selected, perform tests of applications controls.

4. Reconcile the results of application tests to expected results or
to supporting systems and identify any exceptions.

5. Discuss any application test exceptions with responsible users or
information systems personnel, as appropriate, and obtain their
agreement with any test procedure assumptions as well as with
anticipated audit test results.

6. Document audit test results, consider the control and application
risk implications of those test results, and consider the need to
expand the scope of testing or perform other procedures.

Figure 5.5. Steps to gaining a general understanding of an application
(continued)

This type of documentation is important. It allows the auditor to return to the user or systems personnel originally interviewed and ask them if the auditor has described adequately the application. Examples of such application documentation are in Figures 5.7, 5.8, and 5.10 as part of the application review case examples.

Identify Application Control Points

The auditor next identifies key control points within the application being reviewed. In older and simpler batch-oriented systems, this task was fairly easy. The auditor looked for input data acceptance controls, computer decision points, and output data verification controls. Since there may have been only a few programs associated with such an older system, this control identification process was accomplished with minimal analysis.

Identification of application control points is now more difficult. A more modern system may use on-line updating capabilities, be closely integrated with other applications, or use sophisticated techniques such as object oriented programming. Other factors which make this identification of control points difficult include the following:

Data input to the application may be generated by other computerized applications

Controls once performed by data input personnel now may be built into programs

Modern optical scanning input devices and output documents which consist of bar codes make visual inspection difficult

Database files may be used in one application and shared with other applications

The application may make extensive use of telecommunications and thus appear to be "paperless"

There are numerous other reasons why the auditor may have difficulty identifying control points in a modern data processing application. However, the auditor's own documentation helps identify some of these controls. As a rule of thumb, the auditor looks for points where decisions are made within an application. These decision points include such things as checks on the completeness of transactions or on the accuracy of calculations. Often, they are the key controls within an application.

Figure 5.6 lists typical application controls. These are oriented to both classic batch applications and more modern integrated, on-line applications. The auditor should rely on the application description and a general understanding of the application gained through interviews to determine which controls may be applicable.

Test Key Application Controls

Once the auditor identifies control points within an application, the final step is to test those controls to determine whether they appear to be working adequately. This is normally called "control risk assessment" or "compliance testing."

A compliance test is a limited examination of the evidence generated by an application, such as an inspection of an output invoice in a billing system, to determine that the data and computations are correct. A compliance test also may consist of a recalculation or reperformance of some application functions. An auditor may manually recalculate some key application formula to determine whether it is reasonable. A control risk assessment or compliance test is limited in nature and does not cover all transactions or functions within an application.

The purpose of a control risk assessment or compliance test is to determine whether application controls appear to be working. If the auditor is interested in reviewing all transactions in an application processing cycle or all data on an application data file, substantive testing procedures or tests of financial statement balances should be used. These are discussed in Chapter 6.

Normally, the auditor does not perform compliance tests for all controls identified with an application. The extent of testing depends upon the objectives of the applications audit. The external auditor tends to perform compliance tests on internal accounting controls. The internal auditor tends to perform compliance tests over areas such as data processing efficiency or administrative controls.

The following examples illustrate various types of compliance tests available to an auditor and how they might be performed:

Examinations of processing evidence. The auditor wants to ascertain that transactions entered into a system are being processed correctly. When reviewing an on-line manufacturing application, the auditor might record several materials transactions as they are entered on shop floor terminals. After the overnight processing, the auditor verifies that those transactions have adjusted inventory records correctly and that work-in-process cost

Typical Data Processing Application Controls

NOTE: The following describe some general input, processing, and output controls which might apply to a given application. Based on the auditor's general understanding of the application, these controls should be expanded and tailored to the particular application under review.

A. Controls Over the Correctness of Input Data

1. When data is input in a batch mode, there should be batch controls using key field totals, item counts, and hash totals to verify that all data submitted has been input.

2. Key codes or values, such as manufacturing units of measure, should be checked against table files to verify the correctness of inputted data.

3. Check digit schemes should to used for key input fields, such as account or part numbers, to ascertain that the sequence and values of numbers are inputted into a data field correctly.

4. Echo procedures, which return confirmations of correct data to terminal users, should be used where practicable.

5. Input errors should be identified, held for subsequent correction if appropriate, and reported to error sources for action.

6. File control totals and other completeness checks should be used to verify that all files and input data are as complete as expected.

B. Logic and Reasonableness Processing Controls

1. Input tables should be checked periodically to determine that tabled values are understood and are correct.

2. Reasonableness tests should be built into program logic, where appropriate, to identify values which appear beyond pre-established bounds.

3. Label and logic checks should be built into the job control language procedures, the application, and individual programs, as appropriate, to ensure that correct files and versions of the database are being used.

Figure 5.6. Typical data processing application controls

4. Controls should be built into program logic to ensure that calculated values are rounded correctly and use the right precision.

5. Run to run controls should exist between programs to ensure that all data has been processed correctly.

6. Breakpoint and recovery controls should exist such that processing can be resumed in the event of an unexpected interruption.

C. Controls Over Applications Outputs

1. All outputs should have control totals to allow visual or mechanical verification of output data.

2. Outputs should be structured such that transaction and data audit trails are complete and that all output transactions and data can be traced to their source.

3. Distribution of application outputs should be controlled such that only authorized persons have access to output reports or screens.

Figure 5.6. Typical data processing application controls *(continued)*

reports have been updated properly. This verification takes place by reviewing the output reports generated by the system or running special retrieval reports against data files.

Inspection of exception reports. This compliance test is similar to the examination described above. The emphasis here, however, is on testing the error verification routines within the application. The auditor selects transactions input to an application that appear invalid and traces them through the application to determine whether they were reported properly on exception reports. The auditor also can consider submitting test error transactions to a system to verify that they are rejected by the application.

Reperformance of application functions or calculations. This compliance test can be applicable to the automated and the manual aspects of application systems. If a fixed assets application performs automatic depreciation calculations, the auditor manually recalculates depreciation values for selected transactions as a compliance test. The auditor also "walks through" a fixed asset addition or deletion to determine that the application is functioning properly.

Observation of procedures. A common compliance test in a manual environment is an observation of procedures to verify that they are performed correctly. This observation sometimes may be of use when reviewing an automated application. A remote work station receiving downloaded data from a central computer system requires extensive manual procedures to make the download connections properly. The auditor can observe remote work station operations to determine whether these manual procedures are performed correctly.

Use of computer-assisted audit techniques. The use of computer-assisted audit software is discussed in Chapter 6. Often, these techniques are used for compliance tests of automated applications. Audit software can be used to match files from different periods, identify unusual data items, perform footings and recalculations, or simulate selected functions of an application.

Reviews of program source code. In certain limited cases, a review of program source code is used to test selected program functions. For example, the auditor can verify that a certain logic check is performed within a program by verifying the source code. However, this compliance test should be used only with great caution. Because of the potential complexity of a program source code, it is easy to miss a program branch around the area being tested. There also are specialized programs available to compare the program source code with the compiled versions in production libraries.

Compliance tests can be powerful methods to test application controls. They allow the auditor to gain a level of assurance that application controls are working. However, the auditor must be aware that the assurance is not absolute. There is a risk that the auditor may test an application control and find it to be working when, in fact, it normally does not work as tested. Conversely, the test may indicate the application is not working when, in fact, it normally is.

If the auditor finds the results of the compliance tests acceptable, it is possible to conclude that the controls tested are working. The auditor then reports to management that the applications controls tested *appear* to be working correctly. Because of risks associated with such compliance tests, however, the auditor always should be careful to qualify the report to management with a comment about the risks of incorrect results associated with such relatively limited tests.

Sometimes, the auditor finds that the controls tested do not appear to be working correctly. This may indicate that the auditor does not understand correctly some aspects of the application system. The auditor should review the application description and identification of controls to verify that they are correct. It may be necessary to revise the auditor's understanding of application controls and then repeat the audit risk assessment procedures.

If the auditor finds, through compliance testing, that application controls are not working, it is necessary to report these findings. The nature of this report depends upon the severity of the control weaknesses and the nature of the review. If the application is being reviewed for external auditors, the identified control weaknesses may preclude placing any reliance on the financial results produced by the system. If the control weaknesses are primarily efficiency-related or operational, the auditor may report them to data processing management for future corrective action.

REVIEWING APPLICATIONS CONTROLS: SOME EXAMPLES

As stated previously in this chapter, the diversity of data processing applications makes it difficult to provide one set of objectives and audit procedures which is applicable for all applications. Applications can be financial or operational. They can be implemented on minicomputers using purchased software, or custom developed applications located on large systems which use extensive database and telecommunications facilities.

While the auditor develops a general approach to reviewing most data processing applications, it may be necessary to tailor that approach to the specific features of a given application. The following paragraphs describe how an auditor might perform a review of two different data processing applications. The first is a microcomputer system using purchased software with telecommunication links to a larger mainframe machine. The second is a mainframe database application with interfaces to several other mainframe applications.

Auditing a Microcomputer Budgeting System: An Example

In the first applications review example, assume the auditor has been asked to review the controls in a microcomputer capital budgeting system. Assume the financial planning department of an organization developed this microcomputer application, using a popular spreadsheet software package, for capital budgeting analysis and decisions. Although built around a purchased software spreadsheet package, the users coded a series of macro instructions for running the programs. Also, once a budget has been approved on the microcomputer system, the system has the ability to transmit summary data to the mainframe budgeting system to eliminate repetitive data entry steps.

Recently, the auditor was asked by management to review general controls over this financial planning microcomputer system. Audit procedures (as described in Chapter 3) were followed, and the auditor found the general controls to be fairly adequate. Users documented their microcomputer applications, made adequate backups of files and programs, limited system use to authorized personnel only, and followed other good procedures.

Some time after that general controls review, the capital budgeting system was implemented. Because this system now provides direct input to the corporate budgeting system, management asks the internal audit function to review its applications controls. After discussing this review request with upper management and data processing management, the auditor concluded that the objectives of this review are to verify whether the capital budgeting system:

Has good internal accounting controls

Properly makes capital budgeting decisions

Provides accurate inputs to the mainframe budgeting system

Promotes efficiency within the financial planning department

The above objectives are stated in general terms but represent the general format that might be developed for this type of application review. Management typically does not state its objectives in these words. It is the responsibility of the auditor to listen to management's requests and translate them to audit objectives, as in the above example.

Review Application Documentation

The auditor first reviews the documentation available for this example capital budgeting system. Since the application is built around a commercial spreadsheet software product, the auditor expects to find some of the following documentation elements:

Documentation manuals for the spreadsheet software package

Documentation for the programmed spreadsheet macro procedures using "spreadsheet auditor" type software which documents spreadsheet formulas

Procedures for uploading capital budget data to the mainframe budgeting application

Operations procedures for accepting the microcomputer input data to the mainframe data processing function

Procedures covering necessary approvals for the capital budgeting projects input to the microcomputer

The auditor may not find all of the above five documentation manuals or procedures. However, there should be documentation covering the software product used, interfaces with other applications, and necessary manual procedures. These materials are reviewed by the auditor to determine that they are complete; this review also helps the auditor to gain a general understanding of the overall application.

Describe the Application

After reviewing the capital budgeting system documentation and discussing the system with financial planning users, the auditor describes the system for audit workpaper purposes. Since the application is built around a microcomputer spreadsheet product, this description primarily covers the manual interfaces within the application.

Auditors often find it convenient to describe such systems in the form of a flow chart. Sometimes, a verbal description is just as adequate. Examples of each are shown in Figures 5.7 and 5.8. The purpose of this description is to

An Example of a Verbal Description of a Microcomputer Capital Budgeting System

1. Capital budgeting preliminary analysis numbers are prepared manually from preliminary requests on a project by project basis.

2. Capital budget data is input to the microcomputer spreadsheet system to extend and calculate properly budget decision summary data.

3. The spreadsheet application makes use of a file of macro procedures for some calculations as well as a data file of past capital budgeting analyses. These files are protected by passwords through the microcomputer's spreadsheet security system.

4. Capital budgeting personnel call up the capital budgeting application on a microcomputer, input preliminary data for each proposed project, and review results on the screen.

5. When all individual capital budgeting data for projects appear correct, capital budgeting reports are run on the microcomputer and total results receive a detailed review.

6. If totals are not correct or if there appear to be problems with input data, all reports are returned to analysts for rereview and a repeat of the process.

7. If the review shows that results are correct, the microcomputer extract program is run to create a file to transmit data to the mainframe capital budgeting system.

8. Once the control totals from the extract file appear to be correct, the capital budgeting data is transmitted to the mainframe system through normal channels for computer communication.

Figure 5.7. An example of a verbal description of a microcomputer capital budgeting system

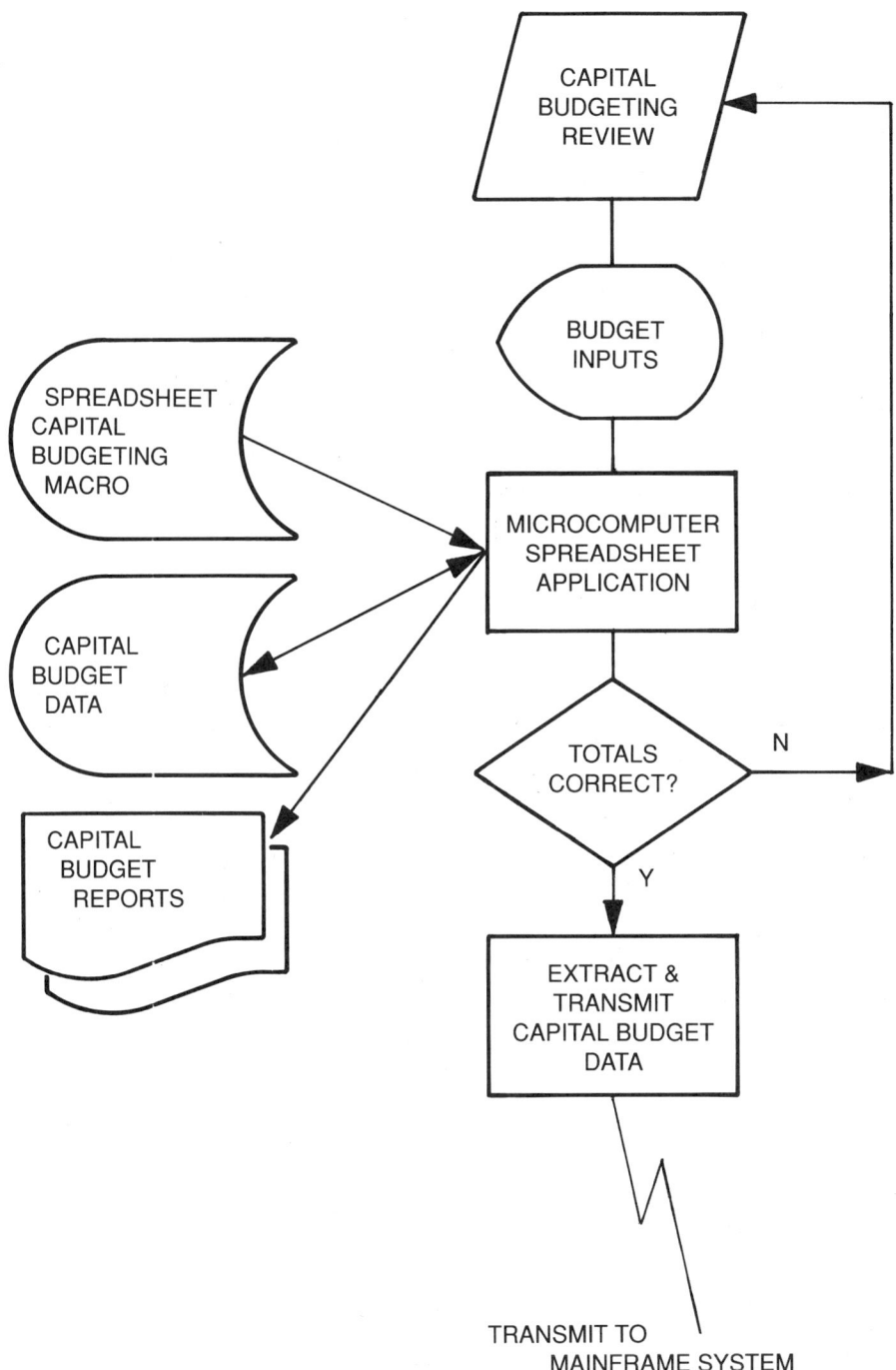

Figure 5.8. An example of a flow chart description of a microcomputer capital budgeting system

provide the auditor with workpaper documentation of the application and a basis for the identification of control points.

Identify Application Controls

Although a rather simple and compact application, the example capital budgeting system has some critical control points. If the spreadsheet macro procedures calculate capital costs, present values, and such related factors incorrectly, management may make incorrect investment decisions. If this data is transmitted incorrectly to the mainframe budgeting system, official budget records may be incorrect. If the microcomputer application is not documented properly, a change in personnel in the financial planning department may result in an inoperable system.

Based on the auditor's understanding of this example system, key application controls should be defined and documented. In this example, the auditor performed a general controls review recently, so it is not necessary to include interdependent controls in the list of controls to be documented.

The controls should be stated in terms of control objectives as was done for the general controls reviews described in earlier chapters. The audit procedures then can be built around each objective. An example of such a set of application control objectives and audit procedures is shown in Figure 5.9. Such a table can be constructed for most application reviews.

Perform Compliance Tests

The final step in the application review is to perform compliance tests of some of the audit controls, as defined in the objectives and procedures set out in Figure 5.9. Depending upon management's and the auditor's relative interest in the application, it may not be necessary to test all controls as listed. Many are related to one another. If no problems or weaknesses are identified in one control area, the auditor may pass on related control areas.

Following the example, some of the tests of application controls might include:

Reperformance of computations. Capital budgeting is based on specific computations such as estimation of the present value of future cash flows based on discount factors. Using another spreadsheet tool or a desk calculator, the auditor could recalculate one or several present value computations generated by the system, to determine the reasonableness of system processes. Any major difference should be resolved with user personnel.

Control Objectives and Audit Procedures
for Reviewing a Microcomputer Capital Budgeting System

Note: These objectives and procedures are designed for a review of the microcomputer capital budgeting application described in Figures 5.7 and 5.8. Similar tables of objectives and procedures would be constructed for other applications.

Objective 5.9.1.   All inputs to the capital budgeting system should be complete and correct.

Procedure 5.9.1.1.   Review the spreadsheet macro program listing and data parameters over spreadsheet cells to determine that all input data appears to be handled correctly. Consider the following:

Field sizes and decimal positions should be significantly large to allow proper present value and other capital budgeting computations

Formulas should be written in proper notation according to the spreadsheet package

There should be controls to screen improper data inputs

Procedure 5.9.1.2.   Determine that documentation exists to describe the process of inputting capital budgeting projects into the spreadsheet application.

Procedure 5.9.1.3.   Observe procedures for inputting data to the spreadsheet application and assess whether controls and data input checking appear adequate.

Objective 5.9.2.   Microcomputer spreadsheet program macro files as well as data files should be protected from improper access attempts.

Procedure 5.9.2.1.   Determine if a "DOS Shell" or some other password based utility program exists on the microcomputer system to protect against unauthorized access attempts.

Procedure 5.9.2.2.   Review procedures for backing up microcomputer programs and files and assess their adequacy.

Procedure 5.9.2.3.   Determine that key table values or other

Figure 5.9. Control objectives and audit procedures for reviewing a microcomputer capital budgeting system

spreadsheet data items are "hidden" from unauthorized review, subject to the capabilities of the spreadsheet software.

Objective 5.9.3. Computations for capital budgeting projects should be mathematically correct and should follow organization assumptions and constraints.

> **Procedure 5.9.3.1.** Review any overall organization rules for capital budgeting projects, such as interest rate and cost of capital assumptions, and ascertain whether the microcomputer model follows those assumptions.

> **Procedure 5.9.3.2.** Select one or more capital budgeting projects input to the microcomputer system for computation and manually reverify the computation.

> **Procedure 5.9.3.3.** After the computation verification of the selected item, print out a listing of selected budget data file items and verify that the selected item was correctly recorded.

Objective 5.9.4. Output reports from the capital budgeting system should provide an audit trail back to input documents.

> **Procedure 5.9.4.1.** Trace output reports from a selected processing cycle to determine that computations correctly reflect the input data.

> **Procedure 5.9.4.2.** Review organizational procedures for verifying the correctness of summary reports from the system and assess their adequacy.

Objective 5.9.5. Control totals should exist to allow for verification of capital budgeting items transmitted to the mainframe system.

> **Procedure 5.9.5.1.** Review procedures for extracting and transmitting data to the mainframe system and assess their adequacy.

> **Procedure 5.9.5.2.** Determine that procedures are in place to assure all transmissions are complete and received correctly by the host system.

> **Procedure 5.9.5.3.** Trace a selected cycle of microcomputer capital budgeting totals to the mainframe application to verify that all data was transmitted correctly.

Figure 5.9. Control objectives and audit procedures for reviewing a microcomputer capital budgeting system *(continued)*

Comparison of transactions. The auditor should select several sets of microcomputer budget schedules and trace them to the mainframe budget system to determine that they have been transmitted correctly.

Proper approval of transactions. Before any budget schedule is transmitted to the "official" mainframe budget system, proper management approvals should be obtained. The auditor can select several schedules transmitted to the mainframe to verify that they have been approved properly and the approval has been documented.

There are numerous similar compliance tests that can be performed for such a microcomputer system. The imaginative auditor performs these depending upon the nature of the audit and the objectives of management. Control weaknesses should be reported to management for corrective action.

Auditing a Mainframe Database Application: An Example

Our second example for an application review is a database-oriented, on-line purchasing system with interfaces to materials requirement planning, receiving, and accounts payable systems. This example system is implemented on a larger mainframe computer at a highly automated manufacturing organization. The auditors have reviewed general controls within the data processing organization and found them generally adequate. Now the internal auditors plan to perform an applications review of the automated purchasing system including its interfaces with other systems.

Although this system was implemented several years ago, it was not reviewed while it was under development. The system is of major accounting significance, but internal audit department scheduling problems prevented a pre-implementation review such as described in Chapter 7. In addition, the application's title of "New Purchasing System" as it appeared on department development schedules did not attract audit attention. Shortly after the application was implemented, however, the external auditors encountered it. They attempted to "audit around" the application in the first year and now have asked the internal auditors to review the application's controls. The system's "paperless" features attracted their attention.

The example organization is an electronics assembly company that purchases high volumes of many small parts and components. Its purchasing department issues "blanket" purchase orders with many of its vendors to supply periodic shipments of parts with quantities specified by a "just in time" automated manufacturing system. The purchasing department makes price and terms agreements with vendors in advance such that unit prices drop as total quantities purchased increase. To promote operational efficiency, the system was designed to minimize paperwork.

A flow chart describing the system is shown in Figure 5.10. The system works as follows:

1. Vendor purchase orders with price and terms agreements are input to the purchasing database. "Blanket" purchase orders are entered annually while others are input as required.

2. The automated materials requirements planning system determines parts requirements. These are input to the purchasing database along with any other manual inputs.

3. The system generates purchase orders on a daily basis. After review, these are mailed or transmitted to vendors. At the same time, these open purchase commitments are input to the automated receiving system.

4. When goods are received, the open purchase order data is called up from the receiving database. The quantities received are entered, and the date received is entered automatically by the system.

5. The material receipts data is input to the inventory system and also input back to the purchasing database. If the receipt is in compliance with purchase order terms, the receipt is set up for payment.

Parts vendors are encouraged *not to send invoices* to the company. For parts receipts in compliance with purchase order terms, the purchasing system will send a record to the accounts payable system authorizing a check to be issued. If the shipment arrives early or incomplete, the check authorization will reflect this. If the vendor sends an invoice, essentially it is ignored.

The automated purchasing system described in this example is a complex, paperless application system not uncommon in larger data processing organizations. Because this application has tight ties to other applications, including materials requirements planning, receiving, and accounts payable, an applications review presents the auditor with the dilemma of where to draw the boundaries of the review. For example, should the receiving or the accounts payable system be included in any automated purchasing applications review because they are tied so closely to that system?

The auditor will do best to limit the scope of such an applications review. In this case, the auditor should attempt to limit the review to the automated purchasing system and its interfaces with other applications. For example, the auditor can verify that the issue of a purchase order correctly initiates a

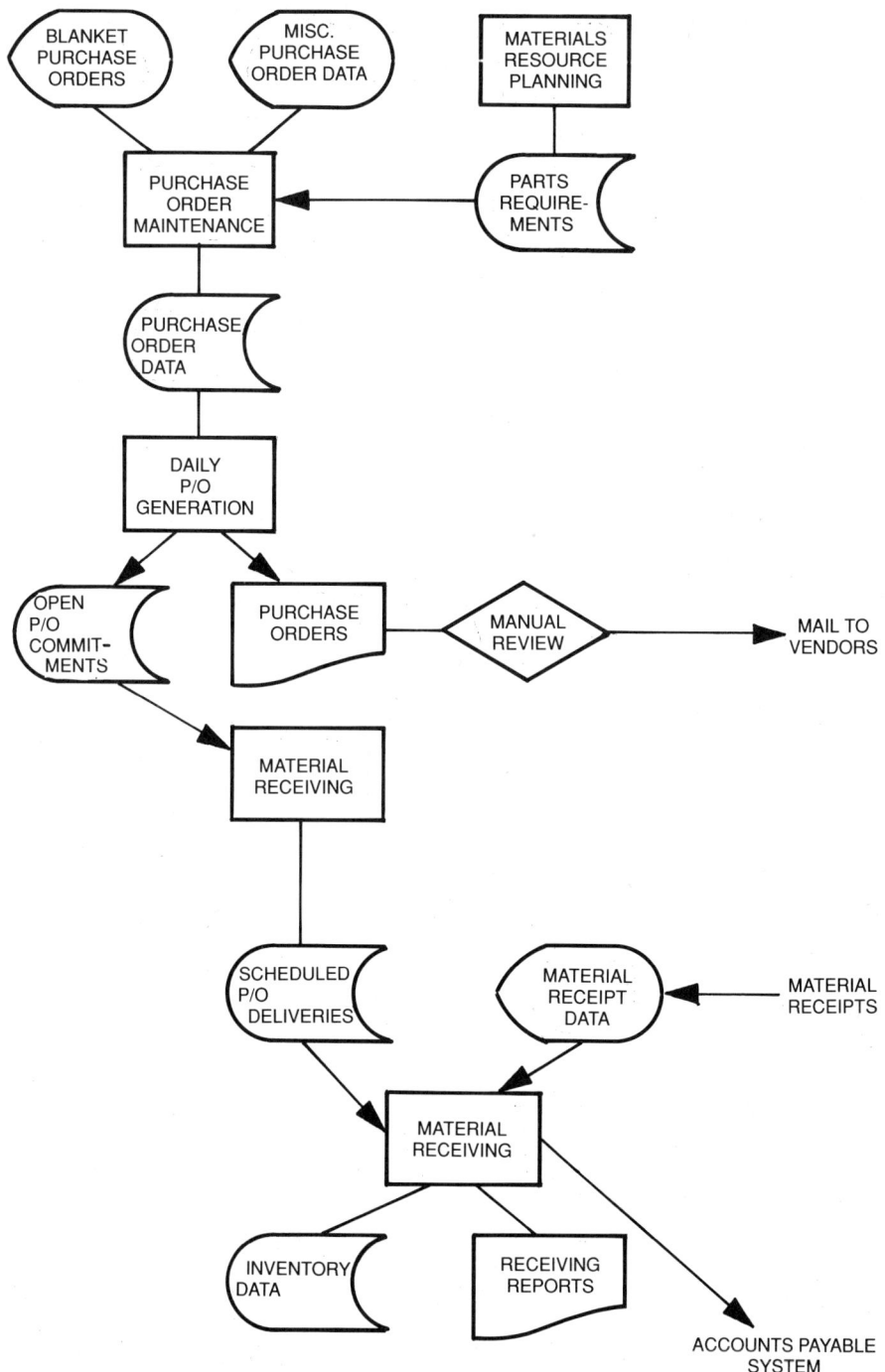

Figure 5.10. An example of a flow chart for a mainframe purchasing system

transaction to the purchasing database and that when the receiving system indicates a receipt, a transaction is received by the automated purchasing application. However, the controls within the purchasing database can be the subject of another applications review.

The auditor's objectives in a review of this sort are many and varied. Primarily, external auditors are interested in gaining an assurance that a receipt of goods is recorded properly as an inventory and an accounts payable item. A more technically oriented internal auditor may want to assure that the automated purchasing system database is maintaining vendor terms and conditions properly.

Before starting on such an applications review, the auditor should carefully define the review objectives and discuss them with all interested parties. An abbreviated set of control objectives for a review of this automated purchasing system might include:

There should be good controls over inputs of purchase order terms and vendor data

The purchasing database should be accurate and complete.

There should be controls to prevent improper purchase orders from being created and transmitted

There should be adequate audit trails over transactions to or from material requirements planning, receiving, and accounts payable

Only transactions for authorized and correct payments should be sent to the accounts payable system

Review Application Documentation

As with the microcomputer application system example, the auditor's first step is to gain a good understanding of this application through discussions with systems designers and users and a review of available documentation. This task can be extensive in a system as comprehensive and integrated as this example. The auditor may review purchasing department procedures, special control procedures over the use of the purchasing database system, and the application's systems and other user documentation.

The auditor may also review workpapers from any past audit department applications reviews of related systems and general controls reviews in such areas as the database administration function. If these interface applications have not been reviewed in the past, the auditor still may postpone any detailed review of them at this time. Any potential control questions regarding these interface systems can be documented on a "to do" list which would be prepared in conjunction with this automated purchasing system review. The auditor would want to expand the review to these other systems only if there appear to be potentially significant control problems.

Describe the Application

The auditor's next step is to describe the system and its controls. This is a much more extensive task than was necessary for the previous microcomputer example.

The auditor's workpaper description of the automated purchasing system can follow the same verbal and pictorial flow-chart format as was discussed previously. However, the verbal descriptions and flow charts should be broken down into smaller units to help the auditor understand the application controls and better manage the application audit. The auditor's description of the example system's procedures for setting up new vendor price and terms agreements is shown in Figure 5.11. Similarly, a flow chart describing how the system interfaces with accounts payable is shown in Figure 5.12.

Identifying Application Controls

There are many potential internal controls issues associated with a large database application such as this example. Many of these internal control issues deal with the interaction of this application with such systems as receiving and accounts payable. Other control issues deal with newer technologies such as the use of a database for maintaining purchase records and telecommunications for the paperless transmission of purchase orders.

The auditor can develop a set of control objectives and audit procedures based on the preliminary set of objectives that were developed at the start of the review and on the understanding of the application gained from the documentation review phase. These application control objectives and audit procedures should be developed along the same general format as the microcomputer capital budgeting objectives and procedures shown in Figure 5.9. However, because this is a much more comprehensive application, there are more control points to be identified. Figure 5.13 lists control objectives and audit procedures for the portion of the example application that interfaces with accounts payable.

The auditor should develop a tailored set of objectives and audit procedures for each portion of the automated purchasing system. Because each application and its subsystems have unique aspects, it is difficult to construct a generic

Auditor Documentation Steps

Note: This figure illustrates documentation that the auditor might use to describe the procedures for setting up new vendor price and terms codes as part of an overall application review of the automated purchasing application described in Fig. 5.10 and discussed in the text.

1. As part of a quarterly update process, a special program is run to identify all vendors with missing prices or cash discount terms in the vendor file. This list is sent to Purchasing for detailed analysis and followup.

2. Also as part of the update process, reports are produced from the vendor file listing total purchases and vendor terms codes; the list is produced in descending order by total vendor purchases and is sent to Purchasing for analysis.

3. Purchasing reviews quarterly lists to determine areas where vendor data is missing or where it may be possible to secure better terms from higher volume vendors.

4. Transactions to run the quarterly update to the vendor file are run in a batch mode; data collection sheets are submitted to the information systems data entry function for input with batch controls.

5. Output from the quarterly updates include a transaction and error report. This report is returned to the originators at Purchasing for review and correction of any errors through a subsequent update.

6. The updated vendor file from the quarterly update is not placed into production until all corrections are complete. That vendor file is then placed into production in place of the earlier version.

7. A written request is prepared by Purchasing and sent to computer operations specifying when the updated file should be moved into production.

8. Particular attention is given to outputs when any updated vendor file is placed in production to determine that it has been added to production as intended, and that it is producing correct results.

9. Miscellaneous transactions are input to the transaction file on an "as required" basis. These include transactions to add new vendors to the file, to change vendor terms or conditions, or to delete data

Figure 5.11. An example of auditor documentation to describe system procedures to set up new vendor price and terms agreements

from the file. These inputs are processed through a regular, weekly update with transaction data collected through an on-line terminal for weekly batch update.

10. Statistical reports covering the contents of the vendor file are run on a weekly basis for distribution to Purchasing for ongoing analysis.

Figure 5.11. An example of auditor documentation to describe system procedures to set up new vendor price and terms agreements *(cont.)*

chart of control objectives and procedures for an application review. However, following the general formats of Figures 5.9 and 5.13, the auditor should be able to construct tables for most data processing applications once the auditor has a general understanding of the application and its specific control characteristics.

When constructing control objectives and procedures for an applications review, however, the auditor may consult with technical personnel if appropriate. In this example, the application receives inputs from a material requirements planning system. If the auditor is unfamiliar with such applications, their unique characteristics should be discussed with other members of the audit team or user personnel who can provide some help.

Once controls have been identified, the auditor should confer with management and others, such as external auditors. This review may prevent the auditor from attempting to perform controls tests which would be difficult or would not achieve significant audit objectives.

Perform Compliance Tests

The final audit step for the review of the example automated purchasing system would be to perform a control risk assessment, or a compliance test, of those controls identified in Figure 5.13. As previously discussed, compliance testing is a "walk-through" approach to testing that will not give absolute assurance that the application controls are working as tested. Based on a test of one or two transactions, the auditor may conclude that a control is working when generally it is not, or that the control is not, when it is.

Depending upon the overall scope of the applications audit, the auditor may not want a compliance test of all controls identified in the applications overview. This decision depends upon the audit budget and the criticality of the application. In any event, the auditor should attempt to perform a walk-through compliance test on what appear to be the more significant controls.

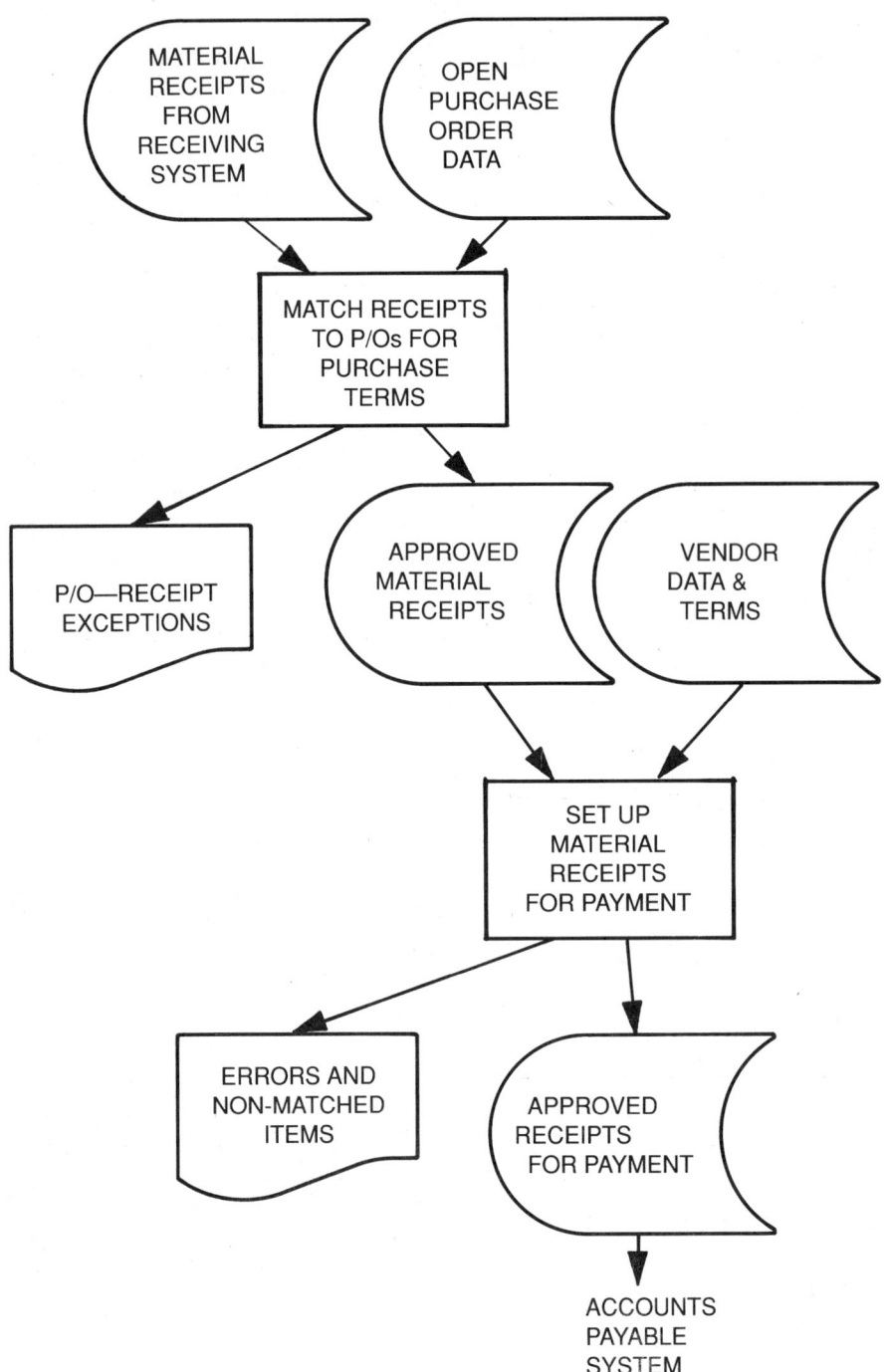

Figure 5.12. An example of auditor documentation of a purchasing system
interface with an accounts payable system

Control Objectives and Audit Procedures for Reviewing an Automated Purchasing System

Note: These objectives and procedures cover the auditor's review of a portion of the purchasing application used as an example in the text. They cover the interface between that application and the accounts payable system. The auditor should develop a similar set of objectives and procedures for the other parts of the application.

Objective 5.13.1. Only valid material receipts should be paid, according to pre-established terms, by the automated accounts payable system.

Procedure 5.13.1.1. Select a sample of material receipt transactions scheduled for payment and determine that they have been processed by the accounts payable system at the scheduled date and with correct terms.

Procedure 5.13.1.2. Review material receipts and vendor file error reports over a period of time; ascertain that all exceptions have been cleared on a timely basis.

Procedure 5.13.1.3. Review control total and run to run controls between the accounts payable and the purchasing systems and determine that they are being reviewed and reconciled on a regular basis.

Objective 5.13.2. The generation of accounts payable checks should be based on quantity and price terms from the purchasing system, payee data from the vendor system, and other data such as inspection results from appropriate departments or systems.

Procedure 5.13.2.1. Select a sample of accounts payable items where checks have been generated and trace them back to determine that quantity and price calculations are correct and that scheduled cash discounts have been taken.

Procedure 5.13.2.2. Review procedures for determining when to take vendor cash discounts and assess whether optimal cost of money considerations are used properly in making such decisions.

Procedure 5.13.2.3. Determine that the application has some

Figure 5.13. Control objectives and audit procedures for reviewing an automated purchasing system

form of reasonableness test such that an exception report or other vehicle is used to flag payable checks that exceed specified parameters.

Objective 5.13.3. Control totals and date stamps should be used such that accounts payable data can be traced back to the originating purchasing system documents.

Procedure 5.13.3.1. Determine that some form of date stamp is used to identify accounts payable transactions.

Procedure 5.13.3.2. Select a sample of transactions from the accounts payable system and trace them back to the material receiving system to determine that items correctly authorized for payment are recognized.

Objective 5.13.4. Application controls should exist to avoid duplicate payments.

Procedure 5.13.4.1. Review application documentation and portions of program code, if appropriate, to determine that controls exist to prevent potential duplicate payments to vendors.

Procedure 5.13.4.2. Using generalized audit software (see Chapter 6) or a report generator, sort accounts payable data by amount due for several periods and match on amount, vendor number, and purchase order number to detect potential duplicate payments.

Objective 5.13.5. Material receipts and purchase order exception items should be reconciled on a regular basis.

Procedure 5.13.5.1. Select a sample of recent error reports generated by the system interface and determine that all reported exceptions have been reconciled.

Procedure 5.13.5.2. Review the system's approved material receipts file, which represents items authorized for payment but not paid, and determine reasons why they remain outstanding.

Figure 5.13. Control objectives and audit procedures for reviewing an automated purchasing system *(continued)*

When deciding which controls to test initially, it is proper to base the testing strategy on comments received from users or data processing personnel on potential control problems. These individuals frequently make comments during an auditor interview such as, "The system would work fine if it wasn't for. . . . " Such comments often point the auditor to potential control weaknesses.

As discussed in the microcomputer application review example, compliance tests involve reverification of computations and comparisons of transactions. Figure 5.14 contains potential compliance tests for the example application, based on the control objectives and audit procedures identified in Figure 5.13.

Detailed compliance testing can be time consuming if a large number of application controls are to be evaluated. The experienced auditor may want the help of an assistant to perform detailed testing procedures.

SUMMARIZING AND EVALUATING APPLICATIONS CONTROLS

The final steps in an applications review are to summarize audit findings, report these findings to management, and plan for any further detailed applications testing. The auditor should document the findings and recommendations in workpapers and, if appropriate, in a formal audit report. These provide both data processing and operations management guidance for taking corrective action to improve applications controls.

When summarizing and evaluating such an applications review, the auditor should be aware of the risks of concluding a control to be working when it is ineffective, or of concluding it is not working when it is. If the auditor has doubts that results of a compliance test are representative, an effective auditor retests the controls or expands the scope of the test. In particular, if management questions results of any limited compliance test, the auditor should consider retesting the particular control.

An application controls review and its related compliance tests provide the auditor with a basis for placing reliance on the results of that application. The auditor may conclude that the application reviewed appears to be operating as intended with generally good controls and procedures. The auditor may stop the review at this point and go on to review other data processing applications. In some instances, however, the auditor wants to perform more detailed or substantive tests. These will be discussed in Chapter 6.

Application Compliance Tests

Note: These compliance tests might be performed for the purchasing and accounts payable application interfaces described in Figures 5.12 and 5.13 and the text. The auditor typically might not perform all of these tests for any given audit but should perform tests in areas where the risk of error or misstatement is greatest.

1. Procedure 5.13.1.1 calls for taking a test of material receipt transactions scheduled for payment and determining that they have been paid by the accounts payable application. Using generalized audit software, an approved material receipts file could be selected and matched to a payment history file several periods later with an exception report for audit analysis.

2. Procedure 5.13.2.1 calls for tracing items which have been paid to determine that quantity and price calculations are correct. Using generalized software, or even a local microcomputer system, the auditor could select a sample of items paid, match it to the material receipts files to extract quantities and prices, and recalculate accounts payable amounts.

3. Procedure 5.13.2.2 calls for a review of cash discount procedures to determine if optimal discounts have been taken. The auditor could design this test by building a microcomputer spreadsheet decision matrix which annualizes vendor cash discount percentages and days with the average cost of capital to determine optimal rates.

4. Procedure 5.13.4.2 suggests a test, using generalized audit software, to identify duplicate payments. This is a compliance type of test.

5. Procedure 5.13.5.2 calls for a review of items on the approved material receipts file which have not been paid to determine reasons they remain outstanding. Generalized software could be used to extract all items on this file prior to a given date to help ease investigative compliance testing.

Figure 5.14. Example of compliance tests for a purchasing and accounts payable application

REVIEWS IN SUBSEQUENT YEARS: A SYSTEMS TEST APPROACH

After performing a detailed applications review, often, auditors are asked to review that same application again after a period of, perhaps, one or two years. Usually, the reason for this subsequent review is to verify that the controls improvements previously recommended have been implemented or that subsequent systems changes have not altered the overall controls environment. The auditor need not rereview the entire application when performing such a follow-up applications audit.

The advantage of an applications systems test is that it is unnecessary to retest all applications controls if some control procedures have been tested in a prior review and were working properly at that time. The auditor need only review and test those aspects of the system that have changed. Such a review can be performed following a four-step approach. First, review the application to determine which control procedures have remained unchanged and which have been modified. This can be accomplished by discussion with key data processing and user personnel, using the workpaper application description as a guide. Next, develop an understanding of any application changes. This usually can be accomplished through a review of systems development documentation materials. Third, update workpaper documentation to reflect applications changes and identify any new controls as might be appropriate. Finally, perform any compliance tests necessitated by applications modifications. Depending upon the nature of these changes, it may be necessary also to retest some existing applications controls.

This approach may avoid reevaluating an application every time it is reviewed. By developing an understanding of any changes that occurred after a given point in time and by reviewing and testing primarily the controls associated with those changes, the auditor can feel relatively confident that the application continues to operate as understood. This approach helps the auditor to do a more efficient job in performing repetitive applications reviews.

THE IMPORTANCE OF REVIEWING DATA PROCESSING APPLICATIONS

The effective auditor places major emphasis on reviewing data processing applications when performing audits of the overall system of data processing controls. This is particularly true because good general or interdependent data processing control procedures often are in place in many established, modern data processing organizations. However, because applications may be developed through a series of compromises with users or without proper quality assurance reviews, individual applications controls may not be particularly strong.

When compared to reviewing a fairly sophisticated data processing application, such as the automated purchasing system discussed in this chapter, a general controls review often provides the auditor with less challenge. To properly evaluate data processing applications controls, the auditor needs a good understanding of both data processing procedures and specific control and procedural characteristics of each application area.

Nevertheless, the effective auditor should spend a substantial amount of audit effort reviewing and testing controls in specific data processing applications. Such reviews benefit general management by providing assurance that applications are operating properly, data processing management by assuring that their design and control standards are being followed, and financial auditors by allowing them to place greater reliance on the output of such applications.

C H A P T E R 6

Evidence Gathering and Testing Applications

INTRODUCTION

A fundamental requirement of auditing is to obtain evidence of the validity of the accounting treatment of transactions and balances and to evaluate the potential for errors and irregularities. Whether a manual or an automated system, this evidence gathering requires more than a review of an application's internal accounting controls. Detailed or substantive testing of the transactions processed by the automated system usually are required to gather evidence properly.

This chapter describes approaches to testing and gathering detailed evidence about automated applications, including one-time tests utilizing methods such as computer assisted auditing procedures as well as continuous audit monitors built into an application. Continuous audit monitors are particularly useful when the auditor has an on-going interest in reviewing exception condition transactions in large, complex systems.

While application evidence gathering techniques usually are used for reviews of financial applications to support external auditors, the same methods are useful for testing and monitoring many non-financial applications.

This chapter will describe how evidence gathering and applications testing can be used for both financial and non-financial types of automated applications.

THE NEED FOR APPLICATION EVIDENCE GATHERING

Many auditors fail to give proper attention to the need to gather evidence when reviewing computer applications. Understanding an automated application and evaluating its internal controls generally is an interesting and challenging audit task, but detailed confirmations of account balances or other evidence gathering tests sometimes are less interesting and more time-consuming. However, evidence gathering procedures can provide the auditor with an opportunity to utilize creative auditing techniques.

An example might better illustrate why applications evidence gathering procedures are important. Assume the auditor performed a detailed internal controls oriented review of a large fixed asset, capital budgeting application where transactions are initiated from a variety of subsidiary systems and where the application eventually provides general ledger and financial statement balances. The auditor tested the systems controls and concluded they were adequate. The auditor also manually recalculated depreciation expense for several selected transactions and found each to be correct. Can the auditor conclude that the fixed assets and accumulated depreciation numbers produced by this sample system are accurate?

In a large organization where fixed assets represents a substantial portion of the balance sheet, the auditor may decide there is too great a risk in relying solely on this internal controls review. The several transactions selected for recalculation as a compliance test may not be representative of the entire population, and there may be an error in certain classes of these transactions. Although application to application controls appear proper, some types of transactions may be assigned to incorrect account groups. Without detailed testing of this fixed assets system, these errors could go undetected.

Auditors should have an understanding of when it is appropriate to perform detailed tests of data processing applications to verify the correctness of transactions or account balances. Some circumstances in which the auditor will want to do this more detailed evidence gathering and testing include:

When there is a perception that the risk of relying solely on internal controls is too high

When the results of limited walk-through or compliance types of tests are inconclusive and suggest more detailed tests

When certain internal controls are weak or difficult to identify

In many instances, the decision of whether to rely just on internal accounting controls and compliance testing or to perform detailed tests of transactions will be a decision of the entire team working on the audit rather than of the computer audit specialist alone. For example, the computer audit specialist may feel that internal controls surrounding a financial application are accurate. However, external auditors who must assess control risk and attest to financial statement balances may insist on detailed testing to better support their conclusion.

Audit Risk When Relying Upon Internal Controls

Auditors generally perform reviews with the objective of detecting errors which could be large enough, individually or in the aggregate, to be significant or material. This is true whether the auditor is performing a financial review or an operational application review. There is always a risk that the auditor's procedures will not detect significant errors or irregularities.

When relying upon application controls only, the auditor often faces a risk that errors, irregularities, or system weaknesses will not be detected. If that risk appears sufficiently high, the auditor will expand procedures to perform detailed testing in addition to placing sole reliance upon an application's internal controls. For a computerized application, those expanded procedures often require detailed application testing and evidence gathering.

The American Institute of Certified Public Accountants (AICPA) has done a good job of defining risk in its General Auditing Standards (GAS). Although most frequently used by external auditors in performing their attest audit work, an understanding of these risk concepts is useful for internal auditors as well. The AICPA has broken down risk into the following three elements:

Inherent risk. Even if there were no related internal control considerations, there is a greater risk of error in some transactions than in others. For example, there is a greater inherent risk of errors or irregularities in an accounts receivables system which handles cash than in a manufacturing scrap reporting system. Similarly, an application which performs complex calculations is more likely to be misstated than a simple one.

Control risk. There is always a risk that the internal accounting controls for an application will not prevent or detect an error or irregularity on a timely basis. Even though the auditor may find good controls over a given application, it usually is appropriate to do some detailed evidence gathering and testing to minimize control risks. Some control risk always exists because of the inherent limitations of any system of internal accounting controls.

Detection risk. This is the risk that the auditor's procedures will not detect a material error when one indeed exists in a given application under review. There is always some detection risk even when the auditor examines 100 percent of the transactions in an application. Inappropriate auditing procedures may have been used or the results may have been misinterpreted. However, detailed sampling and evidence gathering of an application's transactions and balances reduce that detection risk.

The auditor should consider all three risk elements when deciding whether to perform detailed tests of a data processing application. These risk factors are not easily quantifiable and require the auditor's judgment. The nature of the automated application, the quality of the controls surrounding that application, and the type of evidence gathering procedures to be used all enter into this assessment.

While this risk evaluation approach is normally used by external auditors when planning their procedures, it is equally appropriate for internal auditors performing operational reviews of non-financial applications. Management may ask the internal audit function to review controls for a new manufacturing shop floor scheduling system. Based upon the above risk factors, the auditors may decide that detailed testing of that application is necessary before advising management that the controls in that automated application are adequate.

Substantive Testing Due to Weak Internal Controls

There are many instances when an auditor reviews internal controls over an automated application and finds significant potential control problems. While an auditor can publish a report outlining these control weaknesses and recommending a series of systems improvements, management often wants more detailed information to substantiate the auditor's findings. This calls for detailed or substantive testing of the application.

As an example of this situation, assume that the auditor reviewed an on-line order entry system and found a variety of potential control weaknesses. One is that the order entry application does not check customer credit files when authorizing new orders. While management wants this weakness corrected, it is interested also in the extent of the problem at present. To supply this information to management, the auditor could use computer audit software to match the open order file with the credit file to identify potential credit risks.

There are many instances where real or perceived application control weaknesses require the auditor to do detailed testing and evidence gathering. There are other situations where detailed testing will be needed because the internal application controls cannot be evaluated easily. An example of the latter is in the use of a purchased software package with complex calculations. Because documentation other than user manuals may not be available for the microcomputer package, the auditor may decide that detailed testing is the only practical approach to evaluate the application.

The Need for Continuous Audit Monitors

Some applications are so large and complex that there is a need to gather evidence about them on a continuous basis. Assume that a large, multi-division organization uses a centralized accounting system that allows numerous accounting transactions between divisions. Management is concerned about certain unauthorized transactions and asks the auditors to review systems controls. As a result of the review, the auditors conclude that controls are adequate, but due to the number of users and volume of transactions, there is a continuing risk of unauthorized transactions. To check for such activity, the audit department may arrange to install a continuous audit monitor into the application. Continuous audit monitors are tools that allow the auditor to perform ongoing evidence gathering and application testing.

GENERALIZED AUDIT SOFTWARE PROCEDURES

A generalized audit software package is an application designed to help the auditor perform certain data processing functions such as selecting the data from computer application files, printing audit extract reports, and performing other audit related procedures. Some generalized audit software products are designed specifically for auditors, but other general software retrieval products can also be used to perform many desired audit procedures. Some type of audit software is a necessary tool for virtually all effective internal audit departments when operating in today's highly computerized environments.

Background and Reasons for Generalized Audit Software

In the early days of computer auditing, most applications were written in compiler languages, such as COBOL, or in assembler languages. Auditors often did not have the technical skills or the time to write their own retrieval programs, and the many, easy to use software retrieval tools in use today generally were not available. However, by the early 1970s, auditors often found a need to foot, sample, or extract from computerized files. To acccomplish this, they often had to request help from the data processing department to write the necessary retrieval programs. In many instances this department was also the auditee.

This lack of audit software was first solved by the major public accounting firms in the 1970s. Many developed simple audit retrieval programs which allowed audit specialists to run programs against clients' application files. In addition to easy to use retrieval capabilities, other common audit functions were built into the software, such as sequence number gap detection. Soon, these generalized audit software packages became commercially available and now are used widely by many internal audit departments.

Generalized audit software offers auditors the following advantages:

Reduced reliance on data processing. The auditor often needs to operate independently of the data processing department responsible for maintaining the application being audited. Audit software allows the auditor to perform tests of an application without asking the data processing function to write the necessary retrieval software.

Increased audit efficiencies. Audit software can introduce efficiencies into the audit process. For example, external auditors must independently confirm accounts receivable balances. When these accounts are recorded as computerized records, it is more efficient to select records and produce the confirmation letters using computer audit software. In addition the cost of the software often can be spread over several years of use.

Opportunity to observe other controls. By using an independent set of programs, but operating through the auditee's data processing operations, the auditor will have an opportunity to develop a better understanding of other controls within that function. This often points to other areas for subsequent audit work.

In the modern data processing environment, the auditor has many options for audit software. In addition to the standard, generalized audit software, auditors can use software retrieval tools or the fourth generation languages available for many computer systems. The latter are discussed in Chapter 12. Discussed here are considerations for selecting an appropriate audit software package, examples of the types of applications that can be developed, and guidance on developing a computer assisted audit procedure.

Examples of Generalized Audit Software Applications

Generalized audit software can be used to review and analyze data from essentially any computer readable file. The types of applications that can be developed depend upon the audit objectives and the auditor's creativity. Some of the basic types of tasks which are appropriate for audit software applications include the following:

Examining records based on auditor specified criteria. Auditors often need to review large volumes of data for "can't happen," or exception, situations. They may examine inventory status records for negative balances or review accounts receivable balances for accounts over the credit limits. Generalized audit software will produce an output report or file with records selected from criteria defined by the auditor.

Summarizing, resequencing, and reviewing data. There is often an audit need to total all transactions on a file, such as the physical inventory records, or to resequence records, such as an aging of accounts receivable records. Also, audit software is useful for analytical review techniques such as calculating and summarizing turns in a retail inventory.

Testing or reperforming file calculations. The auditor can use generalized audit software to test the accuracy of computations recorded on data files. The auditor may want to recalculate accrued interest on a bank loan file or recalculate depreciation on a fixed assets file. This work can be done on all file records or on a sample.

Comparing data on separate data files. Frequently, auditors need to compare or match separate data files. For example, an auditor may want to match an accounts payable voucher file against an approved vendor name and address file to determine that all data is valid. In addition, the auditor may want to compare the inventory files from two periods in time to identify potentially obsolete items.

Selecting and printing audit reports. Besides producing audit reports based on some of the above criteria, the auditor may be required to produce special reports during the progress of the audit, such as a worksheet report to aid in observing the taking of physical inventory.

Auditors generally combine a variety of the above procedures to build an audit software application for a given type of review. For example, if the auditor is reviewing a financial institution application, the audit software applications might include:

A recalculation of accrued interest on a loan file to verify financial totals

An edit of loan files to detect any unusual rates or terms

A statistical sample of loan balances to produce letters for account confirmation

A match of loan files with personnel files to determine any unusual employee loan activity

A total of loan files with an analysis by date, loan type, and other selected criteria

The above is an example of the types of computations and applications that can be developed through the use of audit software. All internal audit departments operating in a modern, computerized environment should consider the use of some type of audit software.

Figure 6.1 shows the procedures to be followed in using computer audit software, using as an example an application developed to test balances and interest rates for a financial institution loan file. A common brand of audit software was used for this example. Most audit software packages require similar amounts of input and produce similar output reports.

Selecting the Appropriate Audit Software Tool

In a general sense, any computer program used by the auditor is a type of audit software. However, generally, the auditor does not want to develop programs in a compiler language such as COBOL because of the development time requirements. Rather, the auditor wants to use some type of generalized audit or retrieval software. The decision of which approach to use depends upon the computer hardware and software environment, the current availability of software tools, and the auditor's objectives.

No matter what type of audit software is used, the auditor should arrange to have access to a terminal. In a large systems environment, this generally means a terminal where jobs can be entered or placed into the input queue by the auditor. Such a terminal may be a microcomputer, permitting the auditor to download selected file data for detailed audit analysis.

Example of the Use of Computer Assisted Audit Software

This example shows the procedures an auditor might follow to develop an audit software application file to test balances and interest rates for a financial institution loan file.

1. Determine Audit Procedures.

The auditor must first develop a general idea of the types of tests to be performed. These may be determined through discussions with financial auditors who may have an interest in the application. In this example, the auditor is interested in footing or pulling a total for the entire loan file and analyzing certain special conditions such as loans above or below specified parameters.

2. Gain an Understanding of Data Files.

The auditor must gain a general understanding of the application in order to determine which files contain desired data elements to be tested. Often this can be accomplished through a review of documentation and discussions with information systems personnel.

In this example, the auditor probably would decide that the loan file provides the necessary data. The auditor would then secure a file layout or COBOL description of the file, which might appear as follows:

```
01 LOAN-FILE.
   02 ACCOUNT-NO      PIC 9(6).
   02 CUSTOMER-NAME   PIC X(20).
   02 LOAN-AMOUNT     PIC 9(8)V99.
   02 LOAN-DATE       PIC 9(6).
   02 LOAN-TERM       PIC 999.
   02 LOAN-RATE       PIC 99V99.
   02 LOAN-BALANCE    PIC 9(8)V9.
```

An actual COBOL file description, of course, would be much more complex with many more data fields.

3. Program the Audit Procedures and Run Preliminary Tests.

Using the generalized computer audit software package, the auditor would input the description of the data file to be tested and the processing steps required as follows:

Figure 6.1. Example of the use of computer assisted audit software

```
READ LOAN-FILE.
ADD LOAN-BALANCE.
IF LOAN-RATE IS GREATER THAN X%
    OR LESS THAN y%,
    PRINT ACCOUNT-NO, CUSTOMER-NAME,
        LOAN-TERM, LOAN-RATE.
```

The auditor might code the procedure and run a test against the loan file to determine if the audit procedure is working.

4. Run Production Application and Document Results.

The auditor's final step is to run the generalized audit software under controlled conditions using the correct version and the date of the loan file. Output results should be used for subsequent audit analysis and the procedure should be documented in auditor workpapers.

Figure 6.1. Example of the use of computer assisted audit software (*continued*)

Generalized Audit Software Packages

In a larger data processing environment, generalized audit software is often the first or preferred alternative. It offers the following advantages:

Fast learning curves permit auditors with limited data processing skills to build applications quickly and easily.

Preprogrammed audit functions such as sequence number gap detection usually are built into the software.

Limited coding requirements that permit the preparation of retrieval reports with very few program coding steps.

As can be expected, there are some severe disadvantages to the use of generalized audit software, primarily because most audit software products were developed originally for larger, mainframe computer systems using batch oriented, simple file structures. Some of these disadvantages are:

Limited equipment and software availability. Usually, generalized audit software packages are not available for non-IBM larger computers nor for most mini and microcomputers. In addition, the software often cannot directly access many types of database systems.

Relatively high cost. A generalized audit software package may cost an audit department from $15,000 to $20,000 with an annual maintenance fee of several thousand dollars. If only limited usage is planned, this may be too expensive.

Limited transportability. Auditors in many larger organizations are faced with multiple computer sites having differing operating environments. Equipment differences or software license restrictions may prevent transporting the audit software package to multiple sites.

If the auditor's equipment environment appears appropriate, and if sufficient audit procedures are planned, generalized audit software should be considered for testing audit applications and gathering audit evidence. Internal auditors may get help in identifying an appropriate generalized audit software package from their external auditors.

Computer Retrieval or Query Languages

As discussed previously in this chapter, generalized audit software began to be used at a time when data processing departments used only COBOL for their application development efforts, and end users relied upon data processing to develop those applications. This total dependency on COBOL has changed in the modern data processing department, which now also uses some type of retrieval language for producing special request reports or performing special file manipulations. These same languages often are used by end users through an information center. These languages can be used by auditors in place of generalized audit software.

Auditors should inquire as to the types of generalized retrieval languages available at the data center under audit. Usually, they are available for most types and sizes of vendors' hardware, from larger mainframes to micro or minicomputers. Some of these are sufficiently general to be used for most applications. Others are designed to operate only as a query language for a given type of database or for a vendor's application software package. The most general and flexible of these products go by various product names under the general title of fourth generation languages. These are discussed in more detail in Chapter 12.

Generalized data processing retrieval software has the disadvantage of lacking such built-in audit software functions as statistical sampling or sequence number gap detection. However, it generally works quite well for item selection, recalculations, file matching, or data reporting purposes. The specialized audit functions generally are coded into the retrieval language with little difficulty. In addition, if the data processing department has installed such a package, the auditor can share in its use without the need for a separate purchase.

The main disadvantage with these retrieval packages is that the auditor, in a larger organization, often will be required to learn multiple products. There may be one or several installed on the main computer with additional ones at divisional minicomputers. Although the learning curve for these products typically is quite short, an auditor going from one location to another usually finds it more efficient to use only one audit software product. When faced with different equipment types, a copy of a data file can be made for processing at a central, off-site location.

Downloading Audit Data to Microcomputers

Auditors frequently use microcomputers to perform audit tasks. Internal audit departments may have a microcomputer for maintaining staff schedules, producing audit reports, or other administrative purposes. Similarly, external auditors increasingly use portable or transportable microcomputers for financial auditing schedule preparation. These same microcomputers can be used for analyzing application data downloaded from other computer systems.

There are numerous hardware and software tools available which allow data from larger system files to be transferred or downloaded to microcomputer files. In some instances, a hardware device can be installed in the microcomputer to make it look like an intelligent mainframe terminal. Other times, a microcomputer can use a modem and telecommunication facilities to dial into the mainframe to download file data. Either approach requires some coordination with the data processing operations function, but usually there are no technological difficulties in the modern computer center.

The auditor typically initiates a job on the mainframe or minicomputer to transmit the desired data to the micro. Once it has been transmitted to the auditor's disc file, it can be used for such functions as data analysis, recalculations, or selected edits. Full function audit software is just becoming available on microcomputers. The auditor can also copy the downloaded data into a spreadsheet or database package. It is then easily analyzed or manipulated. This microcomputer downloading approach to audit analysis is illustrated in Figure 6.2.

The major disadvantage to downloading files is that only limited amounts of data can be processed conveniently on a smaller microcomputer. The microcomputer must be equipped with a hard disc which, until recently, was limited

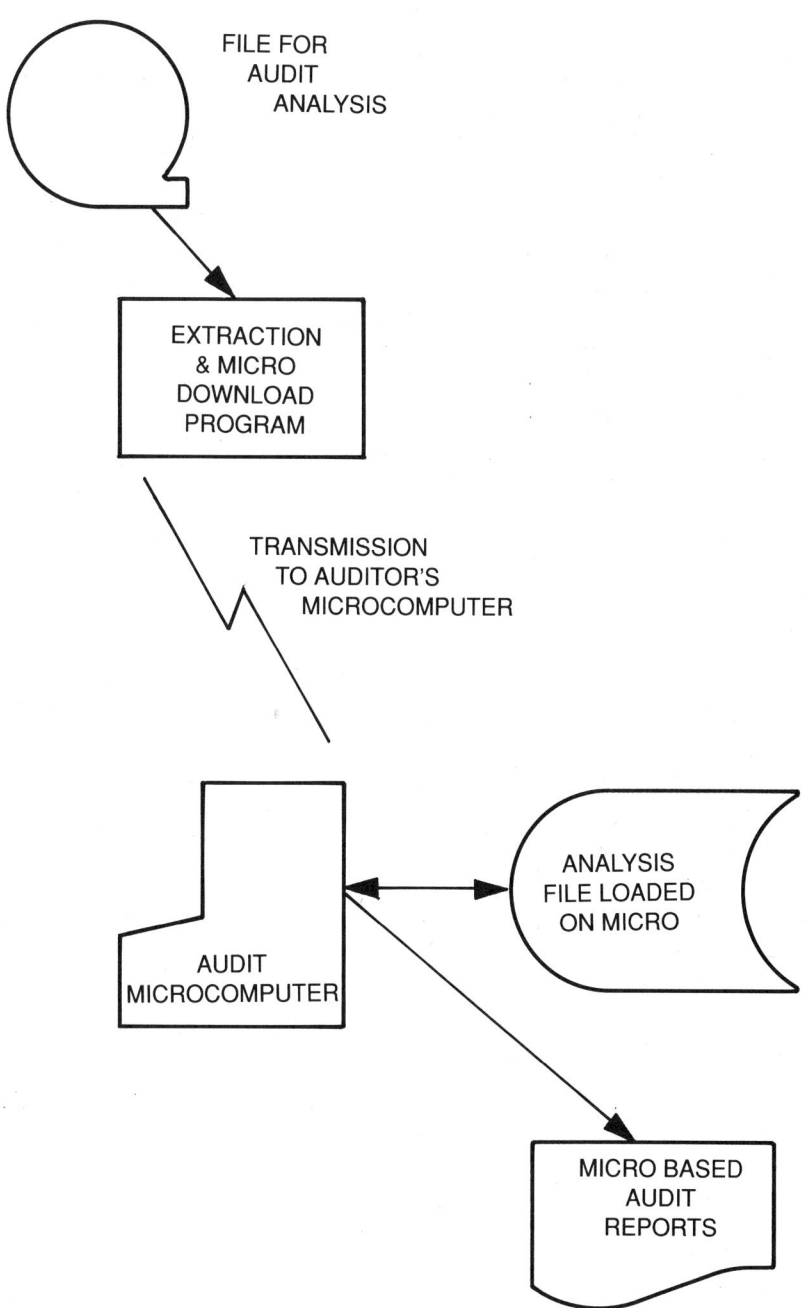

FILE FOR
AUDIT
ANALYSIS

EXTRACTION
& MICRO
DOWNLOAD
PROGRAM

TRANSMISSION
TO AUDITOR'S
MICROCOMPUTER

AUDIT
MICROCOMPUTER

ANALYSIS
FILE LOADED
ON MICRO

MICRO BASED
AUDIT
REPORTS

Figure 6.2. Microcomputer downloading approach to audit analysis

263

to 10 or 20 megabytes of data. Although many newer microcomputers are equipped with much higher capacity hard discs, very large data files cannot be realistically transmitted down to a microcomputer. If nothing else, it takes too long to process such files on smaller microcomputers.

Data formatting presents the auditor with another set of problems when attempting to download data. Usually it is necessary to reformat the mainframe files, which may be in a complex database format, into a simpler format understandable by microcomputer software packages. Despite these problems, the downloading of data from other computers to auditor microcomputers is becoming a useful option for application testing and data gathering.

Developing Computer Assisted Audit Applications

When developing computer assisted audit applications, the auditor should use the same approach whether using generalized audit software, a retrieval language, or data downloaded to a microcomputer. The steps that should be followed are similar to those of a Systems Development Methodology (SDM), discussed in Chapters 2, 3, and 7, except that the auditor may develop an audit retrieval application for a one-time effort. In addition, the auditor will not want to involve other users in the systems development process if the auditor is performing a confidential audit.

Computer assisted audit applications should be developed using the following five-step approach:

Step 1: Determine the objective of the audit test. An important first step for the auditor is to define the objective of the audit test to be run using the audit software. If the desired objective is defined clearly, the identification of testing procedures is a much easier task. An application documentation file or workpaper also should be started with this step.

Step 2: Analyze the application being audited. Once the auditor has defined the objective of the computer assisted test, file layouts and systems flow charts of the application being tested should be obtained to select the appropriate data sources for testing. It is possible the auditor will encounter technical problems at this point which might impede further progress.

Step 3: Design the computer assisted application. The auditor should first define the overall program logic as well as any output report formats. It probably will be necessary to discuss any special codes or other special data characteristics with persons responsible for the computer application. Consideration must be given at this phase to proving the results of any audit tests, for example by balancing to production application control totals. These matters should be outlined in the documentation workpapers.

Step 4: Program and test the application. Once the application has been coded, the auditor should arrange to test it on a limited population of data. The results of this test should be verified for both correctness of program logic and achievement of audit objectives, and recorded in the documentation workpapers.

Step 5: Process and complete the audit application. Making arrangements for processing the computer assisted application requires some schedule coordination between the auditor and data processing operations. Often, the auditor is interested in a specific generation of a data file, and it is necessary to coordinate access to it. During the actual processing, the auditor should take steps to assure that the file or application being tested is the correct one. Depending upon the nature of the audit test, the auditor should prove the results and follow up on any exceptions as required. The audit application workpapers also should be completed at this point. Suggestions to improve the computer-assisted audit procedure next time should be recorded.

Computer-assisted audit procedures are powerful tools which can be used by any auditor. They should not be the responsibility of only the "computer audit specialist" in the audit function. Just as end users make increasing use of retrieval tools for their own report needs, all members of an audit department should understand the available audit tools.

"TEST DECK" APPROACHES TO TESTING APPLICATIONS

The term "test deck" is an old computer audit related term that dates to the earliest days of data processing when applications operated in a batch mode and used punched cards as input and output media. When auditors were interested in the correctness of a computer applications process, the auditor would develop a series of transactions which would achieve known results.

To test a payroll system through the use of a test deck, for example, the auditor might develop a transaction for a known employee showing 40 hours of work, another showing some overtime hours, and a third showing an excessive number of hours that should trigger an error report. Then, a special run of the payroll system would be arranged with the test transactions submitted. The auditor subsequently could verify that the pay was computed correctly, the files were updated correctly, and any necessary error reports were prepared correctly. All audit test transactions would then be purged from any updated files. Through this test deck, the auditor could gain a level of assurance that the system was working correctly.

Test deck approaches fell into disuse by auditors as systems ceased to be batch oriented and became more complex. However, new techniques can still make the test deck approach a viable tool for evidence gathering and testing modern data processing applications.

Test Data for the Modern Data Processing Application

"Test decks" or, more properly, test data approaches can be useful evidence gathering audit tools for many types of applications. Instead of a deck of test cards, for the modern computer application the auditor will use on-line transactions as the "test deck." The auditor inputs a series of test transactions into a system to achieve the following objectives:

To develop a general understanding of the logic and program interrelationships associated with a complex system

To determine that valid transactions are processed correctly

To determine that invalid or incorrect transactions are identified and flagged by program controls

The auditor should recognize that there are some limitations in this testing approach. If a given transaction type is not tested, the auditor will not be able to say that the application works correctly in respect to that transaction. If the documentation is incomplete or incorrect, the auditor may miss a key type of test because the auditor did not understand a transaction.

Manual test data exercises can be developed by tracing user initiated transactions through a normal production cycle or by inputting a series of audit test transactions through a special test run of the application. There are advantages and disadvantages to each of these approaches.

Tracing User Initiated Transactions

The auditor sometimes will be asked to review and gather evidence of transaction controls for a complex, on-line data processing application. The auditor may be asked to verify that an on-line, manufacturing material control system is operating with proper controls. Such a system may have numerous control programs to receive materials into the plant and place them in stores, retrieve these same materials from stores and assign them to work orders, and then add labor and other parts to complete finished assemblies. Because such a system may have numerous transaction types which update the materials files through multiple programs, it is difficult to design tests to verify the correctness of all transaction processing.

When faced with such an application, often, the auditor will be unable to identify any single point for the use of a computer assisted audit procedure. A formal test data approach where the auditor sets up a separate system process also is difficult due to the overall complexity of the system. The most reasonable approach to testing such a system may be to trace user initiated transactions through the production application.

To accomplish this test, following the example of a manufacturing materials control system, the auditor first identifies key control points in the overall system and then observes and records transactions being entered at each control point so that they can be traced to the appropriate screen or reports. As part of this observation the auditor also asks users to input certain invalid transactions to see if they are rejected properly.

Figure 6.3 illustrates this testing approach for an on-line manufacturing system. It shows how the auditor records transactions at one point in the systems flow and then traces them through subsequent steps. This type of transaction tracing approach sometimes is the only way to verify the correctness of processing in a complex data processing application.

The tracing of user initiated transactions is similar to the transaction walkthrough approach described in Chapter 5, except that the auditor captures many more transactions for tracing and verification. This approach often is combined with the use of computer assisted audit techniques at key points in the application to gain a better understanding of the application's processing procedures and its controls. In the example, the auditor could trace inventory transactions input through on-line screens as described. In addition, a computer assisted procedure could be used to compare beginning and ending period inventory status files to highlight differences caused by user initiated transactions.

Tracing user initiated transactions often is not the best way to gain positive assurance that an application is working with proper controls. However, it is an effective approach to gaining a level of assurance that a complex application appears to work with no obvious errors. While individual programs can be tested in some detail, this may be the only way to test an entire operational application.

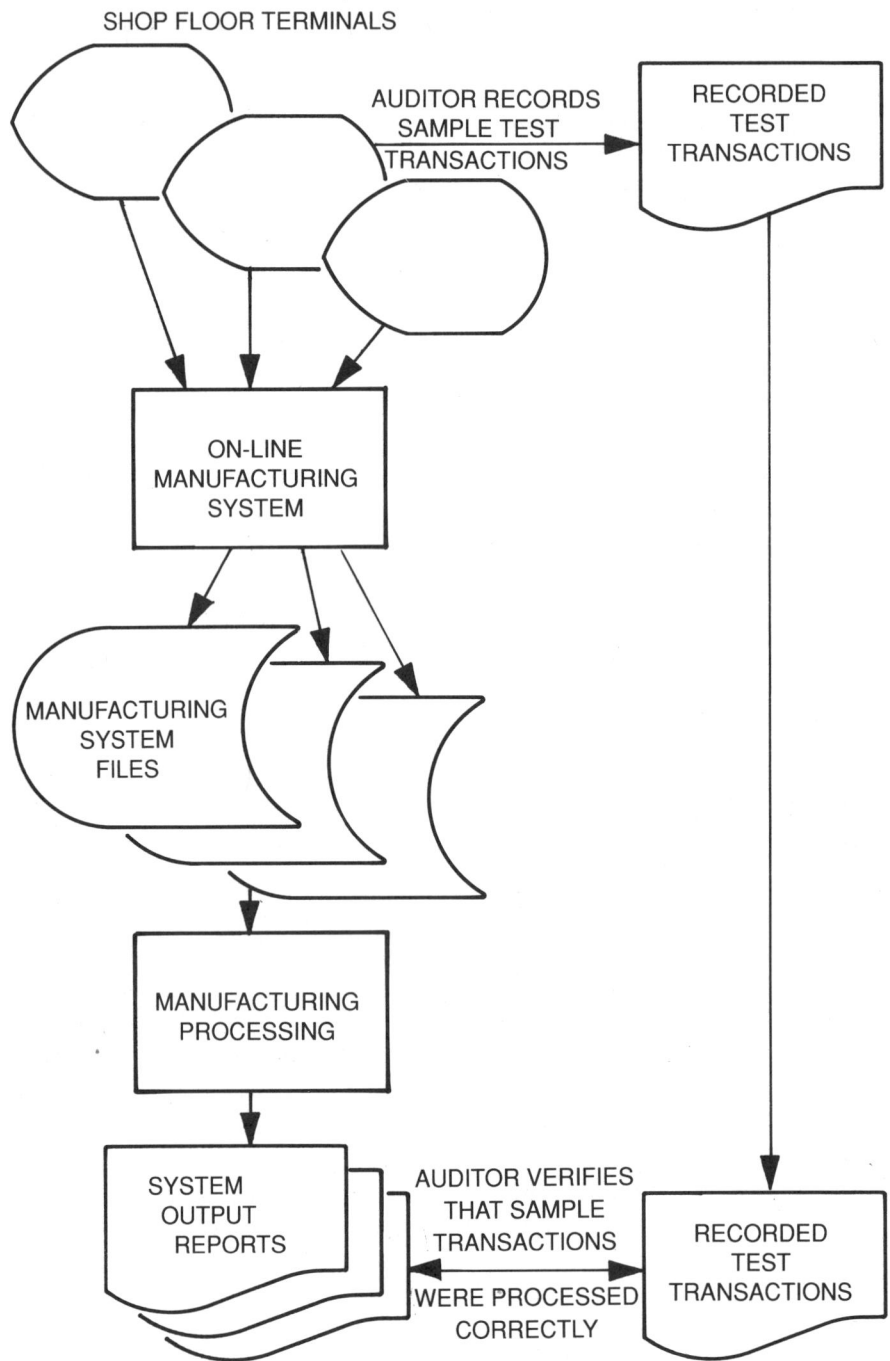

Figure 6.3. An example of a test data approach using user initiated transactions

Using Auditor Initiated Test Transactions

If an application has only one key input point, inserting a set of auditor initiated data through that point for processing in a special run may be an appropriate way to test an application. Assume that an organization has an on-line labor system where workers input their hours and assign that time to various projects using a single-screen format. The auditor may be interested in verifying the integrity and correctness of such a system.

An approach to testing this system is to build an audit file of representative transactions to input to a special test run. These transactions represent a variety of valid entries and error conditions, allowing the auditor to test as many conditions as possible. The auditor then arranges for a special run of the application under review. Copies of live files are used for all processing. The auditor's file of test transactions is input to this test run, and the auditor subsequently verifies the results. This type of test data approach to application testing is illustrated in Figure 6.4.

This test data approach can be an effective way to gather evidence about smaller, self-contained data processing applications. It allows the auditor to run a comprehensive series of test transactions with live data files to verify the correctness of application processing. The approach does have the following limitations:

Data processing operations personnel often object to processing a special run of a production application, fearing that the test data will become intermingled with live data

Because it disrupts data processing operations to run multiple cycles or periods, generally only one cycle of test data for an application can be run

The approach is cost effective only for self-contained applications since it is difficult and costly to design test data for multiple input points

The preparation of a comprehensive set of test data is time consuming

Despite the above limitations, a test data approach is often a useful way to test certain types of applications. It is particularly useful for applications implemented at multiple locations within the organization. By developing a standard set of test data and using it at each location, the auditor can verify, among other things, that there have been no unauthorized changes to the application programs at any location.

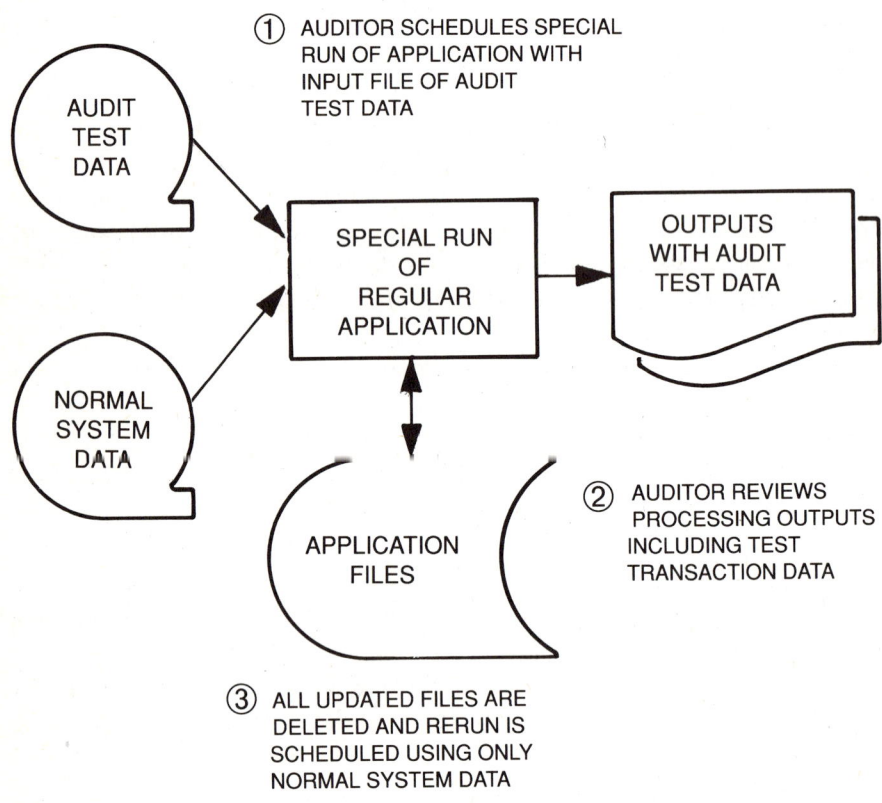

AUDIT
TEST
DATA

① AUDITOR SCHEDULES SPECIAL
RUN OF APPLICATION WITH
INPUT FILE OF AUDIT
TEST DATA

SPECIAL RUN
OF
REGULAR
APPLICATION

OUTPUTS
WITH AUDIT
TEST DATA

NORMAL
SYSTEM
DATA

APPLICATION
FILES

② AUDITOR REVIEWS
PROCESSING OUTPUTS
INCLUDING TEST
TRANSACTION DATA

③ ALL UPDATED FILES ARE
DELETED AND RERUN IS
SCHEDULED USING ONLY
NORMAL SYSTEM DATA

Figure 6.4. An example of a test data approach using auditor initiated transactions

Integrated Test Facilities

The test data approaches described above provide the auditor with a method of testing applications at specific points in time. An integrated test facility, or ITF, on the other hand, is a test data facility built into a data processing application which allows more continuous testing over time. An ITF built into a key application provides auditors with a method to review that application on an ongoing or random basis.

An example will explain how an ITF might be constructed and how it is used. Assume that the auditors are reviewing controls over a central financial system covering a large, multi-division organization. That central financial system receives many transactions from its operating entities for intra- and inter-division financial transactions. Each division carried on the financial application files is identified by its own division code. In addition, the central or corporate accounting function initiates transactions which affect all of the operating entities. The auditors are interested in the controls over these various transactions as well as monitoring them for any improper transactions.

To review this application on an ongoing basis, the audit department could set up a "dummy" division on the system, labeled "Division 99." All system reports produced for this "Division 99" would then be routed to the audit function. Auditors could input test transactions against their audit division from time to time to test the accuracy of the system and verify program integrity. In addition, if another transaction was initiated which impacted all divisions, the auditors would see it on their "Division 99" reports. Figure 6.5 illustrates how such an ITF might be constructed.

An ITF can be an effective and continuous test approach for data processing applications. Normally, a special test run is not required, as auditor transactions are entered with other live data. Just as with other test data processing, the ITF transactions can be compared to predetermined processing results to gather evidence about the application being tested.

The Decision to Build an Integrated Test Facility

The auditor can construct an ITF using one of two approaches. First, the basic data processing application can remain unchanged, with auditors inputting transactions only against the designated code, such as the Division Code 99 example above. It will then be necessary only to purge these transactions from the system after testing has been completed. For an accounting application, this can sometimes be as simple as entering reversal transactions to the system. However, this reversal process generally is more complex. For example, many

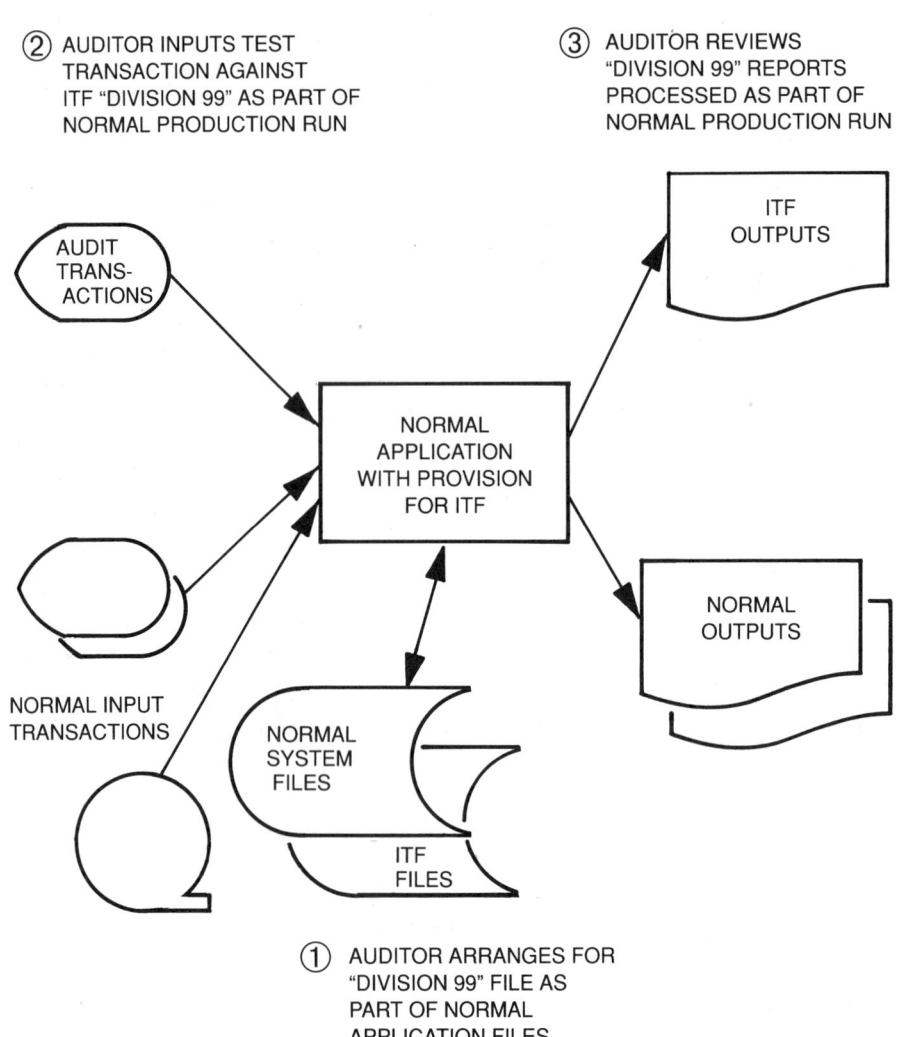

② AUDITOR INPUTS TEST
TRANSACTION AGAINST
ITF "DIVISION 99" AS PART OF
NORMAL PRODUCTION RUN

③ AUDITOR REVIEWS
"DIVISION 99" REPORTS
PROCESSED AS PART OF
NORMAL PRODUCTION RUN

AUDIT
TRANS-
ACTIONS

ITF
OUTPUTS

NORMAL
APPLICATION
WITH PROVISION
FOR ITF

NORMAL
OUTPUTS

NORMAL INPUT
TRANSACTIONS

NORMAL
SYSTEM
FILES

ITF
FILES

① AUDITOR ARRANGES FOR
"DIVISION 99" FILE AS
PART OF NORMAL
APPLICATION FILES

Figure 6.5. Example of an integrated test facility

accounting systems have allocation processes which affect all balances in the system; it is important that other account balances not be altered by the ITF test data.

A second approach to building the ITF is to add a program or modify existing programs to filter out any audit transactions. With this approach, the auditor's test transactions will be transparent to most users of the system. The problem with this method is that it requires special programming as well as the services of the data processing department to build the filter. If data processing is one of the functions being examined, the integrity of the tests could be compromised.

In either event, an ITF is an approach to testing and gathering evidence about a total system, including the manual aspects surrounding the automated application. Because of the program coding and logistics required to establish an ITF, the technique should be used only when the auditor has an ongoing interest in reviewing the application. In addition, the auditor should understand all aspects of the application before starting the ITF. Otherwise, the auditor's ITF transaction activities could have unintended effects, such as initiating action.

Several other areas of caution should be considered before constructing an ITF. In some regulated industries such as financial institutions, auditors may be in violation of state or federal regulations for establishing fictitious accounts in financial applications. The auditor should discuss any ITF plans with the organization's counsel if appropriate. The auditor must take care also that only a limited number of personnel are aware of the ITF. If it is known generally, for example, that Division Code 99 is the auditor's number, anyone making improper transactions will bypass that code.

Constructing an ITF: An Example

An example will explain how the auditor can construct an ITF to test a total system including the data processing application. Assume that the example organization is a distribution company where orders are telephoned in to an on-line order entry desk. The system, after checking credit and inventory stock records, sets up the order for picking and shipment and updates accounts receivable files for customer invoicing. This system is a key strategic application for the company.

The internal audit department is interested in reviewing the financial and operational controls in this order entry application to test for the following:

Are order entry operators properly following procedures in accepting orders, suggesting replacements when items are out of stock, and requesting payment when credit limits are exceeded?

Are inventory status screens properly updated with current availabilities and prices?

Does the order entry system properly update inventory records to indicate that items have been allocated for picking, and credit records to show new orders?

Does the stock-picking documentation properly show the location of items being shipped?

Are all shipped orders properly updated on accounts receivable records?

The above questions have a combination of financial and operational control implications. An auditor could use computer assisted audit techniques to test various portions of the system. An inventory transactions file could be matched to the accounts receivable system files to determine that all shipments have been given to the invoicing system. However, with such a comprehensive and dynamic automated application, a computer assisted procedure would test only one portion. To develop an overall understanding of such an application, an ITF might be appropriate.

The first step is to define clearly the auditor's review objectives. The above questions representing areas of interest should be translated into definite audit objectives. The auditor might restate the last of the above questions as follows:

Determine that all product shipments properly decrement quantities in item status files, update shipment status locator files, and update customer accounts receivable files.

After a detailed review of the system's functions, the auditor should be able to construct a detailed list of such audit objectives.

The auditor's second step is to gain a detailed understanding of the application and its interfaces with other applications. This is necessary to allow the auditor to construct an overall testing strategy. In this example, the ITF strategy involves running dummy orders through the system for audit verification and compliance testing. General approval of the testing strategy should be obtained from management.

This ITF works by having the auditor set up several dummy customer numbers in system files. Orders are placed against those customer numbers and result in the merchandise being picked and shipped to the dummy customers. A program filter should be added to the accounts receivable system to assure that these dummy orders are not invoiced and do not flow into financial records. Also, manual procedures should be established in the shipping department to assure that dummy orders are not shipped. Otherwise, the ITF should be constructed so that most personnel are unaware of it.

Once the auditor determines testing objectives and develops a good understanding of the application, the necessary systems changes are made to operate the ITF. Program filters are constructed (with the cooperation of systems and programming staff, as appropriate) and dummy customers set up in customer files. Finally, an audit test plan is developed. This prevents the auditor's ITF from appearing to be a "fishing expedition" and helps the auditor structure tests to achieve audit objectives.

Figure 6.6 illustrates how this example ITF might operate. The audit department arranges to call in orders for known items, posing as the dummy customers. The orders are placed against known stock items to verify status and prices. In addition, the auditor gains an understanding of the quality of customer service provided by order entry operators. Once placed, the auditor uses actual application output reports and screens to verify the correctness and efficiency of order processing. If appropriate, computer assisted techniques are used at selected points in the applications process.

The above example illustrates how an ITF might be constructed to test a complex, integrated application on an ongoing basis. It takes considerable audit resources to construct and perform such testing. Thus, such an ITF is practicable only if there is an ongoing interest in monitoring the application. This example also tests the human or manual aspects of an application. The auditor should first clear such planned procedures with appropriate members of management and remember to respect the privacy of personnel who unknowingly will be part of the ITF testing.

While this example describes the use of an ITF for a large system with automated and manual aspects, the technique also is appropriate for smaller applications. The two key requirements for using an ITF are: first, the tests should be constructed so that dummy test items can be purged or filtered out, and second, there is sufficient ongoing interest in the application to justify ongoing testing. Given these two requirements, an ITF can be an effective tool for evidence gathering and application testing in the modern data processing environment.

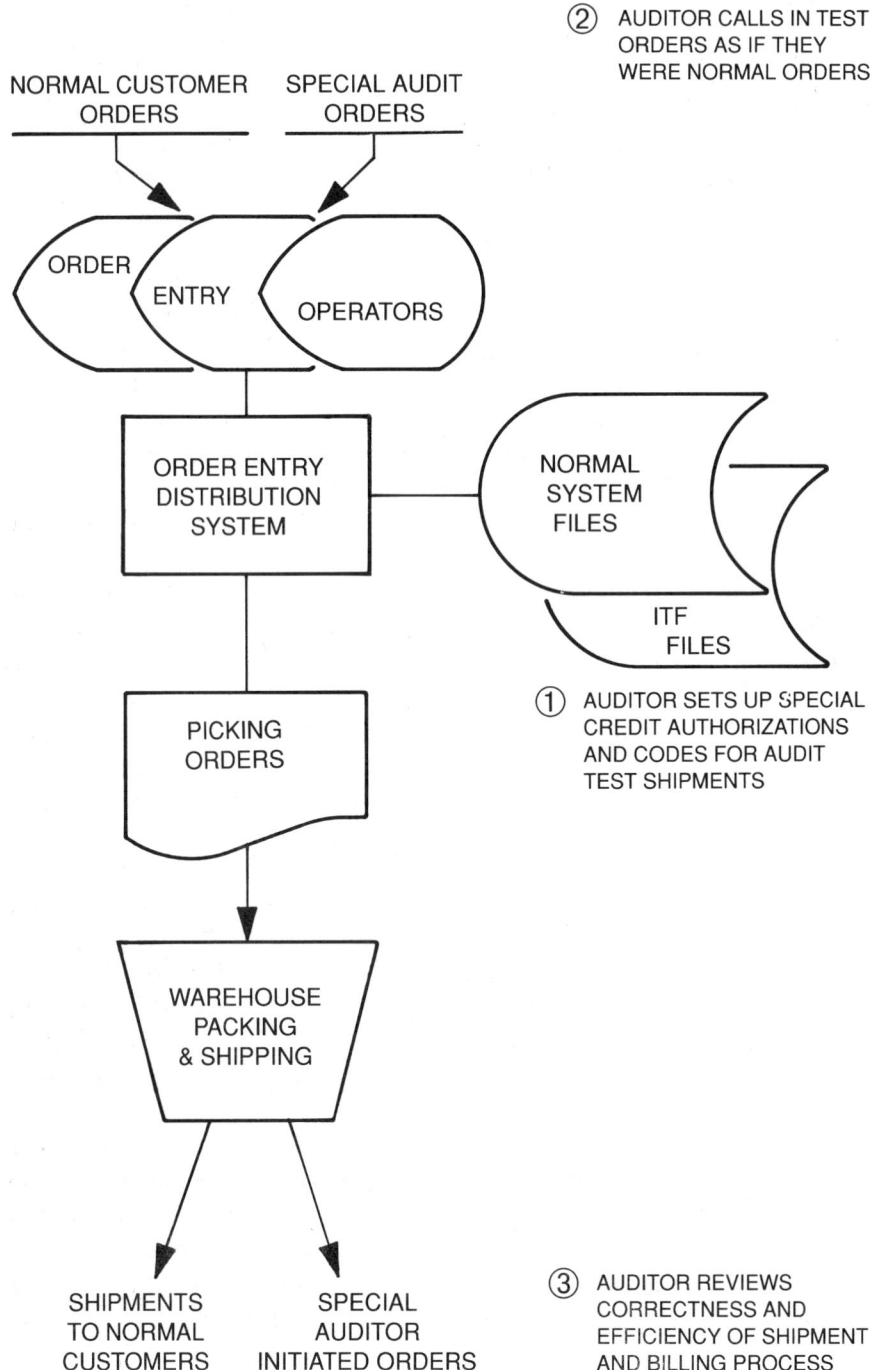

Figure 6.6. Example of an integrated test facility for a distribution company

CONTINUOUS AUDIT MONITORING APPROACHES

The evidence gathering and application testing procedures discussed in previous sections of this chapter generally required the auditor to initiate some action to start the testing process. In many complex systems, however, the auditor also should be interested in monitoring exception transactions within an application on an ongoing basis. A continuous audit monitor is an element of embedded software which monitors an application for certain activities of audit interest and gives that data either to a log file or to a report for later audit analysis. Some earlier literature has referred to this approach as the System Control Audit Review File (SCARF) method; it will be referred to here as the continuous audit monitor method.

In many respects a continuous audit monitor is similar to the error or exception reporting mechanisms built into many applications. Also, it is quite similar to the log file approach used for computer operating systems. The differences are that exception reports usually log only problems, and system log files record only activities. A continuous audit monitor only logs and reports items of predetermined audit interest.

Continuous Audit Monitor Design and Implementation

The purpose of a continuous audit monitor is to gather evidence about transaction exceptions or potentially unauthorized items which may require auditor follow up. Such a monitor does not allow the auditor to perform detailed or substantive tests of an application but allows tests of compliance with preestablished procedures. As with the ITF described previously, a continuous audit monitor should be installed only in an application where there is an ongoing review interest.

An example will better explain how a continuous audit monitor can be constructed. Assume that an auditor is working in a multi-branch financial organization where there are numerous transactions between branches. The auditor has reviewed the financial applications and feels that internal controls generally are adequate. In addition, in support of the external auditor's year-end work, the auditor has used generalized audit software to test key elements of the financial application. However, the auditor is interested also in monitoring and following up on certain exception transactions which may be initiated by various branch users from time to time.

The example application has a large number of exception reports for user follow-up. However, the auditor may be interested only in reviewing certain inter-branch transactions above a specified dollar limit. While regular operational personnel should follow up on such transactions too, the auditor may be interested in the nature of such transactions and the level of follow up activity. A continuous audit monitor allows the auditor to monitor such exceptions.

After gaining a detailed understanding of the data flow within the application, the auditor identifies where inter-branch transactions can be isolated. Arrangements are then made with data processing to insert a monitor program or program code into the application to capture all transactions of interest onto a protected log file for later audit review and analysis. Figure 6.7 illustrates how such a continuous audit monitor might be constructed.

Because a continuous audit monitor often is an embedded item of program code, its parameters cannot be changed in an easy manner. The auditor, therefore, should carefully design the objectives and selection criteria associated with the monitor. Properly constructed, it is an effective tool to monitor applications where the auditor has an ongoing interest.

Continuous Audit Monitor Limitations

While the continuous audit monitor is a powerful tool to gather certain evidence about an application, there are some severe design limitations to this approach. The prime one is that the auditor generally is not able to implement such a procedure independently. The automated audit procedures discussed previously in this chapter often are established independently by the auditor. Given an understanding of file structures, the auditor uses generalized audit software to perform various tests. Similarly, many ITFs can be established independently by an auditor. This is not true for the continuous audit monitor.

Often, it is necessary for the auditor to work with the data processing and programming functions to define the requirements of such a monitor. They then incorporate it into their system design. The most efficient time to suggest the implementation of such a monitor is when a new system is being developed. For example, the auditor might suggest that it be installed during the detailed design phase of a new application, as discussed in Chapter 7.

Generally, it is more difficult to have a continuous audit monitor installed in a system that is operational. Unless there is strong management pressure to install such a tool, the auditor will have difficulty securing the necessary data processing priorities. In addition, once such a system is installed, it could be bypassed or otherwise compromised by subsequent system modificaitons.

The auditor should recognize also that if the data processing and programming functions install such a monitor, they will be aware of it and could bypass

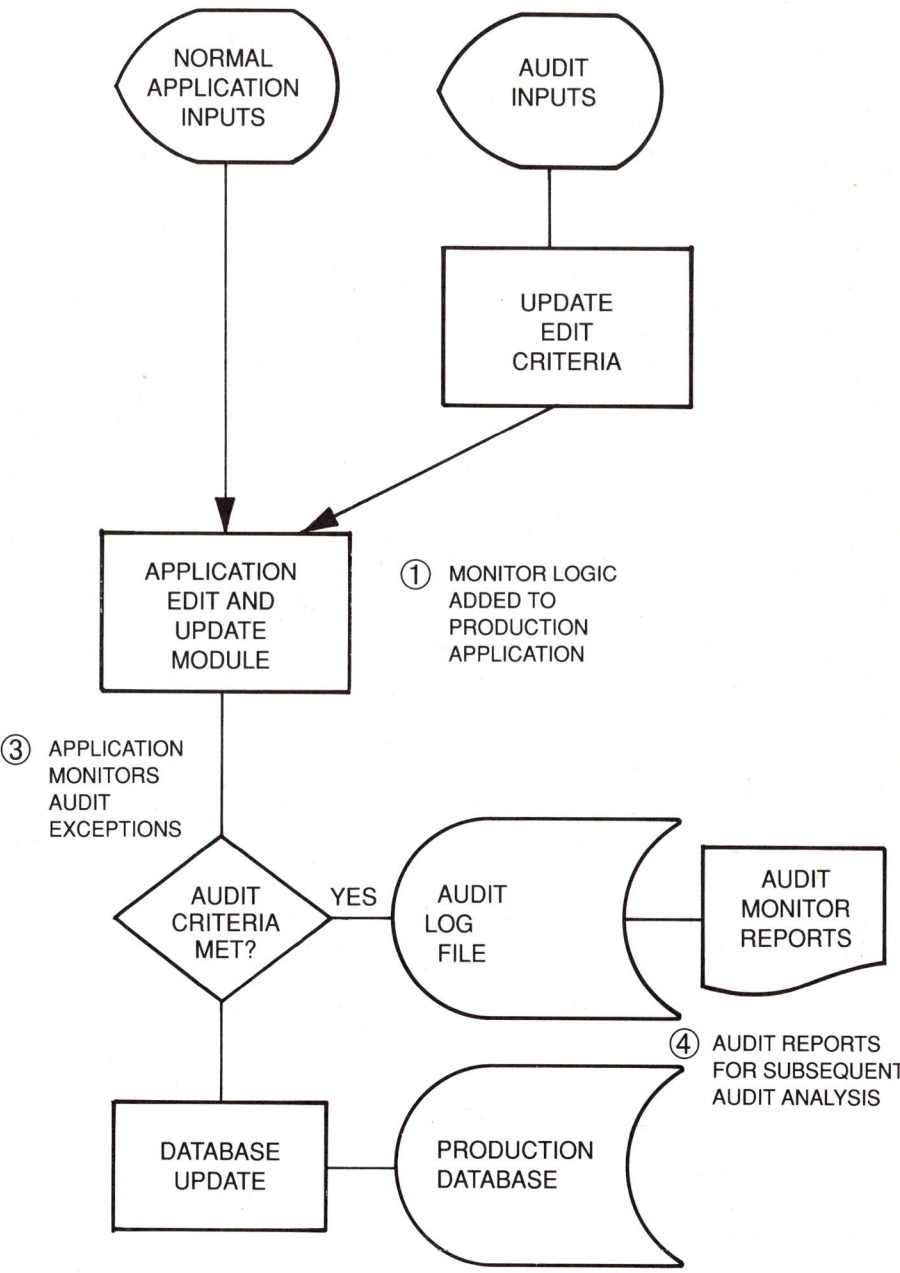

Figure 6.7. Example of a continuous audit monitor

its functions. Nevertheless, a continuous audit monitor is a powerful tool to review certain exception transactions associated with key or critical transactions. Although the example cited was for a financial application, the procedure is applicable also to non-financial applications. In a manufacturing organization, for example, the auditor may install such a monitor to log and report to the auditor all scrap disposals above a specified limit.

EVIDENCE GATHERING IN A PAPERLESS ENVIRONMENT

The auditor's need to build and use computer assisted audit tools will increase in the modern data processing center as more applications become paperless. The traditional paper trail that financial auditors have used to trace and validate transactions will be reduced or eliminated in the more modern computer system. Increasingly, audit tools ranging from the use of generalized audit software to continuous audit monitors will be the only options available for auditors to test and gather evidence about these paperless systems.

There are many examples of paperless systems in the modern data processing organization. The on-line purchasing system used as an example in Chapter 5 had paperless elements. Another example is a financial institution that makes extensive use of automatic teller machines, or ATMs, which often have very limited paper trails. New technologies such as digital image processing will increase the number of paperless systems.

Frequently, the auditor may encounter paperless or limited paper trail systems in a microcomputer environment. For example, some organizations with a network of retail sales representatives have sales personnel input order activity to a portable or laptop microcomputer for subsequent transmission to the central microcomputer or mainframe for order processing. In other implementations, field personnel may use bar code scanners to read stock labels to allow a field microcomputer to set up orders.

The auditor should exercise a level of creativity when gathering evidence regarding these newer, paperless applications. Many techniques described in this chapter apply to these newer systems. For example, the microcomputer field order system could be tested through an ITF at the home-base machine. Auditors could use a test machine to transmit orders into the ITF on the central computer system. A continuous audit monitor could be installed in such a system to record and report to the auditor all unusual order activity outside predefined audit parameters.

Auditors should develop a good understanding of the tools and techniques for the evidence gathering and application testing of all types of automated applications. These procedures help the auditor to support compliance testing of applications controls and attest to the correctness of application financial balances.

CHAPTER 7

Reviewing New Applications Under Development

INTRODUCTION

General management and many data processing departments now recognize that it is more efficient to audit or review the adequacy of an application's controls while the application is being developed than after it has been completed and placed into production. In this role, the auditor acts in the manner of a building inspector reviewing a new construction project. If the building inspector were not invited to inspect the construction project until after the building was completed, it would be difficult to implement recommendations. The inspector would be under pressure to avoid identifying problems requiring significant portions of the building to be rebuilt. Rather, the building inspector should identify problems during the course of the construction and suggest corrections as the building is being completed.

When reviewing new applications under development, the auditor reviews portions of the application under development and suggests corrective action along the way. It is easier for the data processing department to implement these changes during the development rather than after the application is completed and released to users.

The auditor must be careful not to take *responsibility* for the new application's controls, just as the building inspector who points out problems does not take responsibility for implementing solutions. When reviewing new applications under development, the auditor should point out control weaknesses to the developers but should not be responsibile for implementing any corrective actions.

Application development groups, user management, and auditors agree that auditors should play a role in reviewing new data processing applications under development. However, it is often difficult to implement and execute a program of effective pre-implementation audits. This chapter discusses some pitfalls the auditor can encounter and effective approaches to reviewing new applications under development.

OBJECTIVES FOR APPLICATIONS DEVELOPMENT AUDITS

Auditors often consider an applications development—or pre-implementation—audit to be just a review of the controls being built into the new application under development. However, the effective auditor should expand this audit scope to cover other areas as well. The following audit objectives are recommended for pre-implementation reviews of new applications.

An evaluation of controls prior to implementation. This is the most common objective of a pre-implemention review. In the early days of computer auditing, the auditor often waited until a new application was implemented before reviewing controls and procedures. This review often accomplished little because the applications development project team had moved on to other projects and did not have time to implement recommended changes. The most important objective of pre-implementation auditing is to identify recommended controls improvements early enough to be installed during the development process.

An evaluation of project definition and justification. The auditor should not assume that a new data processing project is a given and limit the audit to reviewing its controls. The auditor should also review the justification and definition of the new development project. For example, a new application may have been requested to satisfy the needs of one component of a larger organization without considering the needs of other divisions or deparments. The auditor should point out such definition-related deficiencies to management as part of the pre-implementaion review. Similarly, the auditor should review the justification for the project request with skepticism and verify that the project should be undertaken.

An evaluation of controls over the project development process. Applications development projects often are major efforts that consume substantial time and resources. There should be a project management system in place with a development plan and a means to measure progress against the plan. For major projects, the auditor can render an important service to management by evaluating the controls being used to manage the development effort.

An evaluation of the new system for subsequent audit tests. The pre-implementation audit of a new application provides the auditor with an excellent opportunity to gain a documented understanding of the application so that later audit tests can be designed more easily. In addition, this is the best time to define and suggest automated, embedded audit modules such as integrated test facilities. These evidence gathering approaches were discussed in Chapter 6.

In addition to the above, some internal auditors face statutory requirements for reviewing new applications under development. At least two states have legislation requiring that all new significant state agency applications be reviewed by internal audit departments for controls prior to implementation.

PRE-IMPLEMENTATION REVIEW PROBLEMS

While many agree that it is a good idea for the auditor to review new applications under development, serious problems can be encountered in implementing a pre-implementation review program. Some common problems are discussed below. In spite of these problems, however, pre-implementation auditing can be a worthwhile effort. Too often auditors are accused of "joining the battle after the action is over, to shoot the wounded." Pre-implementation auditing allows the auditor to play a proactive role in the applications development process.

Too Many Review Candidates

Internal auditors sometimes make the mistake of announcing an intention of reviewing and signing off on *all* new applications and major modifications prior to implementation. While this seems attractive for control purposes, the auditor often faces too many requests to review. In a larger organization, there may be dozens or even hundreds of user requests for applications projects generated every week. An internal auditor that attempts to review and sign off on all such requests can be reduced to performing a rubber stamp function. The auditor finds there is no time for detailed or comprehensive reviews and finds only time for nominal approval signatures.

Selecting the Right Application to Review

Even if an auditor decides to review only selected applications, it still is difficult to select applications of *audit significance*. This often is a value judgment for the auditor unless a structured selection method has been established.

A typical applications development department may be working on applications A, B, and C. Given limited time and resources, the auditor may decide to review only B. However, if significant post-implementation problems appear in C, the auditor can be criticized for skipping that system. The auditor reviewing new applications under development needs a consistent approach to selecting applications based on audit criticality and risk.

Determining the Auditor's Role

The auditor can become overly involved in the applications development project. Often, major new information systems are developed through extensive team efforts with numerous user and analyst design review meetings. This is particularly true of applications developed through the use of prototyping methods. Frequently, the auditor performing the pre-implementation review is asked to participate in these design review meetings. If the auditor is involved in the typical meetings when compromises are made during an application development project, it will be difficult to comment on the compromises later as audit points. However, if the auditor does not participate in such design meetings, it will be difficult to perform the review.

To be effective in reviewing new applications under development, the auditor's role must be carefully defined. This requires both negotiation and the understanding of management.

Managing the Audit Review Process

One of the more difficult tasks facing the auditor when reviewing new applications under development is determining the purpose of the audit. It is common for an auditor, after announcing to the information systems department that a certain application will be reviewed, to receive hundreds of pages of requirements studies, general design review documentation, meeting minutes, and other materials. Unless an auditor enjoys reviewing such technical specifications as bedtime reading, this mass of applications development material presents the auditor attempting to cover only the significant portions with a serious problem. The auditor needs an approach based on clearly defined control objectives and audit procedures, to effectively perform such a review.

The Use of New Development Methodologies

Many information systems today are developed with the help of automated, structured design tools to chart and describe application and program flows. This approach is referred to as Computer Aided Systems Engineering (or CASE). Mainframe or microcomputer system design tools allow an applications analyst to define a computer application system and its data file elements through a series of highly organized process flow charts. When some element or relationship changes in the design, it is necessary only to make that one small alteration to the automated design and all other relationships will change. Structured design approaches will be discussed in greater detail later.

These structured design tools and approaches make it easier to build and modify complex applications. However, they also cause problems for the auditor performing the pre-implementation review. Because it is easy to make a modification to an application design, the auditor may be faced with a "now you see it, now you don't" situation when reviewing such an application under development. In addition, the auditor needs to develop an understanding of the new structured tools and approaches before actually performing the review.

FORMALIZED SYSTEMS DEVELOPMENT METHODOLOGIES

Most data processing organizations, large or small, have a formal set of procedures for initiating and developing new information systems. Often, this is called the Systems Development Life Cycle procedure (SDLC) or the Systems Development Methodology (SDM). To effectively review new applications under development, it is essential that the auditor understand the SDM process used by the data processing organization being reviewed.

Formal SDM procedures evolved in response to problems arising out of the informal manner earlier information systems were developed. In the earlier days of data processing, a user might request new applications work very informally by giving the information systems manager a brief note, for example, "to redesign the accounts receivable system." Then, the two might discuss the requirements for this redesign over a cup of coffee. Typically, data processing would go to work on the request with little subsequent discussion. Some months or even years later, a new, redesigned system would be presented to the user that might or might not meet the user's needs. Even worse, in some cases data processing would initiate a project without any significant user input. They might decide that the application needed improvements of some sort and and deliver the modified version with little opportunity for users to understand the changes.

The lack of a long range planning or priority setting process often caused other problems for some applications development functions. The department frequently reported to a controller or financial officer that arbitrarily set priorities for new development projects. For example, an organization might need a new sales order entry system. However, if the controller in charge of data processing said, "I want a new fixed asset system," it should be clear who would receive development priority. The sales order entry system would be placed in a hold status or moved down the priority list.

The above situation often created chaos among users and the applications development department. Users of computer systems often were uncertain how requested applications would be structured and controlled or when the applications would be delivered. Systems development departments often found that management changed priorities and that users did not adequately define requests. Organizations eventually developed SDMs to attempt to solve some of these difficulties.

An SDM is a formalized process to define, develop, and document new computerized information applications and major modifications. The modern data processing organization, large or small, generally uses some type of SDM for developing applications. Although there are numerous minor variations on the procedural steps or the terminology used, Figure 7.1 illustrates a typical SDM.

Project Initiation and Feasibility Determination Phase

Users typically initiate requests for new applications or major modifications by completing a formal document that defines their needs, sometimes estimates the expected cost savings from the new application, and requests a formal management approval of the project.

The next step in this phase is for the data processing organization to determine whether the requested project is "feasible." This word is in quotes because virtually all such projects technically are feasible. However, a

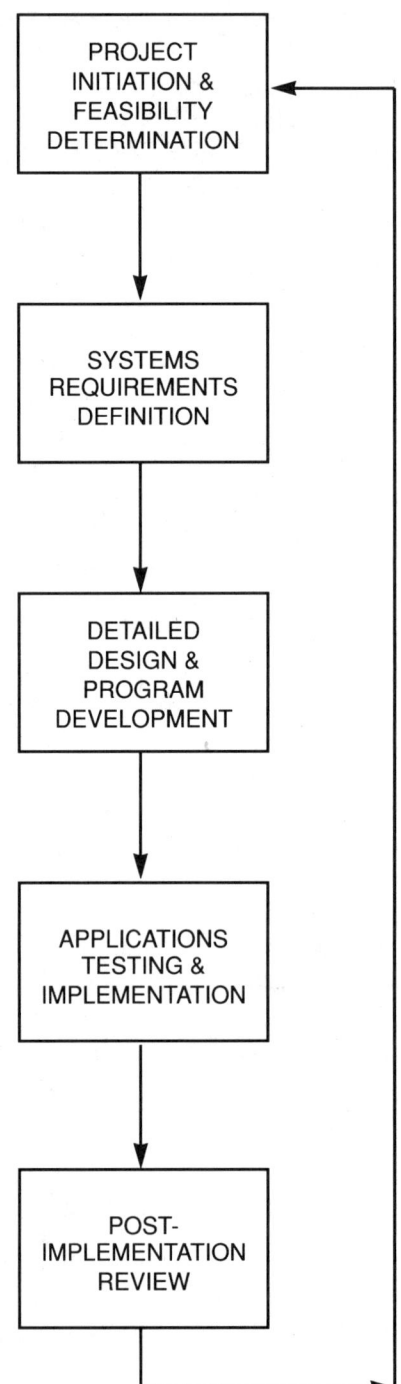

Figure 7.1. Typical systems development methodology (SDM) steps

288

requested project should be evaluated for its practicality and for how it will fit into the overall applications environment. This "feasibility" study or evaluation is documented for subsequent management review.

To complete the project initiation and feasibility phase, the request is either approved for development work by the data processing department or rejected with documented reasons. If accepted and if it is a major project, the request often goes to a management steering committee for priority setting. Such a steering committee should be composed of upper management from major functional areas in the organization. If the request is consistent with their data processing long range plan, the steering committee should compare it to other applications requests and give it a relative priority for action.

Systems Requirements Definition Phase

Once a project has been scheduled for development, the next step is the preparation of a study defining the requirements of the application. Data processing and user personnel work together to define the specific objectives of the proposed new project. These objectives often will be prioritized to determine which are most important to users and most practical to implement.

The requirements definition phase usually results in a fairly detailed general design study for subsequent management review. The study defines prioritized user needs as well as a general description of the technical approach to be taken on the project. If the proposed application requires any new specialized tools, such as a new database package or other purchased software products, the requirements study should identify and define these elements.

The requirements definition study is also the phase where application control concepts are considered and the auditor can express preliminary control concerns before release for detailed design.

Detailed Design and Program Development Phase

Using the final systems requirements study as a starting point, this phase consists of a detailed step by step design of the new application. If new programs are to be developed, they will be documented following data processing program documentation standards. For some organizations, structured design tools may be used in this phase to define the entity relationships, the input-output logic, and data requirements of each program element. These procedures, often called Computer Aided Systems Engineering, or CASE, methods, are discussed later in this chapter.

If the new application consists of purchased software, this phase may be limited to detailed plans for conversion and interfacing with other existing applications. However, this conversion should be documented so that it can be reviewed and approved by the requesting users.

Some newer applications may be developed through prototyping methods as discussed in Chapter 12. Prototyping allows the data processing department to develop a preliminary version using a fourth generation language or other rapid development tool, review that preliminary version with the requesting user, and then make further modifications on an iterative basis until the application meets user requirements. However, if the application is a production system the development still should follow the data processing department's design documentation standards once the application details are formalized.

Application Testing and Implementation Phase

This phase consists of both program level unit testing and overall application testing. Formalized test plans normally are prepared, and users typically are involved in reviewing test results. If problems are encountered in this phase, the project may be turned back for additional programming work. Typically, this is what happens for applications developed under prototyping methods.

After completion of comprehensive applications tests, the users who originally requested the application are asked to approve it for implementation. Following a detailed, documented test and conversion plan, the applications project is able to be implemented. That plan outlines such requirements as the need for new forms, user training, or conversion steps from existing applications.

In a larger data processing organization a quality assurance function often participates in a final review of the program elements of the new application at this point. Quality assurance functions are discussed in Chapter 2. The auditor may look to Quality Assurance to perform detailed reviews for such areas as compliance with departmental programming standards.

Post-implementation Review Phase

The last step in a typical SDM is a post-implementation review to determine if the new application has achieved its original objectives. Normally, this review takes place some time after implementation. If additional changes or improvements are required, they will be documented to initiate a new project and SDM cycle.

Documentation

The above steps describe a typical SDM approach as used by many data processing organizations. Documentation is the key to such an approach.

Users document requirements and developers document design processes. Such documentation helps to minimize misunderstandings and produces better information systems.

Purchased Systems Design Methodology Packages

Some information systems departments implement an SDM procedure through the purchase of a package from an outside vendor. A variety of vendors, including several of the major public accounting firms, offer SDM products with standards manuals, forms, and training materials. While many are similar, their terminology or the number of phases differ slightly. Other information systems departments have home grown approaches, often adapted from one of the published SDM descriptions.

SDM packages vary according to the number of forms or documents required to properly complete a given phase. Some are too paper intensive. For example, one of the purchased SDM packages on the market suggests some 75 "required" unique forms to document properly a new information systems application! That package even suggests that the applications designer complete an SDM form evaluating the level of "synergistic risk" for the new application! Often, when a large number of forms are required or when the purposes of these forms appear obscure, both users and applications developers resist completing them.

Some of the newer SDM packages are tied to automated, structured program design packages. These packages make it easier for systems developers and programmers to define program logic and data element requirements. Some of these structured program design packages have been implemented on microcomputers while others are available on larger machines.

SYSTEMS DESIGN METHODOLOGIES AND THE AUDITOR

To review new applications under development, it is essential that the auditor have a good understanding of the SDM used by the data processing organization. The auditor should understand the SDM's required forms, terminologies, and procedures. Often, this understanding can be gained through reading manuals or attending SDM training classes.

If only portions of a purchased SDM package are used, the auditor might make a general controls audit comment and recommendation regarding the failure to make effective use of the SDM. However, if the information systems department has reduced the number of SDM forms through department policy, the auditor primarily should be concerned with verifiying that all key SDM phases, as outlined above, have been considered.

If the information systems department is not using any SDM, the auditor definitely has a general controls audit finding and recommendation. A modern data processing organization, whether large or small, should be using some type of formalized SDM to implement both in-house developed and purchased software applications. It will be difficult for the auditor to review applications under development without being able to use an SDM as a reference model.

Even if the new application is to be developed through prototyping techniques, the auditor should look for some form of an SDM. The difference here is that there will be a Prototype Development Life Cycle as part of the overall SDM. This life cycle allows the application development to go through a series of iterative cycles as part of the detailed design. Figure 7.2 describes such a prototype development life cycle.

SELECTING REVIEW CANDIDATES

A typical applications development department may work simultaneously on a variety of projects. Some may be major new applications while others may be minor modifications to existing applications. A typical audit department realistically cannot review *all* projects while they are in the pre-implementation status. That would require too many audit resources and, in some respects, would not be the best use of limited audit resources. The auditor needs a method or procedure for selecting which applications under development to review.

Some audit departments attempt to review only major new applications projects and ignore all other potential pre-implementation review candidates. This can be a convenient approach since a typical data processing organization may be working only on one or two major new applications at a given time. However, some seemingly minor system projects or modifications also may be of audit significance and be overlooked.

As a second approach, the auditor may attempt to review *all* new applications requests. However, it is nearly impossible to do a comprehensive review of everything without holding up system development efforts. Time pressures require the auditor essentially to just briefly review and sign-off on most of the new applications request documents.

A third approach for the auditor is to review only new applications where either data processing management or senior management requests audit participation. When taking this role, the auditor often will not be able to review

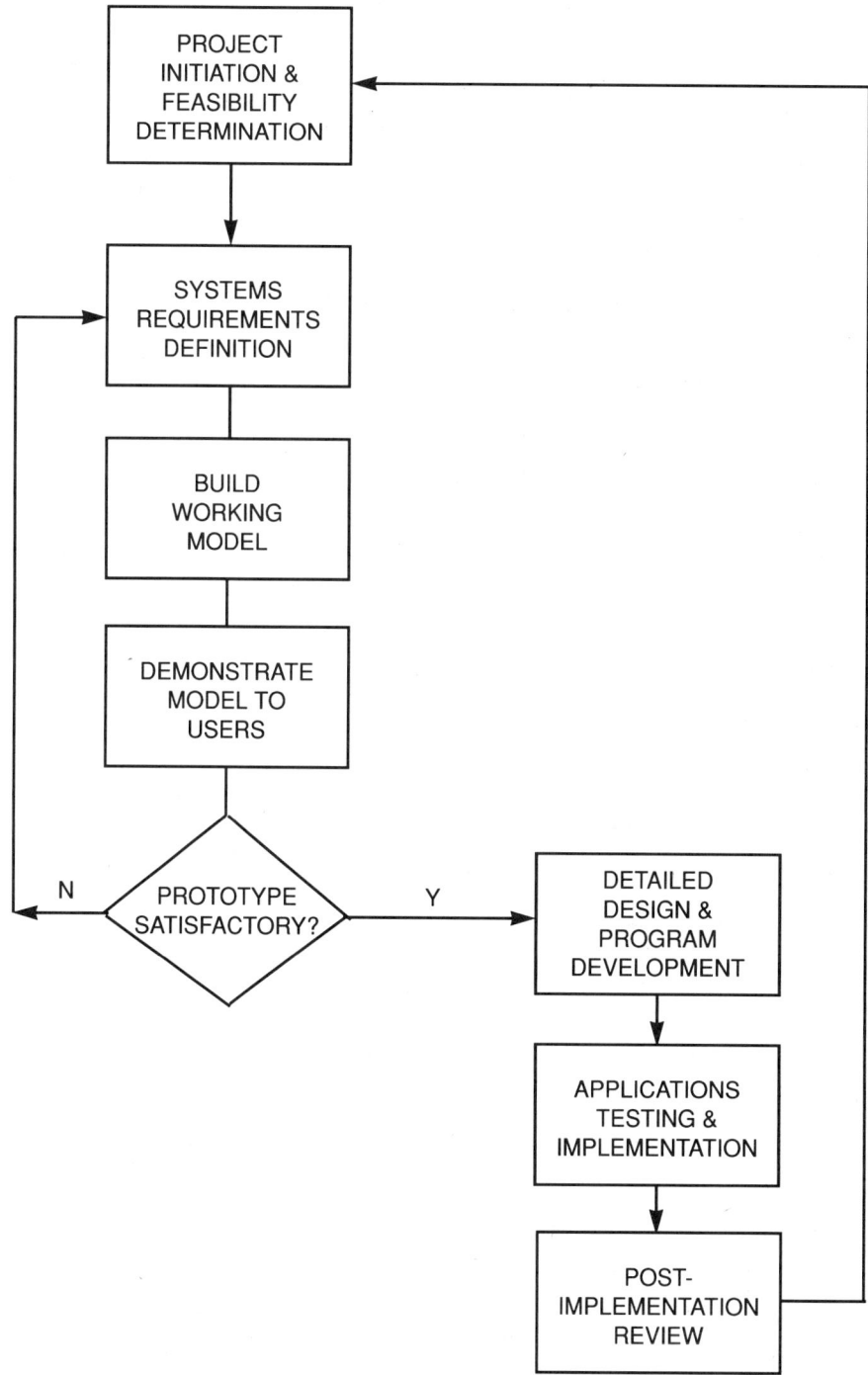

Figure 7.2. Example of a prototype development life cycle

new applications that should have been included and may participate in other reviews that are not necessary, because management does not view new applications development work on the basis of the application's audit risk or controls significance.

A better approach is to develop a formal screening and evaluation approach for deciding which new applications to review. This is often called a risk evaluation or criticality selection approach where the auditor measures the attributes of each potential pre-implementation candidate against a series of weighting factors to select higher scoring projects for review.

New Applications Criticality Selection

To implement a risk evaluation or criticality selection process, the auditor first needs to construct a table of measurement factors for evaluating control risk in new candidates. Essentially, this is a point scoring procedure with projects receiving high scores being selected for potential pre-implementation review. Criticality scoring applications can be tailored to the auditor's own organization and its information systems risk concerns. Criticality scoring is based on the following five broad factors:

1. *Project status*. Project status refers to whether the applications project is a major new effort, a minor new project, or a modification to an existing system. Obviously, major new applications should receive a higher score under this attribute, because major new project efforts represent a greater degree of audit risk to the organization. Perhaps the best way to determine whether a project is "major" or "minor" is to use the total estimated programming hours.

2. *Audit and control significance*. The auditor must determine the relevant control risk of the new applications under development. A sales order entry system, for example, probably will have a higher level of significance than a marketing research information system, because the sales order numbers flow directly to the financial statements while the marketing system may be used for statistical purposes only. The auditor should assign a relative criticality score factor based on an assessment of the audit and control significance.

3. *Technical complexity*. A criticality score factor should be assigned based on the relative technical complexity of the application. A retrieval system based on a fourth generation language probably would receive a lower score than a system written in COBOL because the COBOL system may have a greater chance of programming errors. Similarly, a purchased

software package normally would receive a lower score than an in house developed one.

4. *Interrelationship with other applications.* The auditor should consider the extent the system is interrelated or integrated with other applications. For example, a stand alone application that receives no automated inputs from other applications and supplies none would receive a relatively low criticality factor score. A purchasing system which is part of an overall automated manufacturing system would be assigned a relatively high score.

5. *The impact of application failure.* A criticality factor score should be assigned based on the relative impact of a possible applications failure. If an applications failure would cause the organization to cease operations in a significant area, a higher factor would be assigned. A system failure that only causes inconvenience would receive a lower criticality rating.

Figure 7.3 shows an example of a standard form for a criticality scoring chart. Figure 7.4 is an example of the use of the form for an audit considering the criticality of a new sales and distribution system mandated by the management of a centralized consumer products company with several remote locations. The reader may want to modify some of the factors in the chart, depending upon the reader's organization. A properly assigned criticality score will allow the auditor to justify either selecting or bypassing a candidate for pre-implementation review. Before fully implementing criticality scoring for evaluating pre-implementation candidates, however, the auditor should discuss with management the approach and the assumptions built into the method. In addition, the auditor should evaluate it by applying scores to some applications which were reviewed in the past to determine whether the method appears to be selecting applications correctly.

Using Criticality Factors for Audit Planning

Criticality scoring is particularly useful for deciding which new applications under development should be reviewed as part of a periodic audit planning process. This is a five step process as follows:

1. *Assign a criticality score to all candidates.* The auditor should review all planned new applications or major modifications to assign relative criticality scores. The review can be done as part of a periodic audit planning exercise or as individual applications are authorized. This requires discussing the proposed applications with data processing management and users. The result of this exercise will be a ranked set of review candidates.

Criticality Scoring Chart

NOTE: The auditor should evaluate each application development project according to the following criticality scoring factors. Points should be assigned and then adjusted according to the weighting factors at the end of this chart. Each major criticality factor may receive a total score of up to 50 points. The sum of all the scores for all the factors for each application should not exceed 100 points. The highest scoring applications should be given priority as candidates for pre - implementation review.

		Criticality Scores	
Criticality Factors		**Normal Range**	**Assigned Score**
I. Project Status			
A. *Nature of the application project:*			
New application developed in-house		8–10	_____
Purchased application package		5–8	_____
Major change affecting functionality		6–9	_____
Minor change		0–5	_____
B. *Past history of application change:*			
Significant changes over past two years		6–10	_____
Few changes in past two years		4–6	_____
Two years or more since last change		2–5	_____
New application (no changes)		0	_____
C. *Project development team:*			
Systems contractor, competitive bid		8–10	_____
Systems contractor, sole source		6–8	_____
In-house, remote location development		4–8	_____
In-house development group		2–4	_____
Packaged software with minor vendor changes		1–3	_____
D. *Project management team:*			
User group		8–10	_____
Information systems group		4–7	_____
Joint user and I/S management		1–4	_____

Figure 7.3. Criticality scoring chart

E. *Top management interest in project:*

Project mandated by senior management	8–10	_____
Division or operating unit request	6–9	_____
Project initiated by middle management	5–7	_____
Individual user or department request	2–5	_____
"When time is available" request	0–3	_____
Project Status Score		══════

II. Audit and Control Significance

A. *Type of application:*

Affects financial statement balances	10	_____
Supports financial statement balances	5–9	_____
Supports major organizational operations	8–10	_____
Logistical or administrative support	2–5	_____
Statistical or research application	1–5	_____

B. *Past audit involvement with application:*

Prior audits including recommendations	7–10	_____
Prior audits, limited recommendations	4–8	_____
Audit reviews of related, manual areas	2–6	_____
No audit experience	0	_____

C. *Application control procedures:*

Application-generated internal controls	8–10	_____
Run to run controls with other systems	6–8	_____
User maintained on-line controls	2–6	_____
Batch controls	1–4	_____

D. *Application control responsibilities:*

Controls handled within application	6–10	_____
Remote user control responsibility	5–8	_____
Local user control responsibility	3–7	_____
Parallel, manual control systems	0–4	_____

E. *Computer assisted audit tools capability:*

Candidate for embedded audit facility	8–10	_____
Standard audit software	5–8	_____
Potential use of 4GL tools	2–6	_____
No computer-assisted audit plans	0	_____
Audit and Control Significance Score		══════

Figure 7.3. Criticality scoring chart *(continued)*

III. Technical Complexity
A. *Programming languages used:*

Specialized language such as LISP	8–10	_____
Mixed COBOL and other language systems	6–8	_____
In-house developed COBOL programs	5–7	_____
COBOL produced by application generator	3–6	_____
Report generator or 4GL system	2–5	_____
Purchased application (no programming)	0–2	_____

B. *Database procedures used:*

Distributed database application	8–10	_____
Single database, new software product	7–9	_____
Single database, familiar software	4–8	_____
VSAM or ISAM type data files	2–5	_____
Tape file system	1–3	_____

C. *Systems development methodology approach:*

New SDM procedures being introduced	8–10	_____
Existing SDM procedures used	5–8	_____
Prototyping development	3–7	_____
Purchased software (limited SDM needs)	1–5	_____

D. *Hardware requirements for application:*

New computer system	9–10	_____
New peripheral devices	6–10	_____
New or revised equipment configuration	3–5	_____
Current or existing equipment	1–3	_____

E. *Project team familiarity with technical environment:*

Team new to hardware or software used	7–10	_____
Limited familiarity with environment	4–6	_____
Strong team technical familiarity	0–4	_____
Technical Complexity Score		≡

IV. Interrelationship with Other Applications
A. *Dependency on other applications:*

Inputs from network of polled stations	8–10	_____
Inputs from distributed systems network	6–10	_____
Inputs from local integrated systems	6–8	_____
Inputs through tape or other transfer	2–5	_____
Standalone application	0	_____

Figure 7.3. Criticality scoring chart *(continued)*

B. *Requirements to supply other applications:*

Outputs transmitted directly to network	8–10	_____
Direct outputs through shared databases	6–8	_____
Outputs through file transfers	2–6	_____
Outputs through review and reentry	1–4	_____
Standalone application	0	_____

C. *Communication relationships of application:*

Telecommunications through specialized network	10	_____
Communication through dial-up lines	8–10	_____
Use of standard remote links	5–7	_____
Use of local links only	1–4	_____
Standalone application	0	_____

D. *Interrelationship with end user facilities:*

Uploads and downloads from application files	10	_____
Uploads and downloads from extract files	6–9	_____
End user downloads only	2–5	_____
Standalone application	0	_____

E. *Technical interrelationship considerations:*

Unique hardware and software protocols	7–10	_____
Common hardware—unique software concerns	6–9	_____
Common hardware and software linkages	2–5	_____
Standalone application:	0	_____
Interrelationship Score		=====

V. Impact of Application Failure

A. *Impact of incorrect outputs:*

Potential legal liability	10	_____
Financial statement impact	9–10	_____
Potential for incorrect decisions	4–8	_____
Limited application decision support	1–4	_____

B. *Impact of incorrect files or data:*

Incorrect results passed to other systems	8–10	_____
Corrupted data requiring reconstruction	6–9	_____
Corrupted data requiring reprocessing	1–5	_____

Figure 7.3. Criticality scoring chart *(continued)*

C. *Impact of failure on computer operations:*
    Scheduling problems with related systems   7–10   _____
    Schedule adjustment for reprocessing   4–6   _____
    End user application (minimal impact)   1–5   _____

D. *Impact of failure on project management systems:*
    Requirement to reschedule planned systems   7–10   _____
    Requirement to plan revised application   4–6   _____
    Requirement to plan application fixes   1–5   _____

E. *Impact of application failure on personnel:*
    Need for extra management analysis time   8–10   _____
    Need for extra user clerical time   6–9   _____
    Need for systems or programmer efforts   2–6   _____
    Purchased software vendor support   1–4   _____
        Failure Impact Score   _____

Summary and Weighting

Factor	Score	Weighting Factor	Weighted Score
Project Status	___	0.15	_____
Audit and Control Significance	___	0.40	_____
Technical Complexity	___	0.05	_____
Interrelationship with Other Applications	___	0.10	_____
Impact of Application Failure	___	0.30	_____
Total Weighted Score		1.00	_____

Figure 7.3. Criticality scoring chart *(continued)*

Criticality Scoring Chart

NOTE: *The following standard form has been filled in with score s that might be assigned by an auditor considering the criticality of a new sales and distribution system mandated by the management of a centralized consumer products company with remote locations. The relevant factors are as follows. The application is a major one for the company and will be developed in-house with user participation. The results will feed into financial statement applications, and control s with other systems will be utilized. CASE systems will be used. Pro gramming will be in COBOL. A standard database will be used, and data will be transmitted from remote locations. The existing SDM approach, which was reviewed previously by Internal Audit, will be used to develop the application. New work stations will be installed at remote locations for the first time, with dial-up lines to the central location.*

Criticality Factors	Assigned Score
I. Project Status	
A. *Nature of the application project:*	
*This is a major in-house developed application	10
B. *Past history of application change:*	
*This is a new system	0
C. *Project development team:*	
*An in-house development group will be used	3
D. *Project management team:*	
*Users and members of Information Systems will participate	2
E. *Top management interest in project:*	
*Senior management considers the application a "must" system	10
Project Status Score	25

Figure 7.4. Example of the use of the criticality scoring chart

II. Audit and Control Significance
A. *Type of application:*
   *Sales results will support financial
   statement balances   8

B. *Past audit involvement with application:*
   *This is a new application, but there has
   been audit activity at the remote
   location   3

C. *Application control procedures:*
   *There will be run to run controls with
   other systems   8

D. *Application control responsibilities:*
   *Local users will be responsible for some
   controls   6

E. *Computer assisted audit tools capability:*
   *Standard audit applications are planned   7

   Audit and Control Significance Score   32

III. Technical Complexity
A. *Programming languages used:*
   *This will be an in-house developed COBOL
   application   6

B. *Database procedures used:*
   *A standard database application will be
   used with data transmitted to the central
   location from remote terminals   4

C. *Systems development methodology approach:*
   *An existing SDM, previously reviewed, will
   be used to define the application   5

D. *Hardware requirements for application:*
   *New work stations will be installed at remote
   locations for data collection   9

Figure 7.4. Example of the use of the criticality scoring chart *(continued)*

E. *Project team familiarity with technical environment:*
   *The remote workstations will be new,
   unfamiliar equipment   8

   Technical Complexity Score   32

IV. **Interrelationship with Other Applications**
   A. *Dependency on other applications:*
      *Remote locations will be polled for
      data input   9

   B. *Requirements to supply other applications:*
      *Application outputs will be reviewed and
      then entered in the general ledger system   3

   C. *Communication relationships of application:*
      *Dial-up lines with access controls will be used   8

   D. *Interrelationship with end user facilities:*
      *The local information center may access
      some data   2

   E. *Technical interrelationship considerations:*
      *Common hardware and software will be
      used for the core system   4

      Interrelationship Score   26

V. **Impact of Application Failure**
   A. *Impact of incorrect outputs:*
      *The major risk is that incorrect inventory
      purchase decisions could be made   8

   B. *Impact of incorrect files or data:*
      *Due to remote transmissions, reconstruction
      might be required   7

   C. *Impact of failure on computer operations:*
      *A failure may cause scheduling problems
      with other systems   7

Figure 7.4. Example of the use of the criticality scoring chart *(continued)*

D. *Impact of failure on project management systems:*
   *A major problem might require redesigning
   the entire application   <u>6</u>

E. *Impact of application failure on personnel:*
   *Additional clerical help and systems
   personnel would be needed   <u>8</u>
   Failure Impact Score   <u>36</u>

Summary and Weighting

Factor	Score	Weighting Factor	Weighted Score
Project Status	<u>25</u>	0.15	<u>3.75</u>
Audit and Control Significance	<u>32</u>	0.40	<u>12.80</u>
Technical Complexity	<u>32</u>	0.05	<u>1.60</u>
Interrelationship with Other Applications	<u>26</u>	0.10	<u>2.60</u>
Impact of Applications Failure	<u>36</u>	0.30	10.80
Total Weighted Score		1.00	31.55

Figure 7.4. Example of the use of the criticality scoring chart *(continued)*

2. *Estimate the monthly review time requirements.* The auditor should prepare a month by month estimate of audit time requirements for each potential review candidate. These estimates will be based upon the expected application development timing of the projects as well as the expected review time requirements. The latter will be based on the pre-implementation control objectives discussed later in this chapter.

3. *Merge new candidates with current projects.* Assuming that the auditor is currently reviewing other applications under development, those current plans should be merged with the planning for potential new pre-implementation review candidates.

4. *Determine the hours available for review and make selections.* The auditor should consider the hours available for reviews and select new pre-implementation candidates to be included in the review process.

5. *Review plans with management.* As a final step, the auditor should discuss these pre-implementation review plans with management. This allows the auditor to explain why systems were or were not selected for review. It also allows the auditor to show why audit resource limitations may prevent a review of all candidates.

Problems with Criticality Measurement

The above procedures provide the auditor with an effective approach to select new applications for review. However in the "real world," there always will be some implementation problems. First, the above approach is dependent upon an annual data processing plan. If new project proposals or requests appear randomly over the course of the year, it will be necessary to *continually* adjust the relative criticality scoring. Application implementation delays or schedule adjustments may cause the auditor to pass up or delay reviews of new, critical applications.

The newer structured design or prototyping approaches to applications design may cause the auditor some problems in applying criticality measurements. Such design techniques may be fluid with control concerns poorly defined at the beginning of a project. In addition, the design and implementation may proceed so fast that the auditor will not have an opportunity to evaluate the application. Some approaches to this problem are discussed in the following section.

Also, the auditor may have developed a perfectly proper approach to scoring and selecting new candidates for pre-implementation review, but upper management may decide the auditor should review a given system even though it does not meet criticality standards.

To make criticality measurement selection work, the auditor needs to review all assumptions with management on an ongoing basis. If it is understood that the auditor is performing a comprehensive controls oriented review of the candidates selected, there will be fewer questions about why other applications are not being reviewed.

THE AUDITOR'S ROLE IN PRE-IMPLEMENTATION APPLICATIONS REVIEWS

At the beginning of this chapter, we discussed how the auditor must be careful not to take responsibilty for implementing controls in new applications. The auditor should be the reviewer while the application developers should act upon the auditor's recommendations. However, the auditor still can play several different roles while reviewing new applications under development. The auditor can:

Actively participate with the design team

Monitor applications design and development

Review the application externally at selected intervals

Each of the above roles has advantages and potential pitfalls for the auditor. As a member of the design team, the auditor can be involved in all decisions and compromises that usually go into a comprehensive applications design. This approach gives the auditor the best understanding of the application. However, the auditor can lapse into becoming an integral member of the design committee and lose audit independence. In addition, this mode sometimes can turn into a full time task for an auditor. The only time when a design team role seems to work is when an application is developed through prototyping. In this mode, a version of the application is produced, evaluated, and then modified with little lead time to produce the next version. To be effective in recommending controls improvements when using prototyping, the auditor often needs to play an active role in the design team to keep up with the process.

In the second role, the auditor often attends many of the same design meetings on a regular basis. However, the auditor should make a determined effort to act only as an external, independent reviewer. This role is difficult as the auditor still can lapse into becoming a non-independent member of the design team.

In the third role, the auditor reviews the application only at selected intervals during the applications development process. The auditor might elect first to review an application at the conclusion of the requirements definition phase. The next review might take place early in the design phase to ascertain that any significant control concerns identified in the requirements definition phase have been addressed and that good controls are being incorporated into the preliminary design.

Scheduling is the main problem with this third approach, because the applications design process generally is dynamic with frequent schedule adjustments. When the auditor returns from other projects to review the requirements definition study, it may not be ready, and when it finally is ready for audit review the auditor may be assigned to other projects.

The auditor must exercise scheduling flexibility when reviewing such applications under development. If an application has been rated high on the criticality scale, the auditor may find it necessary to adjust other audit projects to achieve the audit control objectives for the critical application being reviewed.

PRE-IMPLEMENTATION CONTROL OBJECTIVES AND PROCEDURES

To effectively review new applications under development, the auditor needs to follow the same method of defining control objectives and audit procedures that has been used in other chapters of this book. All too often, auditors resist this approach, arguing that applications under development are "different." However, while applications under development are fluid and subject to ongoing developmental change, sets of control objectives and audit procedures still are appropriate for these reviews.

Pre-implementation control objectives and audit procedures have been organized around the same SDM steps discussed earlier in this chapter. The names of various phases may be different from one SDM methodology to another. However, these steps can be tailored to fit the circumstance.

When the auditor selects a given application for review, an important first step is to review the overall audit program with data processing management so that they will understand the planned audit review approach. While some of the detailed audit procedures discussed in subsequent paragraphs may require tailoring to fit a given application, the overall control objectives should be applicable to the pre-implementation review of most applications.

Objectives and Procedures for the Project Initiation and Feasibility Determination Phase

Usually, a new application project has been initiated and some feasibility determination has been made before it is scheduled for an auditor's pre-implementation review. The auditor may have determined criticality and selected such an application for review after this initiation step in the SDM. The auditor's review of this portion of a new application therefore will often be after the fact. Nevertheless, for those projects selected, this is an important review step. When an application project has been selected for pre-implementation review, the auditor should determine that the project initiation request and the feasibility study have been adequately prepared. Figure 7.5 provides objectives and procedures for reviewing this phase of the pre-implementation review. Even though the auditor generally starts the review after this phase has been completed, these procedures should be completed as a first review step.

If the auditor determines that the initiation or feasibility determination process is flawed, the auditor may suggest to management that the given system was a poor choice for development. This can be a difficult role for the auditor. When an application has been initiated without an adequate feasibility determination, often it is because it has a powerful sponsor. The auditor should point out the need for a proper study and evaluation of potential system feasibility.

Objectives and Procedures for the Systems Requirements Definition Phase

In this phase, the auditor reviews the systems requirements study to assess the applications overall control risk. The requirements study provides a general description of the new application including a description of its controls. This is a key point in the pre-implementation review process for the auditor. If the auditor can identify control concerns during this phase of the applications development, it will be relatively easy for system designers to address and correct them.

Figure 7.6 provides a set of control objectives and audit procedures to be performed for the system requirements definition phase of any project. Some of the procedures may require modification if the application under review is composed of purchased software or if it will be a major modification to an existing system. However, the auditor should perform procedures necessary to satisfy all control objectives listed here.

**Control Objectives and Audit Procedures for a
Pre-implementation Review of the Project Initiation Phase**

Objective 7.5.1.   Users requesting a new applications development
project should complete a project work request form consistent
with organization SDM standards.

Procedure 7.5.1.1.   Review project work request forms to determine that the user request:

Adequately describes the purpose and nature of the project

Contains estimates of expected benefits

Has been approved by appropriate levels of management

Procedure 7.5.1.2.   Interview several key application re-
questors to determine that they have a good understanding
of project work request procedures.

Objective 7.5.2.   The user initiated project work request should give
adequate consideration to the costs and benefits of the new applications project.

Procedure 7.5.2.1.   Select several other outstanding project
work request forms and discuss the procedures used for estimating costs and proposed benefits with requesting users.

Procedure 7.5.2.2.   Independently verify the accuracy of user
estimated costs and benefits for the outstanding project
work request under review.

Procedure 7.5.2.3.   Determine if post-implementation reviews
are performed routinely to measure actual application costs
and benefits.

Objective 7.5.3.   The information systems function should review
all user initiated project work requests in a timely manner and
should keep users advised of the request status and disposition of
the request.

Procedure 7.5.3.1.   Review procedures for logging in new project work requests, assigning analysts to perform initial
reviews, and informing users of status and dispositions.

Figure 7.5. Control objectives and audit procedures for a pre-implementation review of the project initiation phase

Procedure 7.5.3.2. Interview several key users to determine their understanding of and satisfaction with information systems procedures over the project work request process.

Procedure 7.5.3.3. Determine that the project work request for the application under review has followed information systems procedures.

Objective 7.5.4. The information systems function should have formal procedures to determine disposition of all new user requests depending upon application size, status, and whether the request is for a new application or an enhancement to an existing application.

Procedure 7.5.4.1. Review procedures for assigning new and application modification requests to proper analysts and determine that more minor requests are processed in a timely manner.

Procedure 7.5.4.2. Review procedures for "emergency" application work requests and determine if these appear adequate given overall SDM documentation requirements.

Procedure 7.5.4.3. Examine project work request files and determine if there are old requests lacking proper disposition action.

Objective 7.5.5. The information systems function should evaluate the approved, user initiated project work requests for technical feasibility, cost effectiveness, and suitability.

Procedure 7.5.5.1. Review procedures for evaluating user requests to determine if formal procedures exist for evaluating feasibility in a systematic manner.

Procedure 7.5.5.2. For the selected pre-implementation application, review the feasibility analysis documentation to determine that:

The request received adequate technical review
Interested users were interviewed to better determine needs
Alternative solutions, if appropriate, were considered
User projected benefits were reviewed and preliminary systems cost estimates were prepared.

Figure 7.5. Control objectives and audit procedures for a pre-implementation review of the project initiation phase *(continued)*

Procedure 7.5.5.3.   Select several other project work requests which have been rejected and determine whether reasons for rejection appear reasonable.

Objective 7.5.6.   There should be a formal evaluation of potential control structure and control risk for all major new application work requests.

Procedure 7.5.6.1.   Review the role of Internal Audit in performing preliminary reviews of new user requests and determine whether this appears adequate.

Procedure 7.5.6.2.   Interview members of the information systems staff responsible for evaluating project work requests and evaluate whether their procedures for reviewing the application control structure and related risk appear to be adequate.

Objective 7.5.7.   The information systems function should prepare formal recommendations on the cost and feasibility of major new project requests for approval by the Steering Committee.

Procedure 7.5.7.1.   Review the documentation for the request being forwarded to the Steering Committee for evaluation and determine that it appears adequate for decision making purposes.

Procedure 7.5.7.2.   Review procedures for submitting larger requests for action to the Steering Committee as well as for handling smaller ones within the information systems function and determine whether these appear appropriate.

Objective 7.5.8.   Project work requests should be evaluated by the Steering Committee in light of overall information systems long and short range plans, and changes or adjustments should be made as appropriate.

Procedure 7.5.8.1.   Review the approved project to determine that it is consistent with long and short term systems plans.

Procedure 7.5.8.2.   If the approved project appears to alter priorities and plans, determine that procedures exist to document those plan changes, to adjust other priorities as appro-

Figure 7.5. Control objectives and audit procedures for a pre-implementation review of the project initiation phase *(continued)*

priate, and to communicate those changes to proper levels of management.

Objective 7.5.9.   The Steering Committee should review all user requests in a timely manner and should set priorities for development work.

Procedure 7.5.9.1.   Review Steering Committee minutes or other records of activity over a limited time period to determine whether the Committee meets with sufficient frequency to evaluate new systems requests.

Procedure 7.5.9.2.   Review Steering Committee priority lists to determine that actions have been taken on all requests that have been reviewed and submitted by Information Systems.

Procedure 7.5.9.3.   Request permission to attend several Steering Committee meetings and determine if adequate attention is being given to user requests approved by Information Systems as well as other user system requests.

Procedure 7.5.9.4.   When projects are rejected or delayed by the Steering Committee, determine that there exist procedures to review and reconsider such requests in light of other information or factors not considered.

Objective 7.5.10.   Formal schedules should be prepared for the approved project work request defining expected start and completion dates.

Procedure 7.5.10.1.   Review schedules of Steering Committee approved projects and investigate the reasons behind any late or otherwise open requests.

Procedure 7.5.10.2.   Determine that procedures exist to inform requesting users of any adjusted or delayed start dates.

Procedure 7.5.10.3.   Through interviews, determine that requesting users understand the scheduled status of the approved project.

Figure 7.5. Control objectives and audit procedures for a pre-implementation review of the project initiation phase *(continued)*

**Control Objectives and Audit Procedures
for a Pre-implementation Review of the Systems Requirements
Definition Phase**

Objective 7.6.1. A project team composed of key information systems analysts and user personnel should be organized to perform a general requirements definition for the new or revised application project.

> **Procedure 7.6.1.1.** Determine that key members of affected user organizations are participating in the requirements definition review team.
>
> **Procedure 7.6.1.2.** Determine that adequate personnel and other resources have been assigned to the requirements definition team to allow them to complete the project properly.
>
> **Procedure 7.6.1.3.** Determine that adequate task plans have been prepared so that each key member of the project team has defined project duties as well as schedules for accomplishing them.
>
> **Procedure 7.6.1.4.** Review initial plans for the requirements definition process to determine that they appear reasonable, meet the objectives outlined in the original application request, and have been approved by management.

Objective 7.6.2. A project management system should be used to monitor the progress of the requirements definition team and to report ongoing status of the proposed application to management.

> **Procedure 7.6.2.1.** Review the project management system to be used by the team and determine whether it appears reasonable for tracking a project of the magnitude of the requested project.
>
> **Procedure 7.6.2.2.** Select another requirements definition project that is in process or recently completed and determine that project management procedures have been followed.

Objective 7.6.3. User interview and survey techniques should be used to reaffirm application requirements and to prioritize objectives.

Figure 7.6. Control objectives and audit procedures for a pre-implementation review of the systems requirements definition phase

Procedure 7.6.3.1.   Review user interview notes or any survey forms to determine that adequate procedures are being used to identify user requirements.

Procedure 7.6.3.2.   If the results of user surveys and interviews appear to point to a different systems direction from the original user request, determine whether appropriate steps are being taken to reevaluate feasibility and application needs.

Procedure 7.6.3.3.   Determine that survey results, if any, have been tabulated and are available for Steering Committee or management review.

Procedure 7.6.3.4.   Determine that procedures exist to route projects back to the Steering Committee if survey results point to different application requirements.

Objective 7.6.4.   Prototyping software tools should be used, if appropriate, to help users define screen and report layouts and to better understand the overall general systems design.

Procedure 7.6.4.1.   Develop an understanding of any prototyping tools available within the information systems organization and determine whether they would be appropriate to the selected application requirements analysis.

Procedure 7.6.4.2.   Review procedures for communicating preliminary report and screen layouts to end users and determine whether these procedures appear appropriate.

Procedure 7.6.4.3.   If prototyping tools are not being used and they appear to be appropriate to the particular application, consider recommending their use for this and related projects.

Objective 7.6.5.   If new or specialized hardware or software tools are to be used, there should be adequate review and selection procedures for identifying the proper products as well as training personnel.

Procedure 7.6.5.1.   Determine whether the general design calls for unique or specialized hardware or software products, such as specialized telecommunications devices or

Figure 7.6. Control objectives and audit procedures for a pre-implementation review of the systems requirements definition phase *(cont.)*

database software, and determine whether adequate attention is being devoted to the review and selection of such products.

Procedure 7.6.5.2. Determine that the costs of any specialized equipment have been factored into the original cost benefit analysis.

Procedure 7.6.5.3. Determine that any specialized hardware or software will interface properly with any existing systems as required.

Objective 7.6.6. The requirements definition and general design should include a detailed analysis of application control points and should identify audit trails.

Procedure 7.6.6.1. Determine that the requirements definition document gives adequate consideration to application controls.

Procedure 7.6.6.2. If the application is intended to replace an existing one, determine that control procedures will remain the same or that key users will understand how the control procedures will differ.

Procedure 7.6.6.3. Interview members of the project team to determine that they understand any control issues surrounding the new application.

Objective 7.6.7. The requirements definition process should result in a general design document, consistent with the SDM, which defines data flows, application inputs, outputs, and processing procedures, and which has been approved by key members of user management.

Procedure 7.6.7.1. Review the general design document and determine that it adequately describes:

Data inputs and interfaces with other systems
General processing logic
System outputs including user approved preliminary screen and report layouts
Application controls and audit trails
Any unique application security considerations

Figure 7.6. Control objectives and audit procedures for a pre-implementation review of the systems requirements definition phase *(cont.)*

A preliminary implementation plan for the detailed design phase

Procedure 7.6.7.2. Interview members of the project team as well as other key users to determine that they understand planned application procedures and concepts.

Procedure 7.6.7.3. Determine that the requirements definition and general design documentation has been approved by appropriate levels of the information systems function, the Steering Committee and user management.

Objective 7.6.8. If purchased software is to be used for the application, a request for proposal (RFP) document should be prepared from the general design document to allow better software vendor selection.

Procedure 7.6.8.1. Review any recent software RFP documents to determine that they are consistent with the general design and provide vendors with sufficient data to prepare proposals properly.

Procedure 7.6.8.2. Determine that adequate procedures have been used to secure proposals from all potential vendors.

Procedure 7.6.8.3. Determine that appropriate procedures for purchasing software applications are being followed.

Objective 7.6.9. Consideration should be given to ongoing application audit approaches during the requirements definition and general design phase of the application.

Procedure 7.6.9.1. Discuss potential ongoing audit approaches with other members of the audit function, both internal and external, and determine whether any unique audit techniques or procedures will be required after implementation.

Procedure 7.6.9.2. If the application appears to be a candidate for an embedded audit technique, as discussed in Chapter 6, develop specifications for such a procedure.

Objective 7.6.10. Audit workpaper documentation should be initiated during the requirements definition phase of the project.

Procedure 7.6.10.1. Develop functional flow charts of the pro-

Figure 7.6. Control objectives and audit procedures for a pre-implementation review of the systems requirements definition phase *(cont.)*

posed application for use as a planning and discussion vehicle with other members of the audit team.

Procedure 7.6.10.2. Begin to prepare detailed workpapers covering the application which will support any subsequent pre-implementation review work as well as ongoing audit activities.

Figure 7.6. Control objectives and audit procedures for a pre-implementation review of the systems requirements definition phase *(cont.)*

During the requirements phase of the review the auditor may decide that special skills are required for the review. For example, if the application involves the installation of a new database management package, the auditor may need to obtain training on the product. Often, classes offered by the vendor to the development staff are appropriate also for the auditor. In larger audit organizations or in an external audit environment, at this phase of the review the auditor may arrange to bring in someone with specialized skills or training. For large projects that may take years to develop and implement, it can be effective to add a specialist to the staff to cover the review of just that project.

The requirements phase of a new applications review also is the appropriate time to determine whether the application will be of such audit significance that an embedded audit module should be included in the design. These techniques were discussed in Chapter 6. If they appear appropriate for the application under review, the auditor should develop requirements specifications for the module and make arrangements for its development and installation.

At the completion of this phase of the review, the auditor should consider writing an informal audit report as discussed later in this chapter. In addition, a workpaper binder should be started to document the new application's controls environment.

Objectives and Procedures for the Detailed Design and Program Development Phase

Typically, this is the longest phase of a new applications development project, and the auditor may want to schedule several reviews during this phase. While each of the periodic reviews should focus on a specific area of the new application development project, the overall purpose should be to answer the following questions:

Does the detailed design meet the general requirements definition?

Do users understand the controls of the new application under development?

Has proper consideration been given to application controls and security?

Are the data processing department's SDM standards being followed?

Have earlier audit recommendations been incorporated into the detailed design?

This phase is an important part of the auditor's pre-implementation review. However, care should be taken not to become buried in detail during this phase. Some data processing organizations attempt to use the auditor as a quality assurance function for the project. However, overall audit effectiveness will be diminished if the auditor's time is spent reviewing details such as compliance with detailed programming standards. Reviews of this sort should be limited to periodic testing.

Figure 7.7 lists control objectives and audit procedures for reviewing the detailed design and program development phase of a new application. The auditor should bring control concerns encountered during this phase of the project to the attention of management as soon as possible so that timely corrective action can be taken.

If the new application under review is a purchased software package, typically, there will be limited in-house design and programming requirements. The use of purchased software is becoming common for many data processing organizations, both large and small, and the auditor should play a role in reviewing controls in a purchased package before it is installed. This chapter discusses some of the unique auditor concerns when performing pre-implementation reviews over purchased software packages.

If the new application is being developed with prototyping methods, the auditor should be certain to review the correct prototype version. Sometimes, an applications developer will decide that a control introduced in an earlier prototype version is cumbersome, and it could be dropped in a later version. Prototyping techniques are discussed below along with other new applications development technologies.

Objectives and Procedures for the Applications Testing and Implementation Phase

This phase of an implementation project normally will include testing the new application, completion of documentation, user training, and conversion of

Control Objectives and Audit Procedures for a
Pre-implementation Review of the Detailed Design Phase

Objective 7.7.1. The detailed application design should be consistent with the general design and requirements definition.

> **Procedure 7.7.1.1.** Review process flow charts and other documentation to ascertain that the detailed design follows the system described in the general design document.

> **Procedure 7.7.1.2.** Discuss any deviations from the general design and requirements definition with the information systems development staff.

> **Procedure 7.7.1.3.** Determine that any deviations between the design phases are documented and are understood by key users.

> **Procedure 7.7.1.4.** Update audit workpapers to reflect any changes between the general and detailed system design, with an emphasis on control related changes.

Objective 7.7.2. The new application should be developed following the information systems' SDM standards and procedures.

> **Procedure 7.7.2.1.** Determine that the detailed design is consistent with the organization's SDM standards for that phase and comment on any exceptions.

> **Procedure 7.7.2.2.** If CASE methodologies are used, apply the controls and procedures outlined in Figure 7.14, Control Objectives and Audit Procedures for Reviewing Applications Developed with CASE Techniques.

Objective 7.7.3. The new application should be developed with proper consideration given to input controls.

> **Procedure 7.7.3.1.** Review the application design to determine that there are adequate procedures to ensure that all input transactions are processed.

> **Procedure 7.7.3.2.** Determine that the detailed design includes adequate procedures for screening input documents for errors and communicating error conditions back to their source.

Figure 7.7. Control objectives and audit procedures for a pre-implementation review of the detailed design phase

Procedure 7.7.3.3.   Review input screen layouts and documents to assess that they follow organization standards and appear to be easy to use and understand.

Procedure 7.7.3.4.   Interview future key users of the new system to determine that they understand their responsibilities for data input and control for the new application.

Objective 7.7.4.   The new application should be developed with proper consideration given to processing controls.

Procedure 7.7.4.1.   Determine that the new application will contain adequate controls to ensure that all transactions received are processed through the application.

Procedure 7.7.4.2.   Determine that the application design gives adequate attention to maintaining audit trails, run to run controls, and log files to ensure ongoing integrity and to allow for application backup and recovery.

Procedure 7.7.4.3.   Identify any key calculations or other processing procedures and determine whether they appear to be correct.

Procedure 7.7.4.4.   Review planned processing controls and procedures with key application users to determine that they understand and approve of how the new application will function.

Objective 7.7.5.   The new application should be developed with proper consideration given to output controls.

Procedure 7.7.5.1.   Determine that the application design contains adequate controls to verify the completeness of output reports.

Procedure 7.7.5.2.   Review procedures for screening and reporting error or exception conditions and determine whether they appear to be adequate.

Procedure 7.7.5.3.   Review planned application outputs to determine that they are consistent with the general design and that they follow standards.

Objective 7.7.6.   There should be adequate project management

Figure 7.7. Control objectives and audit procedures for a pre-implementation review of the detailed design phase *(continued)*

controls to ensure that the new application is developed on a timely basis.

Procedure 7.7.6.1. Review project plans to determine that they have been developed with sufficient detail to identify specific tasks and completion milestones.

Procedure 7.7.6.2. Compare actual detailed design progress to plans and determine whether management has been made aware of any significant deviations.

Objective 7.7.7. The application should be developed in a generally accepted programming language to facilitate future maintenance requirements.

Procedure 7.7.7.1. Consider the software tools to be used for the new application and determine that they are consistent with information systems standards.

Procedure 7.7.7.2. Determine that the application makes use of data dictionary tools, subroutine libraries, and other tools to increase the efficiency and maintainability of the application.

Procedure 7.7.7.3. Determine that all significant programmed procedures are subject to a quality assurance review to minimize errors, assure compliance with organization standards, and promote processing efficiency.

Objective 7.7.8. The application should utilize standard reporting and inquiry tools so that users can retrieve data easily without an understanding of programming concepts.

Procedure 7.7.8.1. Review procedures for linking the application to a fourth generation language, microcomputer system, or other retrieval tool to allow for end user inquiry.

Procedure 7.7.8.2. If users are allowed to input transactions through a microcomputer system, determine that the design contains adequate controls to preserve application integrity.

Objective 7.7.9. Prior audit recommendations made during the general design phase of the project should be incorporated into the detailed application design.

Figure 7.7. Control objectives and audit procedures for a pre-implementation review of the detailed design phase *(continued)*

Procedure 7.7.9.1.   Determine that any audit recommendations made during the general design phase of the application have been incorporated into the application.

Procedure 7.7.9.2.   If the application is to contain an embedded audit module as discussed in Chapter 6, determine that the module is being incorporated consistent with original audit specifications.

Objective 7.7.10.   The application should be developed with proper consideration given to overall application security.

Procedure 7.7.10.1.   Review application security plans for the application and determine whether they appear to be adequate.

Procedure 7.7.10.2.   If the application uses any unique software tools or procedures that do not fit with current information security software products or procedures, determine that steps are being taken to introduce proper security controls.

Procedure 7.7.10.3.   Discuss security requirements with key users to determine whether there are any unique security risks associated with the new application.

Figure 7.7. Control objectives and audit procedures for a pre-implementation review of the detailed design phase *(continued)*

data files to the new application. During this phase the auditor often is able to see whether system controls are working as expected, and can test any embedded audit modules that have been incorporated into the application.

The control objectives and audit procedures listed in Figure 7.8 for this phase are designed to help the auditor determine that the new application is ready for final implementation. However, this phase of the auditor's review often can be difficult if there are system control problems coupled with management pressures to implement the application quickly. Data processing functions often promise to correct control problems in a new application during "phase two," which is to be scheduled later. Frequently, auditors find that because of other priorities, "phase two" never occurs. The auditor should consider the severity of such control problems and either document them for follow up review or make an issue to management about the need for them during the current implementation.

At the conclusion of the application testing and implementation phase, the auditor should prepare a final report on audit pre-implementation review activities. The report should highlight significant control issues identified by the auditor and corrected by data processing. The report should also outline any outstanding control recommendations which have not been implemented. While pre-implementation audit reports up to this point probably have been informal, this final report should follow normal audit department reporting standards.

Objectives and Procedures for the Post-implementation Review Phase

Although the new application is no longer under development, this phase of the audit is still important. The post-implementation review should take place shortly after a new application has been implemented and has had time to "settle down." In other words, the auditor should perform the review after the users have had an opportunity to understand the application and data processing has had time to resolve any final implementation difficulties.

A post-implementation review is different from a normal applications review as discussed in Chapter 5. The prime purpose of this review is to determine whether design objectives have been met and established applications controls are working. Figure 7.9 lists control objectives and audit procedures for such a review. Ideally, the review should be performed by another member of the audit staff to provide an independent assessment of the new application.

Control Objectives and Audit Procedures for a Review of the Application Testing Phase

Objective 7.8.1.   There should be a detailed test plan outlining the application testing steps and the responsibilities for performing the testing.

> **Procedure 7.8.1.1.**   Determine that an adequate test plan has been drafted outlining test phases, objectives, and responsibilities.

> **Procedure 7.8.1.2.**   Determine that one individual or function is responsible for managing the testing effort and monitoring actual progress against testing plans.

> **Procedure 7.8.1.3.**   Interview key application users to determine their understanding of the test plan and their responsibilities to participate in the testing process.

Objective 7.8.2.   The new application should conform with the original design specifications and should satisfy user requirements.

> **Procedure 7.8.2.1.**   Review application input and output documents as well as processing procedures to determine that the application conforms with original specifications.

> **Procedure 7.8.2.2.**   Interview key application users to determine that the application meets current requirements.

> **Procedure 7.8.2.3.**   If there are significant variances between the application version now being tested and original requirements, assess whether these are significant enough to postpone implementation.

> **Procedure 7.8.2.4.**   Determine that any minor application requirements exceptions have been documented and that there is a formal information systems commitment to implement them at an agreed upon later date.

Objective 7.8.3.   Controls as well as all other major components of the application should be working prior to implementation.

> **Procedure 7.8.3.1.**   Review the input, output, and balancing controls over several successive cycles of test processing to determine that they are working adequately.

Figure 7.8. Control objectives and audit procedures for a review of the application testing phase

Procedure 7.8.3.2. Manually balance and reconcile control totals to determine that they are correct.

Procedure 7.8.3.3. Interview key users to determine that they understand the application controls and balancing procedures.

Objective 7.8.4. Testing should include the use of invalid transactions and unusual conditions as well as normal, expected transactions.

Procedure 7.8.4.1. Working with users and application developers, develop a grid of possible, correct and incorrect transaction conditions and determine that these conditions have been tested.

Procedure 7.8.4.2. Review transaction error reports or logs to determine that all error transactions are being corrected properly.

Procedure 7.8.4.3. Review procedures for correcting rejected transactions and determine that the application is handling them properly.

Procedure 7.8.4.4. Determine that the testing includes unusual conditions as well as year end processing cycles.

Objective 7.8.5. If the application is expected to process high volumes of transactions or other data, testing should attempt to simulate those application volumes.

Procedure 7.8.5.1. Determine that application testing procedures give adequate consideration to expected transaction volumes.

Procedure 7.8.5.2. Review capacity management procedures with information systems personnel to determine that the application will be able to accommodate expected volumes.

Objective 7.8.6. Users should receive adequate training covering all significant application functions.

Procedure 7.8.6.1. Review training plans to determine

Figure 7.8. Control objectives and audit procedures for a review of the application testing phase *(continued)*

whether such training is comprehensive and covers all affected users at local and remote locations.

Procedure 7.8.6.2 Observe several training sessions to determine that the sessions are complete and comprehensive.

Procedure 7.8.6.3. Interview selected users to determine that they understand the application after their training.

Objective 7.8.7. Technical and user documentation should be complete prior to application implementation.

Procedure 7.8.7.1. Review technical documentation covering the application to determine that it follows information systems departmental standards and that it includes:

Process flow charts covering major application functions
Program source code listings of all application programs
Detailed descriptions of critical computations or processes
System flow charts describing files and other program requirements

Procedure 7.8.7.2. Review user documentation and determine that it describes the application adequately including the proper processing of error transactions.

Procedure 7.8.7.3. If the documentation includes on-line "help" screens, determine that they are comprehensive and cover all major application functions.

Procedure 7.8.7.4. Interview various classes of application users to determine that they understand application functions and that they are comfortable with application user documentation.

Procedure 7.8.7.5. Test the application documentation by posing several error or exception situations and determining that the documentation adequately describes these error situations.

Objective 7.8.8. If the new application involves a conversion from an existing system, parallel testing should be performed, where practicable, and all conversion linkages should be tested.

Procedure 7.8.8.1. Determine whether parallel testing should be performed prior to application implementation.

Figure 7.8. Control objectives and audit procedures for a review of the application testing phase *(continued)*

Procedure 7.8.8.2. Review any parallel testing plans and determine whether they appear adequate given the criticality, functionality, and requirements of the new and old applications.

Procedure 7.8.8.3. Review the results of a parallel test and determine that the old and new applications provide similar results within the constraints of the new application requirements.

Objective 7.8.9. Control procedures should be in place to ensure that any manual or automated files and data are converted properly to the new application.

Procedure 7.8.9.1. If automated procedures are used for converting existing files, review conversion program controls to determine that the process is taking place properly and that all data is being converted to the new application.

Procedure 7.8.9.2. If applicable, review procedures to convert manual data and determine whether the procedures appear adequate.

Procedure 7.8.9.3. Review data conversion processes and determine that there are adequate controls over this process.

Objective 7.8.10. There should be a formal applications acceptance and approval step prior to actual implementation.

Procedure 7.8.10.1. Review user acceptance procedures and determine that requesting users formally have approved the new application.

Procedure 7.8.10.2. Review procedures for turning the new application over to computer operations and determine that all documentation is complete and that the application has been converted properly to a production application.

Procedure 7.8.10.3. Review pre-implementation workpapers and determine that all outstanding material control concerns have been addressed and the application is ready for implementation.

Figure 7.8. Control objectives and audit procedures for a review of the application testing phase *(continued)*

**Control Objectives and Audit Procedures for a
Review of the Post-implementation Phase**

Objective 7.9.1. The application should operate in accordance with the general design objectives and should satisfy user requirements.

> **Procedure 7.9.1.1.** Review original general design documentation as well as subsequent correspondence to determine that the application meets original objectives.

> **Procedure 7.9.1.2.** If the new application is not meeting original objectives in any material way, determine that plans are in place to make the necessary modifications.

> **Procedure 7.9.1.3.** Interview key users to determine that they are satisfied with the new application.

> **Procedure 7.9.1.4.** Review any program change requests or other documentation concerning the application to determine whether it is operating as originally planned.

Objective 7.9.2. Controls over the new application should be operating as intended and should be adequate.

> **Procedure 7.9.2.1.** Review controls over application inputs and outputs to assess their adequacy.

> **Procedure. 7.9.2.2.** Interview key users to determine that they understand application controls and that the controls appear to be working as intended.

> **Procedure 7.9.2.3.** Review error logs and other reports covering rejected transactions to determine that these procedures are working as intended.

> **Procedure 7.9.2.4.** Interview computer operations personnel to determine that the application is working properly and in a well controlled and efficient manner.

Objective 7.9.3. Application users should be realizing cost/benefit estimates as outlined in the initial project request.

> **Procedure 7.9.3.1.** Review original cost/benefit estimates for the application and determine whether planned benefits are being achieved.

Figure 7.9. Control objectives and audit procedures for a review of the post-implementation phase

Procedure 7.9.3.2. If there are significant negative variances from the original cost savings, determine whether these have been recognized and whether steps are planned to take corrective action.

Procedure 7.9.3.3. Interview the original requesting users to determine whether the application is achieving initial estimated savings.

Objective 7.9.4. The application should be operating consistent with standards established for full production applications by computer operations.

Procedure 7.9.4.1. Determine that the application is being operated according to computer operations production standards.

Procedure 7.9.4.2. Review problem logs or other operations documentation to determine whether there is any ongoing pattern of difficulties with the new application.

Procedure 7.9.4.3. Interview operations and control personnel to ascertain that the application is operating correctly.

Objective 7.9.5. Embedded audit procedures or other audit software processes should be functioning in a cost-effective manner.

Procedure 7.9.5.1. Determine that any embedded audit procedures are working as intended.

Procedure 7.9.5.2. If appropriate, test embedded or other computer assisted audit procedures to determine that they are working and are achieving required objectives.

Procedure 7.9.5.3. Determine that all audit documentation covering the application review is complete and in order.

Figure 7.9. Control objectives and audit procedures for a review of the post-implementation phase *(continued)*

NEW APPLICATIONS DEVELOPMENT APPROACHES AND THE AUDITOR

Earlier, we discussed an approach to reviewing new applications under development following the data processing department's established SDM and a classic applications development life cycle approach. However, new methodologies and technologies are finding their way into the applications development process which allow applications to be designed and developed faster and to be better structured to meet user needs.

This section will discuss briefly two new development methodologies, prototyping and automated development tools. Another key new method, the use of fourth generation languages, will be discussed in Chapter 12.

The new applications development methodologies allow developers to design an application faster, but they do not change the auditor's basic concerns for reviewing new applications under development. No matter what methodology is used, the auditor should be concerned that the application is designed with adequate controls, that it has been defined and justified properly, that the project development process is being managed properly, and that the system is documented properly.

Prototyping

The traditional SDM process described previously works best when users know exactly what they want in a new application and when requirements can be interpreted easily into a new information system by the development staff. All too often, however, users do not define requirements sufficiently, and it becomes necessary to make changes to the applications definition during the course of the design. In some instances, these changes may be caused by the auditor's pre-implementation recommendations.

Systems development functions, using tools such as COBOL, often have problems reacting to ongoing change requests from users, due to the inflexibilities of the development COBOL tools. Thus, design definitions frequently are frozen at some point during the development process. Then, when the application is completed, it does not meet user needs. This situation creates an SDM development cycle that looks like Figure 7.10.

Prototyping is a technique to help avoid these problems caused by improper user definition. The developer gives the user a test or tentative version of the new application for review shortly after the request is made for the application. The user then reviews that test version and suggests changes, and the developer makes modifications to define the design of the application. This may result in

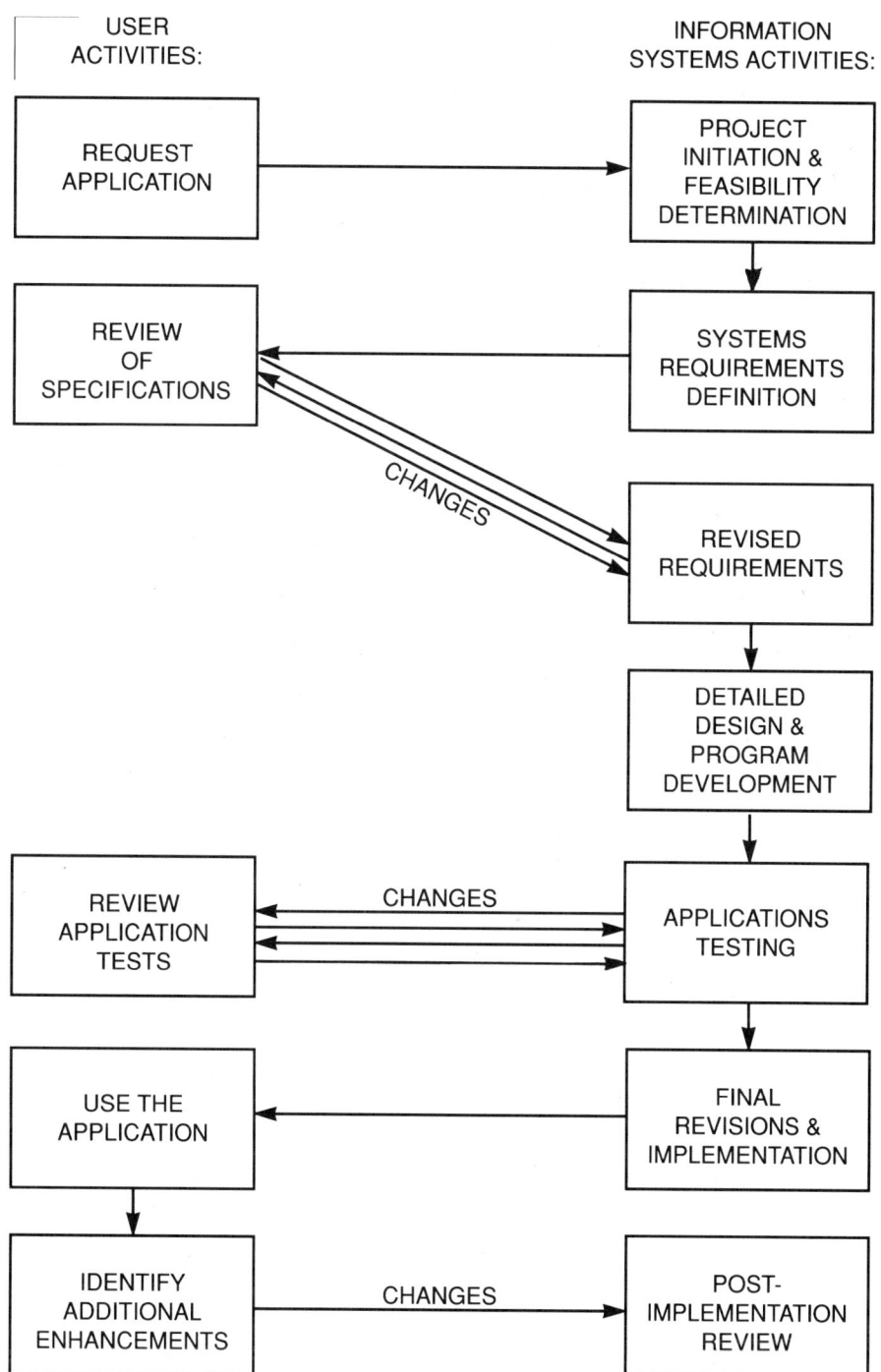

USER
ACTIVITIES:

INFORMATION
SYSTEMS ACTIVITIES:

REQUEST
APPLICATION

PROJECT
INITIATION &
FEASIBILITY
DETERMINATION

REVIEW
OF
SPECIFICATIONS

SYSTEMS
REQUIREMENTS
DEFINITION

CHANGES

REVISED
REQUIREMENTS

DETAILED
DESIGN &
PROGRAM
DEVELOPMENT

REVIEW
APPLICATION
TESTS

CHANGES

APPLICATIONS
TESTING

USE THE
APPLICATION

FINAL
REVISIONS &
IMPLEMENTATION

IDENTIFY
ADDITIONAL
ENHANCEMENTS

CHANGES

POST-
IMPLEMENTATION
REVIEW

Figure 7.10. A development cycle experiencing definition problems

another prototype, more changes, and still other prototypes until the final application is developed. This creates an SDM process as shown in Figure 7.11.

Many applications development groups have found that prototyping is an effective tool for producing new applications which better meet user requirements. Rather than waiting through a long development cycle, users can see a tentative version of the application, suggest changes to the requirements definition, and eventually receive applications which better meet their needs.

Because changes can occur rapidly during a prototype development cycle, auditors reviewing such projects must be more involved in the development process. It does not work, for example, for the auditor to review the detailed applications design specifications and then return later to review the results of applications testing. Because of prototyping, many changes may have taken place.

In addition to the rapid and changing development process associated with prototyping, the auditor faces other issues which should be considered when reviewing such an application under development, including:

Temporary versus final solutions. The various interim versions of an application designed through prototyping are temporary solutions to an applications project designed to show users how the ultimate system will operate. However, often users accept a temporary solution as the final system. Auditors should verify that all prototyping temporary solutions contain all the controls expected in the final version.

Defining needs too narrowly. It is easy for applications developed under prototyping to solve limited, departmental needs but miss major business problems. For example, a given function in an organization may need a certain type of information system which the auditor knows is needed also in other organizational units. In the rush to deliver a prototype to the requesting department, systems developers may miss the larger business problem.

Controls may be lacking. It is easy for applications developers working with prototyping techniques to give less attention to system controls than under more traditional development procedures. Prototype developers and users, for example, might not give much consideration to access controls when working with various versions of a proposed retrieval screen. Auditors reviewing such prototype development efforts must give particular attention to these control concerns.

The move to full production is difficult. Systems development and operations groups often have a difficult time bringing an application development under prototyping into a full production status. This move requires that the

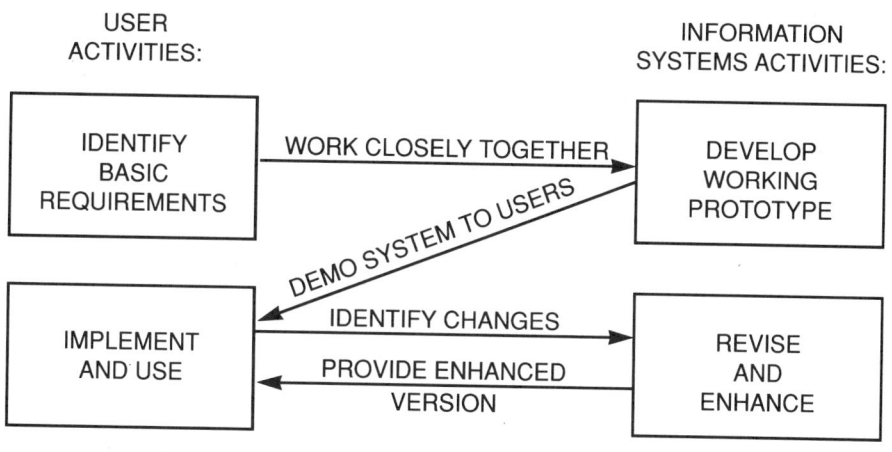

Figure 7.11. A development cycle using prototyping techniques

unstructured development be documented and defined properly. Often, the user who worked with and helped to define the prototype ends up running it in a production status as an end user.

Because of these factors, the auditor needs to treat differently pre-implementation reviews of applications developed through prototyping. Figure 7.12 lists objectives and procedures for auditing new applications under development when prototyping techniques are used. While the auditor's concerns during the requirements definition and post-implementation plans essentially remain unchanged, prototyping does change the auditor's approach when reviewing the detailed design and applications testing phase of the development process.

Automated Development Tools

A large number of "automated development tools" have been introduced to help data processing programmers and analysts develop better applications faster. There are many such products available today with newer, more powerful ones being introduced regularly.

Automated development or CASE tools generally have some or all of the following attributes:

**Control Objectives and Audit Procedures for
Reviewing Applications Developed with Prototyping Techniques**

Objective 7.12.1. Prototyping tools should be used for appropriate application projects where traditional designs are either ineffective or not cost effective.

> **Procedure 7.12.1.1.** Review the general requirements of the application being considered for prototyping development and determine whether that approach appears appropriate rather than developing the application from a set of formal SDM specifications. Factors to consider which favor prototyping are:
>
> > Users lack a complete understanding of requirements
> > A non-static model is needed for ongoing changes to the application requirements
> > There are communications gaps between users and the systems developers
>
> **Procedure 7.12.1.2.** Determine that users understand the capabilities and limitations of the prototyping tool being used. Such tools are usually one of the following types:
>
> > Tools that produce screen formats or images only
> > Tools that produce screens which have some data editing capabilities
> > Tools that can create a basic working model which can create, update, and delete records, and produce output reports
> > Tools that create miniature versions of the full function application
>
> **Procedure 7.12.1.3.** Interview key users to determine that they understand that the resultant prototype application may have to be rewritten into a full production application.
>
> **Procedure 7.12.1.4.** Develop a hands-on familiarity with the prototyping software product being used to assess its capabilities for developing controlled applications as well its potential use as an audit tool.

Objective 7.12.2. There should be a formal Prototype Development

Figure 7.12. Control objectives and audit procedures for reviewing applications developed with prototyping techniques

Life Cycle (PDLC) which defines the sequence of activities necessary to verify the application requirements and implement preliminary and final prototype models.

Procedure 7.12.2.1. Review PDLC procedures with information systems personnel and determine that they include the steps necessary to define and build prototype models.

Procedure 7.12.2.2. Assess whether PDLC procedures effectively link into the organization's overall systems development methodology procedures such that prototype applications can be converted to and maintained in full production status.

Procedure 7.12.2.3. Select several applications implemented over a recent period and determine whether they were built following the PDLC and are implemented following overall production standards.

Objective 7.12.3. There should be a formal procedure, involving key users and developers operating as a team, to define the requirements of applications under development using prototyping.

Procedure 7.12.3.1. Review the structure of the prototype implementation team and determine that all key users are represented.

Procedure 7.12.3.2. Review preliminary requirements for the application and determine whether the proposed prototype solution addresses only a local solution or a more total, organization-wide solution to the problem.

Procedure 7.12.3.3. Attend selected prototype planning meetings and determine that users are explaining requirements properly, and that prototyping analysts are developing a proper conceptual solution which meets requirements and has adequate controls.

Objective 7.12.4. The application prototype process should result in a series of demonstration models which allow users to further define needs in order to develop the final model.

Procedure 7.12.4.1. Attend selected prototype review meetings

Figure 7.12. Control objectives and audit procedures for reviewing applications developed with prototyping techniques *(continued)*

to determine that the prototype developers are modifying the model to meet any revised requirements.

Procedure 7.12.4.2.   Determine that users of the prototype model are giving adequate attention to learning to operate the prototype, to document their suggested refinements, and to prioritize requested changes.

Procedure 7.12.4.3.   Review data control and other prototype procedures to assess whether controls will be adequate.

Procedure 7.12.4.4.   Determine whether the results of prototype meetings, suggested changes, and test results are documented adequately.

Procedure 7.12.4.5.   Determine that an adequate implementation plan has been developed and is being followed for the overall prototype development effort including expected completion dates.

Procedure 7.12.4.6.   Determine that procedures exist to freeze the prototype development process when the application meets general user requirements such that prototyping efforts do not restrict the resources required for other development efforts.

Objective 7.12.5.   The completed application developed under prototyping should meet user requirements, be responsive to future changes, and should have adequate controls.

Procedure 7.12.5.1.   Assess whether there has been adequate testing over the prototype developed application to provide assurance that it will operate in production correctly.

Procedure 7.12.5.2.   Review the conversion plan for the application and determine whether it appears adequate for bringing the application into production.

Procedure 7.12.5.3.   Determine that adequate user and systems documentation has been prepared for the application prior to bringing it into production.

Procedure 7.12.5.4.   Assess whether any computer assisted audit techniques would be appropriate for testing the application and document recommendations.

Figure 7.12. Control objectives and audit procedures for reviewing applications developed with prototyping techniques *(continued)*

Workstation based. A basic concept of automated applications development is that the designer no longer needs paper and pencil to perform systems design tasks. While programming code has been created through on-line terminals for some time, the applications design process has not been as automated. Designers used flowchart templates and descriptive documentation to define their designs. The new CASE tools move much of this work to a microcomputer workstation or a terminal.

Structured design techniques. Structured design has been used since the early 1980s. It is a procedure for defining applications through a series of data or process flow diagrams, showing various relationships, from top level down to detailed design charts. CASE automated development workstations contain the graphics tools to produce such charts and make changes to them. An example of such a structured design chart is shown in Figure 7.13.

Data entity capabilities. In addition to diagramming capabilities, CASE automated development workstations often have a facility that allows the developer to define the data entity relationships that eventually will be used in the application's data dictionary. Changes here will translate into changes in the automated structure diagrams.

Automatic code generators. Many but not all CASE tools contain the facility to generate COBOL code from the automated, structured design. While this probably will not eliminate traditional programmers in the short run, it will greatly relieve them from such repetitive tasks as writing data retrieval programs. While the use of these generators is efficient, the code they generate often is almost unreadable.

While not all data processing organizations today use CASE procedures, the auditor will see their use increase in the near future. Such tools give the application developer the ability to rapidly alter application structures during the design phase. While control techniques have not yet become well established due to the variety of CASE approaches now being attempted, Figure 7.14 contains objectives and procedures that the auditor should consider when reviewing a new application which is being developed through CASE techniques.

PRE-IMPLEMENTATION REVIEWS OF PURCHASED SOFTWARE

Purchased software applications are becoming common in the modern data processing organization. For many routine applications, it is not cost effective to develop an application in house when it can be purchased as a package. For

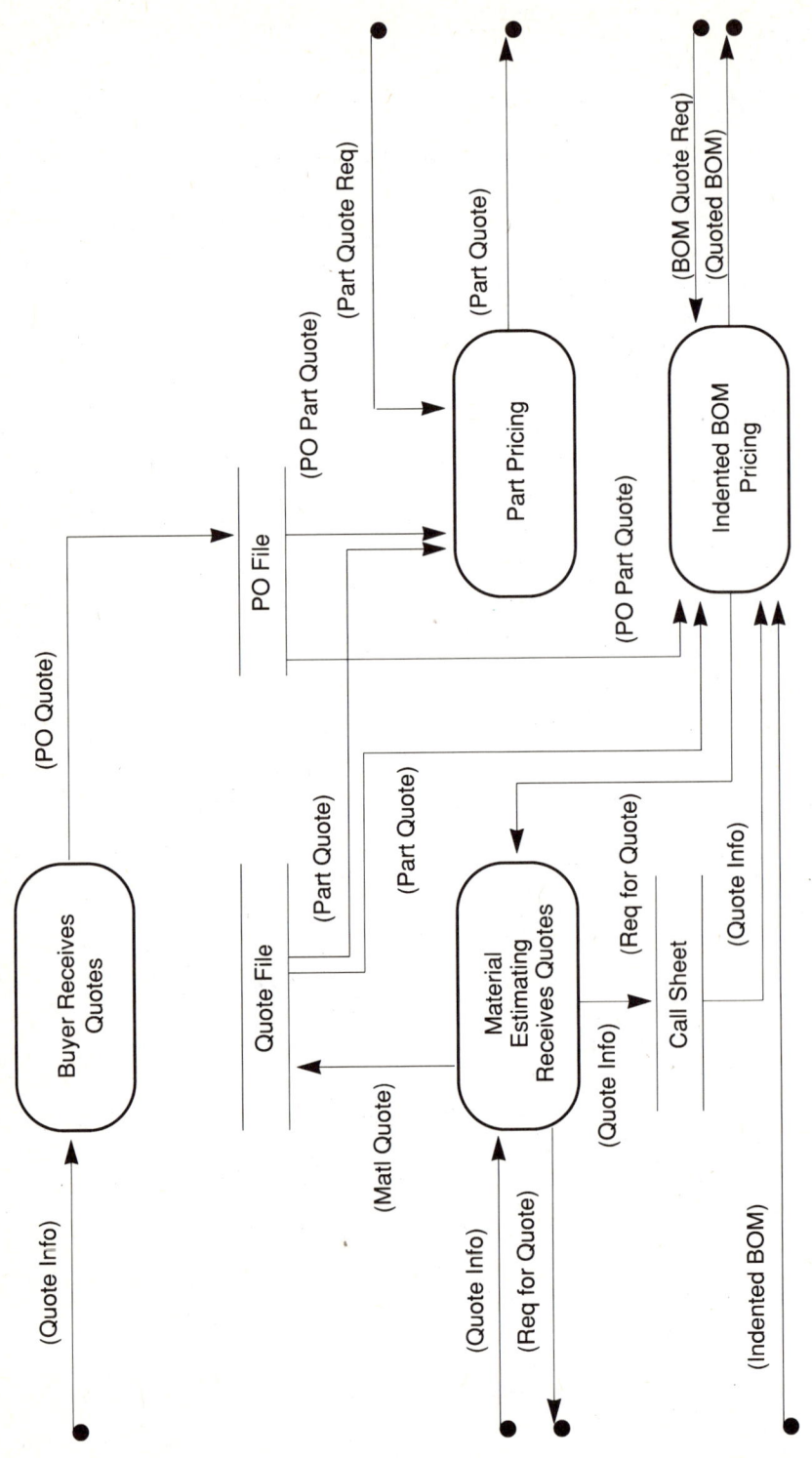

Figure. 7.13.   An example of a structured design chart

338

Control Objectives and Audit Procedures for Reviewing Applications Developed with CASE Techniques

Objective 7.14.1. System developers, users, and auditors should have a good general understanding of the CASE tool in use including its documentation features and linkages to program code generators.

Procedure 7.14.1.1. If the CASE tool is being introduced to the organization for the first time, determine through interviews and discussions that all involved personnel have a good understanding of the product including its strengths and limitations.

Procedure 7.14.1.2. Determine that the users of the application under development understand how CASE procedures will change classic SDM techniques.

Procedure 7.14.1.3. Review the features of the CASE tool being used and determine that systems developers are using its potentially powerful features, such as the ability, through graphical approaches, to explode and implode diagrams and dictionary specifications.

Procedure 7.14.1.4. Determine that the information systems organization has developed standards for performing business analysis and for using and documenting new applications with CASE tools.

Procedure 7.14.1.5. If a program code generator is part of the CASE tool, determine that the program code produced is compatible with existing programmed applications as well as overall programming standards.

Objective 7.14.2. The CASE tools should be appropriate for the application under development and should include appropriate documentation and change procedures.

Procedure 7.14.2.1. Discuss with management the decision to use CASE tools for the application under development and determine whether the tool appears appropriate for the project.

Procedure 7.14.2.2. If the application is a major revision or

Figure 7.14. Control objectives and audit procedures for reviewing applications developed with CASE techniques

rewrite of an existing application, review procedures used to document existing program and data flows into the new CASE tool.

Procedure 7.14.2.3. If appropriate, determine that the CASE tool was used first for business function analysis, then to document processes and data flow procedures, and finally to document program modules.

Procedure 7.14.2.4. Determine that procedures exist to control and document changes to the CASE design.

Procedure 7.14.2.5. If CASE documentation is limited to work station graphical formats rather than paper, determine that users and auditors have access to and understand the use of CASE work station terminals.

Objective 7.14.3. Users of the application under development should have an understanding of the diagramming technique used to define and develop the application.

Procedure 7.14.3.1. Review procedures to present CASE documented procedures and processes to users and determine that users have a proper understanding of these methods.

Procedure 7.14.3.2. If key users do not appear to have an understanding of the CASE developed function and data flow diagrams, determine that steps are being taken to train users in the use of the CASE tool.

Procedure 7.14.3.3. Review the prototyping capabilities of the CASE development tool and determine that it is being used and is understood by key users.

Objective 7.14.4. Data flow diagrams and related CASE documentation should emphasize data description consistency and should document application control points.

Procedure 7.14.4.1. Review the CASE tool and determine that it has the capability to analyze design documentation entered by analysts and to highlight structured methodology inconsistencics.

Procedure 7.14.4.2. Review the CASE tool and determine that

Figure 7.14. Control objectives and audit procedures for reviewing applications developed with CASE techniques *(continued)*

it has the capability to produce properly general design, operations, and end user documentation.

Procedure 7.14.4.3. If there appear to be deficiencies or inconsistencies with the installed CASE tool, review flow diagrams and related documentation to identify potential design inconsistencies.

Objective 7.14.5. The CASE designed project should include adequate levels of quality assurance and user walk through reviews.

Procedure 7.14.5.1. Determine that information systems developers make proper use of application walk through meetings to describe application functions to users.

Procedure 7.14.5.2. If the CASE designed application under review is a reverse engineering or redesign of an existing application, determine that key users have been involved in validating the business rules contained in the older application.

Procedure 7.14.5.3. Determine that a formal team has been assembled to perform quality assurance reviews and that the team has established clear objectives including:

> Reviews of CASE design completeness and clarity
> Reviews of CASE generated program code standards as well as the correctness of interfaces with other applications
> Reviews of systems test results and any exceptions to initial test plans

Procedure 7.14.5.4. Attend selected user walk through and quality assurance meetings covering the CASE designed application and determine that adequate attention is being given to application controls.

Figure 7.14. Control objectives and audit procedures for reviewing applications developed with CASE techniques *(continued)*

example, most payroll applications do not require custom programming and many purchased packages are available that can effectivly perform the function.

Similarly, some applications are difficult to develop and can be purchased quite inexpensively. A microcomputer spreadsheet application is an example. No one would consider developing such a package when a powerful one can be purchased for less than $500!

Auditors should not ignore purchased software packages when considering candidates for pre-implementation review. In many instances, there can be as significant a risk and exposure with such packages as would be found with in-house developed applications. The requirements for the purchase can be poorly defined, the developers may add changes to the package which do not follow good control procedures, or the conversion and testing may be inadequate. In addition, just because an outside vendor offers a software package for sale is no reason to assume that controls in the package are good.

Figure 7.15 lists objectives and procedures for performing pre-implementation reviews of purchased software packages. Purchased software packages play an important role in the modern data processing organization. The effective auditor should participate in the pre-implementation reviews of these software products.

PRE-IMPLEMENTATION AUDIT REPORTS

Many internal audit departments have a formal procedure for issuing audit reports. Draft reports are prepared, auditees prepare responses after discussion and negotiation on the draft, and a final audit report is issued with copies to various levels of management. This format often is inappropriate for reviews of new applications under development.

An individual controls problem with a particular pre-implementation program or output report which is identified by an auditor, can be corrected by the applications developer almost at once. There is little need to discuss such a finding in a formal audit report. However, audit management should understand the use of this type of reporting format for pre-implementation reviews.

Auditors should consider issuing informal memo reports after each phase of the pre-implementation review. These memo reports discuss the scope of review activities and document any audit concerns. If some of the prior concerns have been corrected, the actions taken and current status of controls issues are discussed. Of course, the auditor should develop a workpaper binder

Control Objectives and Audit Procedures for Reviewing Purchased Software Applications

Objective 7.15.1.   There should be a formal requirements analysis and feasibility study for the purchase of a software package.

> **Procedure 7.15.1.1.**   Determine that the request for the new application includes an initial requirements definition, even though a purchased application is contemplated, and apply the control objectives and audit procedures set out in Figures 7.5 and 7.6 of this chapter.
>
> **Procedure 7.15.1.2.**   Review the formal make versus buy analysis for the application purchase decision and determine whether due care was exercised in this analysis process. Factors which would favor a software purchase include:
>
> > An overall management policy that software should always be purchased unless nothing is available
> >
> > An application need for which there are many purchased alternatives available, such as payroll or accounts payable
> >
> > An application area which requires frequent, legally mandated changes, such as a pension application subject to tax law revisions
>
> **Procedure 7.15.1.3.**   Review long range information systems plans and assess whether the purchased application will fit into any overall strategic computer systems plans.

Objective 7.15.2.   All potential vendors of software for the application should be considered to determine those that best meet requirements in a cost effective manner.

> **Procedure 7.15.2.1.**   Determine whether appropriate procedures were followed to identify all potential software vendors. Such procedures include reviews of published reference lists, searches of data processing periodicals, and referrals from hardware vendors.
>
> **Procedure 7.15.2.2.**   If a formal Request for Proposal (RFP)

Figure 7.15.   Control objectives and audit procedures for reviewing purchased software applications

was used to solicit vendor bids, determine that the RFP correctly defines application requirements.

Procedure 7.15.2.3. If Information Systems did not use an RFP for the software selection, interview management to determine reasons for the omission. A valid justification might be a limited number of vendors supplying software in the desired area.

Procedure 7.15.2.4. Review procedures used for preliminary or initial screening of vendors. Appropriate reasons for screening out vendors from consideration include:

Incompatible software or hardware requirements
Critical factors which do not meet requirements
Cost in excess of original make versus buy estimates
Poor vendor experience

Objective 7.15.3. The capabilities of potential application packages and the qualifications of their vendors should be checked prior to entering into contract negotiations.

Procedure 7.15.3.1. Determine that a formal evaluation process is used for the final group of potential packages. This might include an evaluation of:

The age of the product and the number of installed sites
Vendor policy on maintenance and upgrades
Vendor policy on access to source code
The application's control features
The availability of a fourth generation language or other report generator tool
The quality of documentation and vendor training programs
The availability of a user group

Procedure 7.15.3.2. Determine that arrangements have been made for vendor product demonstrations of final candidates with key user participation.

Procedure 7.15.3.3. Interview key users to determine that they understand the final package selections and are satisfied that they meet requirements.

Procedure 7.15.3.4. Review the document outlining the final

Figure 7.15. Control objectives and audit procedures for reviewing purchased software applications *(continued)*

selection recommendations to determine whether it is consistent with the package selection analysis process.

Objective 7.15.4. Potential purchased software packages should be reviewed for the adequacy of control procedures and suitability of operating within the information systems environment.

> **Procedure 7.15.4.1.** Review the initial requirements documentation to determine whether there are any unique control considerations associated with the application.

> **Procedure 7.15.4.2.** Review the final package selection candidates for controls and audit trails; communicate any concerns to information systems management.

> **Procedure 7.15.4.3.** Determine that computer operations and the systems programming function have reviewed the final candidates to assess any unique technical or environmental problems.

Objective 7.15.5. The application software vendor should be able to provide adequate training, documentation, and support.

> **Procedure 7.15.5.1.** Determine that both Information Systems and users have completed an adequate review of documentation and have identified any potential problems or concerns.

> **Procedure 7.15.5.2.** Determine that the software evaluation team has secured the names of several installed sites for the application and has questioned those users through a telephone poll. Review documented polling results.

> **Procedure 7.15.5.3.** Determine that information systems representatives and users visit at least one installed site to observe and determine satisfaction with the application package.

> **Procedure 7.15.5.4.** If timing is appropriate, suggest that a key user attend an application users' group meeting to assess the level of satisfaction and complaints.

Objective 7.15.6. Application software should be leased or purchased through a formal contract which defines the responsibilities and rights of each of the contracting parties.

Figure 7.15. Control objectives and audit procedures for reviewing purchased software applications *(continued)*

Procedure 7.15.6.1. Obtain a copy of the software contract and determine that it properly defines the deliverables, costs, and such key factors as:

Delivery dates

Commitments to maintenance and updates

Policies regarding software modification

Policies regarding transportability of the application package to other computer systems

Codification of any representations or special allowances made by sales and technical representatives

Procedure 7.15.6.2. Determine that the organization's legal department or outside counsel is involved in software contract negotiations.

Procedure 7.15.6.3. Assess the potential long term viability of the contracting vendor and, if appropriate, suggest that the contract contain terms for such matters as holding a version of the source code in escrow.

Objective 7.15.7. There should be a formal project implementation team with the responsibility of moving the new application package into production status.

Procedure 7.15.7.1. Determine that there is a project implementation team that consists of key users as well as information systems personnel.

Procedure 7.15.7.2. Review implementation plans and assess whether they are sufficient for the magnitude of the application.

Procedure 7.15.7.3. Interview members of the project implementation team and ascertain that they are aware of their responsibilities.

Objective 7.15.8. Any local changes introduced to the application software should be added only if necessary and kept to an absolute minimum.

Procedure 7.15.8.1. Determine whether any local changes are scheduled to be introduced to the software package.

Procedure 7.15.8.2. Review contract terms regarding policies

Figure 7.15. Control objectives and audit procedures for reviewing purchased software applications *(continued)*

for local software changes and determine whether any planned changes are in compliance with those terms.

Procedure 7.15.8.3. If significant local changes are included in the purchased application implementation, apply the Control Objectives and Audit Procedures set out in Figures 7.7 and 7.8 of this chapter.

Procedure 7.15.8.4. Determine that informations systems personnel understand their responsibilities in keeping the application current in light of local changes and vendor upgrades.

Objective 7.15.9. There should be a formal program of acceptance testing prior to converting to the new application package.

Procedure 7.15.9.1. Determine that a testing and implementation plan exists and is being followed.

Procedure 7.15.9.2. Determine through reviews of test results and interviews with key users that the application conforms to specifications and satisfies user needs.

Procedure 7.15.9.3. Review application controls to determine that they are adequate and working prior to implementation.

Procedure 7.15.9.4. Determine through interviews that key users are aware of application functions and control procedures.

Objective 7.15.10. Arrangements should be made to receive ongoing help and maintenance in the use of the applications software product.

Procedure 7.15.10.1. Determine that a vendor "help" telephone line exists and that this level of support is sufficient.

Procedure 7.15.10.2. Interview other key users to determine whether they are satisfied with vendor support arrangements, and if not, recommend that Information Systems begin negotiations with the vendor to improve these arrangements.

Figure 7.15. Control objectives and audit procedures for reviewing purchased software applications *(continued)*

covering review activities. This binder documents pre-implementation activities and provides a basis for later applications reviews.

At the conclusion of the review, the auditor can issue a formal audit report following audit department standards. Where appropriate, this report discusses pre-implementation audit findings and corrective actions taken. However, the main function of the final report is to highlight outstanding control issues which still need to be corrected within the new applications system.

AUDITOR SKILL NEEDS FOR PRE-IMPLEMENTATION REVIEWS

For many types of audit reviews, it is of prime importance that the auditor understand computer audit, control, and security concepts. The auditor need not be a skilled data processing specialist. When reviewing new applications under development, however, the auditor does need a higher level of technical skill and a general understanding of the control implications of various new technologies such as prototyping or CASE. In addition, an acquaintance with newer data processing techniques, such as electronic data interchange or artificial intelligence procedures, is needed.

The auditor should have a personal objective of having a strong understanding of the technical concepts used in these new applications. This can be accomplished by attending the same vendor training classes that will be given to data processing personnel. Also, it can be accomplished through outside technical seminars and reading. The greater the auditor's knowledge of these technical areas, the greater the auditor's credibility when dealing with data processing applications development specialists.

PRE-IMPLEMENTATION REVIEWS IN THE MODERN DATA PROCESSING CENTER

Auditors have been performing pre-implementation reviews of data processing applications since the beginning of computer auditing. However, while the concept appears attractive to auditors and applications development management, implementation of this concept often is difficult. If the auditor orients reviews to the organization's SDM procedures and takes the role of an independent observer of controls, these reviews can be effective.

The need for pre-implementation reviews increases with some of the current trends in the modern data processing organization. Many organizations seek out purchased software as a first choice for new applications. All too often, there is a tendency to blame the software vendor for control weaknesses and take minimal corrective action. The auditor can play an effective role in highlighting such deficiencies and influencing corrective actions. This need for pre-implementation reviews also is of particular importance when new applications are being developed through prototyping methods or are being developed by end users.

The auditor in the modern data processing organization should devote a significant portion of review activity time to new applications under development. This results in a more effective audit function as well as better controlled data processing applications.

Section III

SECURITY FOR THE MODERN DATA PROCESSING CENTER

CHAPTER 8

Physical Security Strategies and Risk Evaluation

INTRODUCTION

Auditors have been evaluating and recommending controls for improving computer security since the 1960s when computer systems were first introduced into organizations. These early systems often were placed in glass-walled lobbies for all to see. Few professionals or organizations then considered the vulnerabilities and security exposures associated with business data processing systems. Auditors, however, were among the first professional groups to recognize the importance of computer security controls and procedures. Today, even though formal computer security administration functions have been established in many organizations, auditors still are viewed as the computer security "experts" or advisors.

Computer security concerns have changed much from those early days of glass-enclosed computer rooms with batch-oriented systems. Then the auditor was primarily interested in protecting the expensive computer hardware from damage. Today the auditor is equally concerned with physical security and information security, covering protection of data and programs. This chapter discusses physical security and risk evaluation. Information security will be discussed in Chapter 9.

This chapter explains strategies for installing physical security controls in the modern data processing center. However, before recommending the installation of potentially complex and expensive physical security controls, the auditor should understand the risks associated with *not installing* such controls. Therefore, this chapter also discusses several approaches to computer security risk assessment and suggests an approach that is well suited to the modern data processing center.

Many levels of management today are aware of the vulnerabilities of their data processing systems and the need to install security controls. Often, they are not aware, however, that similar vulnerabilities exist within such areas as telecommunications networks, office automation equipment, and departmental microcomputer systems. Auditors can and should continue to identify and evaluate computer security risks and recommend the installation of effective controls. The auditor's role in evaluating data processing security and recommending improvements should extend beyond the traditional computer center to these other areas of data processing activity.

ASSESSING COMPUTER SYSTEM PHYSICAL SECURITY EXPOSURES

Physical security risks and control procedures often are covered in the auditor's review of general data processing controls. Even before these risks can be considered, however, the auditor should understand potential computer physical security exposures. This chapter discusses physical security exposures while Chapter 9 discusses information security exposures. It is difficult, however, to draw a definite line between physical security and information security in the modern data processing organization. Both chapters will attempt to identify these areas of potential overlap.

When considering physical security exposures in the modern data processing organization, the auditor should consider the entire scope of data processing operations. While they usually include any central computer operations centers, there may be physical security exposures surrounding remote computing devices, telecommunications equipment, office automation networks, and other forms of organization automation. In addition, these computer security exposures should be taken in context with similar exposures within the entire organization. An auditor's concerns about the lack of fire detection and prevention devices in a computer operations function will accomplish little if similar controls do not exist within the rest of the organization's physical facility.

Data processing physical security exposures can be divided into the following categories:

Natural or catastrophic disasters
Power and environmental failures
Communications systems failures
Sabotage, riots, and malicious damage
Unintentional damage

Following are discussions of each of these exposures in the context of the modern data processing organization. Some exposures involve considering "can't happen" situations. By understanding some of the types of exposures which a modern data processing organization may face, the auditor will be in a better position to help evaluate those risks and suggest controls which will provide appropriate safeguards.

The auditor should consider data processing physical security in light of the data processing organization under review. Some exposures which may be appropriate to a larger organization are less applicable to the smaller function. After reviewing the various exposures, the auditor should develop a detailed list of potential physical security exposures which then can be evaluated using one of the risk evaluation approaches discussed in the next section.

Natural or Catastrophic Disasters

Natural disasters refer to floods, earthquakes, or other Acts of God. This is probably the least common type of disaster which a data processing organization faces. Natural disasters also are the most difficult to predict and generally the most devastating when they occur. For example, an organization and its data center may be located on a "hundred year flood plain"—where there is a high probability of a major flood once every hundred years. However, there could be two major floods within a three year period followed by none in hundreds of successive years.

An evaluation of the probable occurrences of such natural disaster risks often indicates a low probability of occurrence. This will be discussed in the section on risk analysis. The auditor should give consideration to the remote probabilities of natural disasters which may affect the data processing facility. Some data can be gathered from recent experiences with similar natural disasters, the local weather bureau, and historical records. The latter has been mentioned because some historical events repeat themselves. For example, weather records and recent minor seismic disturbances predict there will be a significant earthquake in the San Francisco area at some future time. However, one would have to consult history records to ascertain that there was a significant earthquake in the St. Louis, Missouri, area in the early 1800s. There is certainly an exposure in both areas that such events will be repeated.

Should the auditor and management be concerned about such natural disaster exposures? In most instances, there is probably little that can be done to predict or prevent them. Rather than attempt to establish probabilities of such events occurring, the auditor should establish a disaster recovery plan with a backup processing site located in a secure location. Such procedures are discussed in Chapter 10.

Catastrophic disasters refer to fires, environmental failures (such as gas leaks), or other misfortunes where man may play a role. An organization has a degree of control over some types of such catastrophic disasters. The building housing the computer system can be constructed with fire retardant materials and equipped with detection and prevention devices. Although there will be some exposure to such man-made disasters, a proper controls environment can reduce the risks.

There are other types of catastrophic disasters, however, where the data processing organization has little or no control. An area chemical plant or nuclear power plant could have a catastrophic accident that would render the nearby geographic area uninhabitable for an extended period. An accident on a passing train carrying chemicals could cause similar results. Although caused by man, a given organization has little control or power to prevent such catastrophic disasters. Nevertheless, the probability of occurrence is probably higher than the "hundred year" floods discussed above.

Should the auditor and management be concerned about catastrophic disasters? There certainly should be a major concern where management has some control. It is quite easy to prevent and detect fires within the facility. It is more difficult to prevent a major fire which starts in the building next door. It is also difficult to prevent a chemical disaster caused by a passing railroad car.

The auditor helping management identify physical security exposures and develop controls appropriate to those risks should recognize that there is a small probability that natural or catastrophic disasters could destroy or disable the data processing facility, and even the entire organization. There is a physical security exposure here, but other than a good disaster recovery plan, there are few controls that can be installed to limit the exposure.

Power and Environmental Failures

The risk of damage or destruction to electrical power systems or air conditioning systems can be a major exposure to the larger, modern data processing function. Some larger computer systems require water chiller systems to cool the equipment. Most larger systems require conditioned electrical power and have strict air conditioning requirements. These facility requirements represent a physical security exposure to the data processing organization.

Power and environmental failures represent less exposure to organizations with mini or microcomputers. These smaller systems generally do not require specialized air conditioning or electrical power sources. However, there is an obvious need for normal power for the data processing equipment.

The auditor evaluating physical security exposures should survey the power and environmental exposures within the data processing organization. This requires an understanding of the environmental requirements of the computer hardware as well as the current status of backup sources of supply. Figure 8.1 provides a checklist to help the auditor survey such power and environmental requirements. This information should be used to help the auditor identify vulnerabilities in this area of the data processing resource. Once vulnerabilities have been identified, the auditor should prepare a detailed list of probable physical security exposures for the function. This information then can be used for the risk evaluation process and to help identify areas where security can be improved.

Communications Systems Failures

The typical modern data processing organization is dependent upon its communications network. This may be either a telecommunications network with lines connecting various remote facilities throughout the organization's operations or it might be a local network connecting remote terminals within the facility to a central computer center. In addition, modern organizations increasingly have local area networks linking various office automation and microcomputer devices.

The external telecommunications network generally is supplied by an outside telecommunications organization that provides common carrier lines, leased lines, or specialized communications links. While the ability of such a telecommunications carrier to continue in business represents a type of physical data security exposure, the risks typically are fairly low. If an outside

**Security Exposures Checklist for Evaluating
Power and Environmental Failure Risks**

*NOTE: This checklist is designed to help identify potential electrical
power, air conditioning, or related environmental risks within computer
operations. Questions are designed so that a "NO" answer represents
a potential environmental security exposure.*

POTENTIAL SECURITY EXPOSURE	YES	NO	N/A
1. Are computer room power sources and air conditioning facilities independent of other facilities in the building?			
2. Are exterior power transformers and air conditioning units shielded from direct exposure to outsiders?			
3. Are all power lines protected by adequate circuit breakers?			
4. Are master power switches clearly marked and located near each computer room door?			
5. Is there a power conditioner installed to shield against electrical spikes?			
6. Is there an uninterruptible power supply or an emergency power system?			
7. Is the uninterruptible power supply system tested on a regular basis?			
8. Is the computer room equipped with emergency, battery operated lights for use during power failures?			
9. Are any electric door locking devices also tied to the emergency power system?			
10. Are there emergency light and power systems, as required, in any nearby tape vaults or forms storage rooms?			
11. Is the computer room equipped with thermostats and monitors to regulate temperature and humidity?			

Figure 8.1. Security exposures checklist for evaluating power and environmental failure risks

	YES	NO	N/A

12. Are there audible alarms which sound when temperature and humidity exceed defined limits?

13. Are chiller units located to minimize water damage if there is a plumbing failure?

14. Is the computer room located to minimize the danger of overhead leaks from pipes or the roof?

15. Is the floor equipped with drains to avoid water accumulation in the event of a flood or fire?

16. Are water level detectors installed beneath the raised floor?

17. Is smoke and fire detection equipment approved by a recognized authority?

18. Are supplemental heat detectors located under the false floor?

19. Is the computer room equipped with a zone fire control system?

20. Does the fire control system both activate local audible alarms and automatically notify a nearby fire department?

21. Is the duct system designed to exhaust smoke and combustion products to the outside?

22. Are portable fire extinguishers located at every door and at other key points in the computer room?

23. Are ceiling tiles and wall surfaces constructed with a flame resistant material?

24. Are fluorescent light fixtures designed to minimize the danger of melting or dripping?

25. Is the computer room inspected on a periodic basis by competent fire inspection authorities or insurance specialists?

Figure 8.1. Security exposures checklist for evaluating power and environmental failure risks *(continued)*

telecommunications organization made the investment in lines and switching gear, some other organization usually takes over the telecommunications company in the event of an economic failure. There is, of course, a significant risk if the carrier does not have adequate backup facilities. The primary exposure with an outside carrier, however, is in information security of data transmitted over the lines. This is discussed in Chapter 9.

Many of the communications physical security exposures facing an organization exist within that organization rather than over remote lines. These include vulnerability of the networks and switching equipment to failure as well as to malicious damage. Many physical security exposures associated with communications networks are included in the checklist in Figure 8.2 along with other risk assessment considerations discussed in the following section on damage from sabotage, riots, and malicious acts.

The most basic security exposure that data processing organizations typically face regarding communications networks is the potential inability to quickly reconstruct networks following a disaster. Many organizations built internal and external networks on an ad hoc basis with little documentation. It is necessary to have a good understanding of the existing network to reconstruct all or a portion of it on a short term basis.

As part of a review of communications systems vulnerabilities, the auditor should prepare a list of potential exposures. This list should be classified by the various components of the telecommunications network which could fail. These exposures should include just those that are unique to telecommunications and should not be a duplication of other data processing exposures, such as natural disasters, which have been considered separately.

Sabotage, Riots, and Malicious Damage

Data processing management first limited access to computer rooms and placed them in unobtrusive locations due to a concern over potential malicious damage. Auditors expressed concerns and management began to realize their computer systems were vulnerable. A malicious intruder could pull one or a handful of wires from an older mainframe's back panel to render the machine useless. A locked, limited access computer room was necessary to limit this physical security risk.

During the 1970s, many computer centers in the United States and Western Europe became targets of political insurrection, vandalism, and riots. Some terrorists decided that certain computer systems were symbolic of whatever it was they opposed and made them targets for destruction. Although this type of activity has subsided in recent years, circumstances can cause it to increase once again.

Data processing management reacted to such threats of malicious damage or sabotage by establishing physical controls surrounding the larger system computer facility. These controls ranged from tight access restrictions over entry to the computer operations center to placing that computer center in a separate, remote, "blockhouse" facility. Many smaller, minicomputer systems that do not process critical data often are located in secure facilities because of similar management security concerns.

While good access controls have eliminated or reduced security exposures in many larger computer centers, little has been done to reduce the physical security exposures surrounding microcomputers and other office automation devices. The concern surrounding these devices often is less one of malicious damage than of the potential theft of equipment. For example, users with microcomputers in many larger organizations have found boards and other components taken from their machines. This can be a significant exposure as many personnel with the organization often own their own machines and, thus, may be motivated to take these microcomputer boards and other components.

When reviewing physical security exposures for a minicomputer environment or for an information center facility, the auditor should consider these potential risks in that context. There may be a security exposure due to poor access controls over an information center facility. However, it would limit the effectiveness of that information center if tight access controls were to be installed and users not encouraged to use the facility. There should be a trade-off between the importance of having a user friendly environment and the potential of physical security exposures.

Many factors go into the assessment of physical security exposures in a modern data processing organization. Figure 8.2 is a checklist to help identify potential physical security exposures. As discussed earlier, these potential exposures do not mean that a control should be installed. Rather, a risk evaluation should be performed and, if management is not willing to accept such a potential risk, the control for that risk should then be installed.

Unintentional Damage

The auditor assessing physical security exposures also should consider exposures caused by human error or unintentional damage. While they are difficult to identify directly, day to day control and operational practices will indicate potential unintentional damage exposures. For example, unless there

Security Exposures Checklist for Evaluating Sabotage, Riots, and Malicious Damage Risks

POTENTIAL SECURITY EXPOSURE	YES	NO	N/A
1. Is the computer facility located away from high traffic areas in the organization?			
2. Is the computer facility located away from the building's outside walls and without windows?			
3. Are fences and other access controls installed to shield any outside power transformers or air conditioning units?			
4. Are interior walls surrounding the computer room constructed of a heavy material to avoid possible break-ins?			
5. Are secure facilities, with locked doors, used for tape and forms storage rooms?			
6. Is traffic to the machine room restricted to authorized individuals and then through a controlled entrance door?			
7. Are authorized employees required to wear visible photograph identification badges?			
8. Are data processing personnel trained to challenge and report any improperly identified visitors?			
9. Are access controls, such as a cypher lock or a badge reader, used to control computer room access?			
10. Are cypher combinations or badge controls changed on a periodic basis?			
11. Are other doors to the computer room facility used for emergency exits only and equipped with strike locks and audible alarms if opened?			
12. Are there special access control procedures in place for maintenance, customer engineering, and other non-information systems personnel including:			

Figure 8.2. Security exposures checklist for evaluating sabotage, riots, and malicious damage risks

	YES	NO	N/A
Visitor badges?			
Registration logs?			
Policies to escort all visitors?			
13. Are there procedures to limit and control parcels and tapes leaving and entering the computer room?			
14. Are all employees required to sign an agreement that specifies their role in the organization and acknowledges the ownership of data processing equipment, programs, and data?			
15. Is there an individual in the information systems organization assigned overall responsibility for physical security controls?			

Figure 8.2. Security exposures checklist for evaluating sabotage, riots, and malicious damage risks *(continued)*

is a rule prohibiting smoking, eating, and beverage drinking around data processing equipment, there is an exposure that the equipment may be damaged due to smoke or spilled food. Similarly, an otherwise competent employee may allow an unauthorized person to enter a computer operation without checking identification, a failure that may cause physical damage or compromise information security.

Figure 8.3 contains a checklist of unintentional damage exposures to data processing operations. Controls to prevent these exposures are usually more easily justified than the exposures identified in Figures 8.1 or 8.2, where management may decide that the cost of installing an uninterruptible power supply, for example, is sufficiently large that it will accept the risk of an extended power failure. Many controls which will correct Figure 8.3 exposures can be installed with relatively little effort. For example, the auditor may determine that a procedure permitting operators to store backup diskettes in their homes creates a risk of unintentional damage. The auditor can recommend without further risk assessment that a safe deposit box be rented to store backup diskettes.

Security Exposures Checklist for Evaluating Unintentional Damage Risks

POTENTIAL SECURITY EXPOSURE	YES	NO	N/A
1. Does the computer operations organization have an adequate segregation of duties?			
2. Are there job descriptions for all employees and are there procedures to ensure that employees meet the requirements of those descriptions?			
3. Are background checks, within legal limits, performed for all new employees?			

4. Are new employees required to sign agreements covering:
    Nondisclosure of any organization informtion?
    Bonding?
    Conflicts of interest?
    Security clearances (if applicable)?			

5. Are new employees adequately trained before starting new job assignments?			
6. Is there an ongoing program of continuing education for all information systems employees within their areas of technical expertise?			
7. Are operations employees required to complete "Problem Report" logs in the event of any operations failure or problem?			
8. Do operations shift supervisors complete a report at the end of each shift detailing activities and problems?			
9. Are operations shift and problem reports reviewed by information systems management on a regular basis?			
10. Is there a documented procedure, known to employees, covering employee counseling or disciplining in the event of an improper employee action?			

Figure 8.3. Security exposures checklist for evaluating unintentional damage risks

	YES	NO	N/A

11. Do employee resignation and dismissal procedures provide for appropriate:
    Return of access keys and badges and revocation of passwords?
    Return of confidential materials and other documentation?
    Changes to locks and combinations?

12. Are there adequate procedures covering the access limitations and supervision of any temporary personnel working within the facility?

13. Are fire evacuation drills conducted on a regular basis?

14. Are key employees aware of their roles, as documented, in the published disaster recovery plan?

15. Is there adequate off-site storage for critical applications and system media including:
    Copies of the operating system and related software?
    Source and object code versions of all programs?
    Proper versions of backup tapes to allow reconstruction of key applications?
    Database backup files and related operating software?
    Operations documentation and job control language procedures?
    Key forms and other supporting supplies?
    Copies of disaster recovery procedures?

16. Is the off-site storage area in a secure location convenient to the backup processing site?

17. Have alternative transport services to the off-site location been identified in the event of a strike, bad weather, or other problems?

18. Can the computer center remain operational using supervisory personnel during a strike?

Figure 8.3. Security exposures checklist for evaluating unintentional damage risks *(continued)*

365

	YES	NO	N/A

19. Are locked cabinets, movement logs, and related controls used to control sensitive materials within the computer operations area?

20. Are checkwriting signature plates and related materials stored separately from the associated documents?

21. Are logs, identification badges, and other controls used to monitor the distribution of sensitive output reports or documents?

22. Are there restrictions against food, drink, and smoking within the computer room facilities?

23. Is there a lounge or break area near the computer facility for employee use?

24. Are there adequate policies covering personal use of computer facilities by employees, such as for education?

25. Does the information systems organization perform periodic self assessment reviews to evaluate the risk of unintentional damage?

Figure 8.3. Security exposures checklist for evaluating unintentional damage risks *(continued)*

APPROACHES TO DATA PROCESSING SECURITY RISK ANALYSIS

Risk analysis is a formal approach for evaluating computer security exposures by determining the relative potential losses associated with each exposure and the probability of that loss occurring. Risk analysis is really the process that all individuals follow in many day to day decisions. Before an individual driving an automobile exceeds a posted speed limit, he determines the relative risk of being stopped for speeding on that particular stretch of road, the cost of a citation, and the relative probability of being stopped at that point in time.

Data processing physical security often lends itself to a formal process of risk analysis. This section discusses three such approaches to risk analysis. Each of these approaches could be the subject of a complete book.

Quantitative Approach to Risk Analysis

A quantitative approach to risk analysis requires the auditor or security specialist to go through a formal statistical analysis which identifies and values all assets within the data processing function, identifies all threats to those assets, estimates the probability that each of the threats will occur, and ranks expected losses as a result of the threats. This approach has its origins in product safety analysis techniques. For data processing, it was perhaps first presented in 1977 by Robert Courtney when he was with IBM. Since then, there have been many published variations to this same basic quantitative risk analysis approach.

This approach is based on calculating the risk, R, that a security threat will occur. Risk is determined from the probability, P, of a security exposure occurring a given number of times per year, and E, the exposure cost attributed to any such loss. The risk is then stated in terms of probable loss dollars per year according to the relationship:

$$R = P \times E$$

Assigning Exposure Probabilities

The first step in this risk analysis is to assign a probable exposure to all identified physical security risks, stated in terms of expected rates of occurrence annualized per year. The probable exposure is referred to as the Loss Multiplier when expressed as a decimal. If an event is expected to occur once in every three years, the probable exposure is ⅓ and the Loss Multiplier is 0.333. Table 8.1 shows Loss Multipliers for a range of probable exposures.

The auditor should ask data processing personnel and users to estimate the

Expected Loss Probability Range Values

FREQUENCY	ANNUALIZED FREQUENCY	LOSS MULTIPLIER (P_L)
Once in 300 years	1/300	.00333
Once in 30 years	1/30	.03333
Once in 3 years	1/3	.33333
Once in 100 days	365/100	3.6500
Once in 10 days	365/10	36.500
Once per day	365/1	365.00
10 times per day	365/.1	3650.00
100 times per day	365/.01	36500.0

Table 8.1 Expected loss probability range values

probability that an event will occur. For example, the auditor might ask data processing personnel to estimate the probability that one of the system's disc drives will crash once per day, once every ten days, once every 100 days, and so forth. If there are significant differences in models or types of equipment, the analysis should be performed separately for each. The auditor should talk to a variety of technical personnel who can give reliable estimates. Where available, actual maintenance statistics can be used to support these probabilities. When there are differing probability estimates, the auditor should develop an average response to select the appropriate probability category.

Figure 8.4 illustrates how Loss Multipliers might be calculated for several example exposures. To avoid needless speculation, the auditor compiling the probabilities should avoid asking subjective estimate questions about events such as natural disasters where personnel have no expert knowledge. It is reasonable to ask a data processing operations manager to estimate disc head crash probabilities. It accomplishes little to ask that same individual for the probability of a major earthquake.

The assigning of probable exposures in this type of exercise is subjective at best. We tend to base estimates of future performance on recent past history. If one of those "hundred year" floods had taken place the past spring, people generally would give future floods a much higher probability rate than deserved.

The Calculation of a Loss Multiplier for Sample Exposure Risks

1. Dial-up Telephone Line Failure

A computer center uses a regular telephone line for communication with a remote unit. Because this line goes through the main switching unit for the facility, the auditor has been advised that the line fails on the average of two times per day. A review of equipment trouble reports in the computer room supports this estimate.

The annualized loss multiplier is:

Occurrence rate $= 2/\text{day} = 0.5$ days per event

Loss Multiplier, $P_L = 365/0.5 = 730.0$

2. Microcomputer Equipment Failure

A specialized microcomputer system has been installed on the factory floor. Vendor literature for this equipment states that its Mean Time Between Failures (MTBF) is 10,000 hours.

The annualized loss multiplier is:

Occurrence rate $= 10,000/24$ hours $= 416.67$ days per event

Loss Multiplier, $P_L = 365/416.67 = 0.876$

3. Computer Center Power Failure

A computer center is located in a rural area on the edge of two power utility grids. Due to summer storms and other factors, operations estimates a major power failure about every three months. Computer center records support this estimate.

The annualized loss multiplier is:

Occurrence rate $= 3$ months $\times 30$ days $= 90$ days per event

Loss Multiplier, $P_L = 365/90 = 4.06$

Figure 8.4. The calculation of a loss multiplier for sample exposure risks

Delphi techniques, developed by the RAND Corporation in the late 1960s, sometimes are an effective method of obtaining an independent consensus opinion from experts on loss probabilities. The concept requires each member of a panel of "experts" to submit to an independent analyst written answers to a set of questions. Names are removed from the answers and they are tabulated and sent back to the "experts" to revise their original answers based on the anonymous responses of the others. The process is repeated until a consensus is reached.

Delphi techniques can be effective to obtain consensus opinions on such unknowns as risk analysis probable exposure factors. However, the exercise can be expensive in terms of the time and effort required. In addition, the technique appears to work best for more global issues, such as the state of organization automation in the year 2000, as opposed to the expected probabilities of floppy diskette failures due to human mistakes.

Regardless of the method used, this step in the risk analysis process should result in a list of probable exposures for each of the significant risks identified. Although some of these annualized exposures may appear small, they should not be discarded from the list before the expected exposure losses are calculated.

Estimating Expected Exposure Losses

The next step to quantitative risk analysis is to calculate expected exposure costs or losses for each of the exposures identified. The auditor or analyst should calculate an estimate for each of the significant physical security exposures identified. These estimates should include the cost of replacing an asset as well as the more intangible costs of business interruption, extra personnel, and opportunity losses. Figure 8.5 provides a list of some of the factors to consider when estimating the costs of expected losses.

Just as a fair amount of subjectivity was involved in assigning the probable exposures, this cost estimating process is difficult. It is relatively easy to estimate the expected replacement cost for the loss of a mini or microcomputer system due to a fire. However, estimating the cost of business loss due to the unavailability of the equipment is much more difficult. For example, if a model of minicomputer hardware is no longer manufactured, much of the software may have to be rewritten to work on a replacement machine.

The auditor compiling these exposure cost estimates should take care to document all estimates and computations. Once the results of the risk analysis are compiled and released to management, the auditor's results may be

Factors for Estimating the Costs of Expected Losses

*NOTE: The following are alternative factors to consider when esti-
mating expected losses due to a risk exposure. Care should be taken to
assign the most appropriate cost to the identified exposure. For example, it may cost $8,000 to install a microcomputer system as a replace-
ment for one damaged in an industrial accident. However, because the
microcomputer system operates in a key operational area, there may
be a much greater expected loss due to lost opportunity costs.*

1. The current replacement cost of the asset.
2. The costs to repair damage to the equipment less any expected
   insurance recoveries.
3. The costs to replace the equipment including costs for:
   Ordering and shipping
   Installation costs
   Testing and startup processing time
4. The costs of maintaining a backup or recovery capability including the annual costs to test that capability.
5. The costs of hazard protection and business interruption insurance.
6. The costs to continue operations without use of the equipment
   including:
   Lost business
   Operating by alternative methods such as manual procedures
   Legal liability costs

Figure 8.5. Factors for estimating the costs of expected losses

questioned. This documentation will allow the estimate to be supported or modified with additional data.

Using the cost of expected losses calculated here and the annualized probable exposures developed earlier, risks can be calculated for all of the expected exposures using the relationship discussed previously: $R = P \times E$. The result is a series of expected dollar losses per year for each of the exposures evaluated. When calculating these values, it is necessary only to state risks in terms of rounded values. It is of little value to multiply two very subjective estimates together and then present the expected joint probability in terms of many decimal places, implying a high degree of accuracy.

Figure 8.6 illustrates a hypothetical risk analysis calculation using this method. The final chart is organized by the type of exposure. However, it often is organized by descending values. The values in the figure are hypothetical and used for illustration only. This computation is easy to perform using a microcomputer spreadsheet program. The data can be modified easily for changes in estimates or presentation format.

Using Quantitative Risk Analysis Results

The purpose of the risk analysis is to identify the physical security exposures with the highest probable annualized losses. This can be a powerful method for identifying areas where corrective action should be taken to protect the organization from data processing physical security losses. The same method can be used to analyze and rank potential losses associated with information systems as discussed in Chapter 9.

Quantitative risk analysis is a common technique used in many agencies of the federal government and in financial institutions. However, because of the computations required, the exercise can be time consuming and difficult. In addition, the method described here is an abbreviated approach. Many of the formal procedures used in federal agencies, for example, employ much more complex statistical analyses which sometimes raise more questions than they answer.

Quantitative risk analysis can be a useful tool for identifying and quantifying some data processing physical security exposures. However, it should be recognized that the calculated risks are only as good as the underlying estimates of costs and probabilities. Many of these values are difficult if not impossible to estimate, and considerable management time can be spent in attempting to estimate probabilities where there is no historical data.

Risk Analysis Example

Note: The following table presents risk calculations for various data processing exposures. A four step process is followed:

1. *Identify all potential exposures to Information Systems operations using the checklists in Figures 8.1 through 8.3.*
2. *Estimate the probability of occurrence and calculate the annualized loss multiplier for each exposure, as illustrated in Table 8.1 and Figure 8.4.*
3. *Calculate expected losses, using the estimated exposure factors outlined in Figure 8.5.*
4. *Calculate the annualized probable loss exposures as illustrated in the following example and rank the losses to determine the most critical exposures.*

EXAMPLE EXPOSURE	ANNUALIZED LOSS MULTIPLIER P_L	EXPECTED LOSS E $	ANNUALIZED PROBABLE LOSS R $
1. Loss of computer center due to:			
Short term power failure with recovery within four hours	3.04	2,000	6,080
Air conditioning failure causing one day outage	0.33	17,500	5,775
Catastrophic fire causing extended outage	0.03	800,000	24,000
2. Damage to key program library files due to:			
Operator error requiring backup from off-site storage location	36.50	1,000	36,500
Introduction of virus program requiring extensive reprocessing	0.17	20,000	3,400
Deliberate and malicious destruction of many on-site and off-site files	0.06	500,000	30,000

Figure 8.6. Risk analysis example

EXAMPLE EXPOSURE	ANNUALIZED LOSS MULTIPLIER P_L	EXPECTED LOSS E $	ANNUALIZED PROBABLE LOSS R $
3. Failure to process the payroll system due to:			
Tape or media failure requiring an immediate rerun	365.00	500	182,500
Theft or destruction of check documents requiring five days to reconstruct	0.11	12,000	1,320
Malicious damage to files or programs requiring short term manual procedures	0.01	85,000	850

Figure 8.6. Risk analysis example *(continued)*

Simulation Techniques for Risk Analysis

Quantitative risk analysis requires an auditor to develop probability estimates of whether certain risk exposure events will happen. Simulation techniques attempt to test certain data processing exposures to determine what safeguards exist and where addditional controls are needed. Simulation techniques can be particularly useful methods for analyzing exposures due to human failures.

As an example of a simulation risk analysis procedure, assume either the auditors or data processing management are concerned about access controls in a computer operations center. When the control desk personnel responsible for monitoring access are queried about procedures, they supply all of the "correct" answers. With no further testing, the auditor might conclude that access controls are adequate. To test those access controls, the auditors (in conjunction with management) might arrange to have an unknown person attempt unauthorized access through the computer operations entry point. If successful, that unknown person might leave a briefcase or some other evidence of the visit and depart.

Physical security risk assessment simulation techniques can be an effective method of testing physical security exposures in very limited areas. They can be used for testing human-intensive procedures such as physical security access controls, controls over the release of key tapes or reports, or controls over initiating new jobs in the computer center. The functioning of the entire data center can be tested through a simulated emergency as discussed in Chapter 10.

There are many areas, however, where such simulation techniques are not appropriate. They should not be used if the results of the simulation would result in significant disruption to data processing operations or destruction of data or programs. It would be inappropriate to attempt to test information access controls by attempting to break into the system and bring it down through operating system manipulation.

Simulation techniques should be used only when the auditor or management has a strong interest in making a point regarding a known controls weakness or security exposure. Although the test should be performed with total confidentiality, it should be cleared also with proper levels of upper management. It may be a good idea to discuss any simulation test plans with the organization's human resources function as well as its legal counsel. If the results of the simulation were to identify criminal activities, the simulation techniques might be considered entrapment and not valid in any subsequent legal proceedings.

In certain instances, simulation techniques can be useful tools to identify data processing security exposures and vulnerabilities. However, the method should be used with caution. Without proper planning, the method can result in major disruptions to the organization which could create a new set of potential risks. The quantitative technique discussed above or the qualitative methods discussed below often are more appropriate.

Qualitative Approaches to Risk Analysis

The quantitative risk analysis approach explained above allows data processing decision makers to evaluate physical security risks in terms of probable expected losses. Using this information, management more easily can justify installing security controls and safeguards. Qualitative methods, along with simulation techniques, are alternate approaches to risk analysis that avoid the detailed statistical analysis required by the quantitative aproach.

Data processing managers as well as auditors and computer security specialists generally have a good intuitive understanding of data processing physical security risks. They know it would be a major problem if the telecommunications switching unit were to fail, and that it would be worse if the entire main computer system failed. While quantitative numbers would reveal this risk, good intuitive reasoning often does the same.

This section will discuss two approaches to qualitative risk analysis where the auditor plays an important role. One of them is a physical security risk assessment review. The other is a management oriented physical security risk analysis exercise.

Auditor Physical Security Risk Assessment Reviews

Chapters 2, 3, and 4 discussed data processing general controls reviews of large, small, and distributed computer systems. The lists of control objectives and audit procedures in those chapters covered some physical security controls along with other controls. In some instances, however, the auditor may want to perform just a specialized physical security risk assessment review.

A data processing physical security review will be more detailed than a general controls review. For example, Figure 2.4 included audit procedures for testing for the existence of access controls for the data center. Following those procedures, the auditor would inquire whether some type of access control system was in use, ask for a demonstration, evaluate that access control system, and then proceed to the next step in the audit program. In a specialized physical security review, the auditor would perform additional extended access control evaluation procedures. These might include checking the frequency of cipher lock changes or matching the list of issued access key cards to a list of current employees.

Output from a physical security risk assessment review might be a standard audit report highlighting identified physical security risks and suggesting corrective action. The output could also be a less formal report which serves as input to a management oriented physical security risk assessment review as discussed in the next sections.

Figures 8.7, 8.8, and 8.9 contain control objectives and audit porcedures for specialized physical security reviews of large, small, and distributed computer systems. These lists are broken down into general computer system size as was done in earlier chapters. However, these size designations should be used only as guidelines. Some small systems have the attributes of large systems, and the auditor should decide which procedures are most appropriate.

One purpose of a specialized physical security review is to obtain a qualitative assessment of data processing physical security risks. This assessment approach does not include the detailed probability analysis included in the quantitative risk analysis procedures discussed previously. When performing such a qualitative review, the auditor should take care to avoid drawing extreme conclusions. If the auditor finds a relatively minor control weakness, it should not be reported that the entire data processing operation is at extreme risk. Minor risks should not be extrapolated, and the relative risk should be considered when evaluating physical security risk exposures.

Management Oriented Physical Security Reviews

Many data processing managers use educated guesses to determine what types of physical security controls to install and what types to accept. Managers often have a concern that malicious damage could be caused to data processing equipment by a disgruntled employee. Access controls limiting entry to the computer room can be a very effective control. However, if the computer center has not been subjected to an electrical power failure in recent memory, management often does not see the need to install an uninterruptible power supply for that same computer room.

Management often can accomplish an effective non-numerical risk analysis through surveys and detailed discussions. This process requires the data processing management group to survey security exposures and decide on courses of action to limit security risks. The auditor can play an important role

**Control Objectives and Audit Procedures for
Large Computer Physical Security Reviews**

Objective 8.7.1.   Computer equipment should be kept in a secure, environmentally controlled facility.

    Procedure 8.7.1.1.   Tour computer room physical facilities to ascertain security strengths and weaknesses including:

        The general location of the computer room within the overall facility such that it is outside heavy traffic patterns.

        The placement of computer room outside walls and any windows to limit access to unauthorized individuals.

        The general structure of interior walls to determine that they are secure and are constructed from floor to true, not false, ceiling.

        The location of air conditioning units and power transformers to determine that they are properly protected.

    Procedure 8.7.1.2.   Observe the temperature and humidity controls in place and assess the adequacy of procedures for monitoring those controls.

    Procedure 8.7.1.3.   If the computer facility is located on a multi-floor facility, assess the risk of damage from plumbing failures, equipment, or occupants of upper floors.

    Procedure 8.7.1.4.   Review the overall area surrounding the computer room and assess the risks of such potential nearby hazards as:

        Airports or airplane landing patterns

        Chemical plants or other hazardous facilities

        Warehouse buildings or other nearby structures susceptible to fire

        Rivers, historical flood plains, or other possible causes of flooding

    Procedure 8.7.1.5.   Determine that the computer facility is located inconspicuously with no references or direction signs.

Objective 8.7.2.   Access to the computer operations facility should

Figure 8.7. Control objectives and audit procedures for large computer physical security reviews

be limited to authorized persons and that access should be controlled.

Procedure 8.7.2.1.   Observe the type of cypher lock, key lock, or badge system used to control access to the computer room and assess its adequacy.

Procedure 8.7.2.2.   Determine that locks or lock combinations to the computer room are changed on a periodic basis.

Procedure 8.7.2.3.   Review the list of assigned key cards or access rights and determine that all persons on the list are still authorized employees.

Procedure 8.7.2.4.   Determine that logs or special badges are used for visitors to the computer room and assess their adequacy and use.

Procedure 8.7.2.5.   Review procedures for allowing maintenance and other facilities personnel access to the computer room and assess the adequacy of these procedures.

Procedure 8.7.2.6.   Visit the computer room facility on an unannounced basis during a non-prime shift and determine that access control procedures are being followed.

Procedure 8.7.2.7.   Determine that terminated employees are immediately escorted from the computer room and that their access rights are canceled.

Procedure 8.7.2.8.   Observe any supplementary doors within the computer room facility and determine that they are equipped with exit only locks and audible alarms.

Objective 8.7.3.   Appropriate controls should be in place in the computer operations facility to protect equipment, materials, and employees against fire or water hazards.

Procedure 8.7.3.1.   Determine that the computer facility is protected by zone controlled smoke and fire detection equipment, both above and below the raised floor, and that activation of these devices will result in an audible alarm in the computer room as well as automatic notification at the nearest fire department.

Procedure 8.7.3.2.   Determine that the computer room is

Figure 8.7. Control objectives and audit procedures for large computer physical security reviews *(continued)*

equipped with an overall, zone controlled fire control system using Halon or equivalent suppressants as well as appropriate portable extinguishers.

Procedure 8.7.3.3.   Determine that master power switches are located at each major door to the computer room.

Procedure 8.7.3.4.   Observe that fire evacuation charts are posted prominently and determine that evacuation drills take place on a periodic basis.

Procedure 8.7.3.5.   Determine that the computer room facility is equipped with flame retardant, waterproof plastic covers for placement over major data processing equipment items.

Procedure 8.7.3.6.   Determine that the computer facility is inspected periodically by local fire inspectors, and review their last report to identify any open items for corrective action.

Procedure 8.7.3.7.   Discuss the computer room design with the organization's facilities manager to determine that flame resistant materials have been used for floor and ceiling tiles and that the ductwork has been constructed to minimize the risk of fire.

Procedure 8.7.3.8.   Determine that there are drainages under the raised floor to help avoid water accumulation in the event of flooding.

Procedure 8.7.3.9.   In addition to fire evacuation drills, determine that there are published procedures for an orderly shutdown of computer facilities in the event of flooding or a major weather disturbance.

Objective 8.7.4.   Physical security controls should be in place over tape storage and telecommunications controls areas.

Procedure 8.7.4.1.   Visit the prime tape storage facility and determine that access, fire, and other controls are appropriate and consistent with procedures used in the main computer room facility.

Procedure 8.7.4.2.   Review procedures for logging tapes in and out of the tape library and assess the appropriateness of the controls.

Figure 8.7. Control objectives and audit procedures for large computer physical security reviews *(continued)*

Procedure 8.7.4.3.   Visit the telecommunications control area, if outside of the main computer room, and determine that access, fire, and other controls are appropriate and consistent with other computer room procedures.

Procedure 8.7.4.4.   Review cabling of telephone or local network lines from remote devices to the telecommunications facility and determine that these lines are shielded or obscured from view.

Objective 8.7.5.   Computer operations should have adequate insurance coverage consistent with the potential risks and the desires of management.

Procedure 8.7.5.1.   Review related insurance coverage with appropriate individuals in the organization and determine whether there appears to be adequate coverage for data processing equipment and media including:

   Losses from fire or flood damage
   Losses from equipment breakdowns such as sprinkler system leakage
   Losses from theft or vandalism
   Civil commotion or riot losses

Procedure 8.7.5.2.   Review insurance coverage for business interruptions and assess adequacy given the organization's dependence upon key computer applications.

Procedure 8.7.5.3.   Where there appears to be insurance related vulnerabilities, discuss these concerns with appropriate levels of management.

Objective 8.7.6.   An effective and tested disaster recovery plan should be in place for computer operations.

Procedure 8.7.6.1.   Determine that the organization has a disaster recovery plan following the outlines presented in Chapter 10.

Procedure 8.7.6.2.   Interview several members of the disaster recovery team, as designated in the plan, and assess whether they understand their roles as outlined in the plan.

Procedure 8.7.6.3.   Determine that key aspects of the plan,

Figure 8.7. Control objectives and audit procedures for large computer physical security reviews *(continued)*

such as arrangements for the alternate processing site, are still current and operable.

Procedure 8.7.6.4. Determine that the disaster recovery plan is being tested periodically, and review the results of the last several tests including any documented recommendations for improvement.

Objective 8.7.7. Backup copies of files, documentation, and critical forms should be stored in a secure, off-site location.

Procedure 8.7.7.1. Visit the off-site storage location and assess whether security and environmental controls appear to be adequate.

Procedure 8.7.7.2. Review procedures for designating files as candidates for off-site storage and assess whether proper consideration appears to be given to application criticality.

Procedure 8.7.7.3. Select several key applications and determine that key versions of tape files as well as documentation and special forms are stored in the off-site location.

Procedure 8.7.7.4. Secure an inventory listing of all items in the off-site location and determine whether it appears to be complete and current.

Objective 8.7.8. Adequate procedures should be in place to record equipment failures and to maintain the equipment on a regular and emergency basis.

Procedure 8.7.8.1. Review manual and automated logging procedures for recording equipment failures and assess whether they appear to be followed.

Procedure 8.7.8.2. Determine that procedures are in place to obtain appropriate levels of approvals for calling in maintenance personnel during all operating shifts.

Procedure 8.7.8.3. If on-line remote maintenance techniques are used for hardware or software problems, determine that those activities are logged and reported to management.

Procedure 8.7.8.4. Review operations schedules to determine that equipment maintenance is performed on a regular basis consistent with equipment requirements.

Figure 8.7. Control objectives and audit procedures for large computer physical security reviews *(continued)*

Objective 8.7.9. Controls should be in place to monitor the distribution of output reports as well as the introduction or release of tape files.

Procedure 8.7.9.1. If users pick up output reports from the computer operations facility, determine that a lock box or supervised, badge identification system is used such that only authorized persons may pick up their reports.

Procedure 8.7.9.2. If an office courier or mail distribution system is used for output report distribution, determine that procedures are adequate to ensure that reports go only to appropriate recipients and that confidential reports are sealed.

Procedure 8.7.9.3. Review procedures for the production and distribution of key documents, such as payroll checks, and assess whether security procedures appear adequate.

Procedure 8.7.9.4. Review controls over releasing tape files to outside users and determine that adequate levels of approval are required.

Procedure 8.7.9.5. Review procedures for bringing data and program files into the computer system and assess management's understanding of the vulnerability to computer viruses with the introduction of such files.

Objective 8.7.10. There should be adequate physical asset controls over all computer hardware and related equipment.

Procedure 8.7.10.1. Trace selected fixed asset records from accounting records back to actual computer room equipment.

Procedure 8.7.10.2. Determine that all major items of computer equipment, if owned, are labeled with fixed asset tags and that they appear on fixed asset records.

Procedure 8.7.10.3. Using the equipment vendors' monthly lease or maintenance billings, trace a sample of selected items from the bills to actual equipment in the computer room operations area.

Procedure 8.7.10.4. Determine that control procedures are ade-

Figure 8.7. Control objectives and audit procedures for large computer physical security reviews *(continued)*

quate when maintenance and other outside personnel remove circuit boards or other items from the computer room for repair or upgrading.

Figure 8.7. Control objectives and audit procedures for large computer physical security reviews *(continued)*

in this process by helping develop the physical security surveys, performing detailed reviews of larger computer centers, and leading management in security-oriented discussions.

Qualitative risk assessment works best with a security risk analysis team to discuss physical security vulnerabilities, identify major risk areas, and implement the proper controls. Such a team should consist of key members of data processing management, the data security specialists, plant security personnel, and members of the audit department. The team first should gain a general understanding of any potential data processing physical security risks within the total organization.

The typical, modern organization has a data processing function that often is distributed. There may be one central data processing operations center but additional smaller, peripheral sites. In addition, significant data processing applications may be processed at freestanding divisional minicomputers, at departmental microcomputers, or through specialized process control computers. There may be data processing physical security risks at all of these locations. However, the local data processing management often does not have the necessary knowledge of security practices to determine whether the risks are significant.

A first step to understanding any data processing physical security risk is to perform a limited physical security review at each of these data processing sites. This is often an area where auditors can play a significant role by performing a physical security assessment. Control objectives and audit procedures for such reviews of larger data processing operations are found earlier in this chapter. However, auditors generally cannot review all of the data processing sites within a large organization in the time available to do the overall risk analysis. User completed surveys often are also necessary.

A departmental data processing physical security survey allows management to gather information about all computer processing facilities within the organization. The auditor can use experiences gained in developing other audit programs and questionnaires to help data processing management develop such a user physical security survey. An example of a departmental survey

**Control Objectives and Audit Procedures for
Small Computer Physical Security Reviews**

Objective 8.8.1.   Adequate access controls should be in place to prevent theft or damage to the mini or microcomputer hardware or its components.

> **Procedure 8.8.1.1.**   Observe the location of the mini or microcomputer system and determine whether it is in an area of limited traffic.
>
> **Procedure 8.8.1.2.**   Assess the size and criticality of the small computer system to determine whether it should be located in a facility that can be locked during non-prime shift hours.
>
> **Procedure 8.8.1.3.**   Determine that locking devices are used to secure any freestanding microcomputer systems, and assess whether these locking controls are adequate.
>
> **Procedure 8.8.1.4.**   Even if the small computer system is not attached to its work platform, assess whether locks are used to secure the chassis from improper removal of boards and from unauthorized operation.
>
> **Procedure 8.8.1.5.**   Visit selected work areas during off shift hours and determine that the mini- or microcomputer workstations are properly locked, that work areas are clear of programs and diskettes, and that locking keys are properly secured.
>
> **Procedure 8.8.1.6.**   Determine that asset tags or other identification markings are used for all microcomputer equipment.
>
> **Procedure 8.8.1.7.**   Select a limited sample of computer equipment from the accounting department fixed asset records and trace these to identify the actual physical equipment.
>
> **Procedure 8.8.1.8.**   Review procedures used for any portable microcomputers, and assess whether they provide adequate controls.

Objective 8.8.2.   Program and data files, documentation, and supplies should be kept in a secure location to prevent theft or damage.

Figure 8.8. Control objectives and audit procedures for small computer physical security reviews

Procedure 8.8.2.1.   Review procedures for logging and controlling small computer system programs and determine that master copies of program libraries are kept in a secure location.

Procedure 8.8.2.2.   Observe the handling of program diskettes and documentation surrounding microcomputer work stations and determine that the materials are logged properly, controlled, and secured during off shift hours.

Procedure 8.8.2.3.   Select a sample of recent microcomputer software purchases and trace to the actual software located at microcomputer stations to determine that the software is logged and properly controlled.

Procedure 8.8.2.4.   Review procedures in place to discourage unauthorized copying of microcomputer programs and data files and assess their adequacy; such procedures may include:

> Signed employee agreements not to copy unauthorized materials
>
> Maintaining critical data files in an encrypted format
>
> The use of password protection devices to prevent unauthorized file access

Objective 8.8.3.   The small computer system should be protected from the introduction of malicious, computer virus type programs.

Procedure 8.8.3.1.   Determine that a policy, acknowledged by employees, is in place which prohibits the introduction of unauthorized programs to any microcomputer system.

Procedure 8.8.3.2.   Review the extent of dial-up modems attached to small computer systems and assess whether policies are in place to prohibit the downloading of programs from sources such as computer bulletin boards.

Procedure 8.8.3.3.   Determine whether virus detection software has been installed on computer systems and review procedures for use of this software.

Objective 8.8.4.   Backup copies of key files should be kept in a secure offsite location.

Figure 8.8. Control objectives and audit procedures for small computer physical security reviews *(continued)*

Procedure 8.8.4.1.   Survey a sample of small computer systems and assess whether file backup mechanisms, such as "streamer tape" units for microcomputers, are adequate given the size and criticality of the system files.

Procedure 8.8.4.2.   Review backup procedures for a sample of small computer applications and determine that they appear to be understood by users and are being followed.

Procedure 8.8.4.3.   Determine that backup copies are used to reconstruct key files on a test basis in order to assess the adequacy of backup procedures.

Procedure 8.8.4.4.   Review the contents of the offsite storage location and determine that proper backup versions of files and locations are present and properly labeled.

Procedure 8.8.4.5.   If the organization has charged a key employee with the responsibility of keeping backup files at the employee's home, interview the employee to determine that files are properly stored in a safe location.

Objective 8.8.5.   The small computer system should be protected from environmental hazards as well as electrical power fluctuations and failures.

Procedure 8.8.5.1.   Determine that electrical line surge protectors are installed on all microcomputer systems.

Procedure 8.8.5.2.   Assess the use of standby or uninterruptible power supplies and determine that such devices are installed for critical microcomputer systems.

Procedure 8.8.5.3.   Observe a sample of small computers installed in non-office production areas and determine that the units are shielded properly and protected from dust or temperature extremes.

Procedure 8.8.5.4.   Observe a sample of microcomputers in office areas and determine that floor pads in carpeted areas or other devices are used to prevent static electricity.

Procedure 8.8.5.5.   Review electrical power supplies for any minicomputer system and determine that it is on a separate line and protected with adequate circuit breakers.

Figure 8.8. Control objectives and audit procedures for small computer physical security reviews *(continued)*

Procedure 8.8.5.6.   Determine that there are management policies against food, drink, or smoking materials near mini or
microcomputer units.

Figure 8.8. Control objectives and audit procedures for small computer
physical security reviews *(continued)*

questionnaire is shown in Figure 8.10. The auditor should tailor this document
to the needs of the specific organization.

The objective of this survey is to gather security related information from
departmental personnel who are not necessarily data processing professionals.
Therefore, the survey should be non-technical and easy to complete. Another
consideration is possible objections to such a survey by independent data
processing departments. To avoid such organizational disputes, the audit
function can be a neutral third party in circulating and following up on the
survey forms.

When the surveys have been completed, the risk assessment and evaluation
team can analyze them and identify potential areas of risk. After a detailed
discussion of the survey responses, the team may identify some critical
physical security vulnerabilities. For example, they may determine from the
surveys that a single remote operating unit has an older model, freestanding
minicomputer with no backup machine available either within the organization
or in the market. This lack of a backup processing capability can be a severe
security risk.

The risk assessment team often can do an effective job of identifying
potential physical security risks. Through reviews of the surveys and general
discussions, a prioritized list of risks and vulnerabilities can be developed.
Additional outputs from this team are:

A security risk report for management action. The team should develop
a detailed report outlining identified security risks and recommendations for
corrective action. For example, the team may identify the lack of a data
processing emergency backup site and recommend establishing such a
capability.

A series of short term actions to improve security. The team may identify
certain areas where a minor change in procedures or small corrective actions
can improve data processing physical security. For example, the team may
determine that the cipher lock to a machine room has a combination which
is not changed regularly.

388

**Control Objectives and Audit Procedures for
Distributed Processing Physical Security Reviews**

Objective 8.9.1.   Adequate and consistent physical security procedures should exist at all nodes of the distributed processing system.

> **Procedure 8.9.1.1.**   Review physical security procedures and published documentation which covers both the distributed hardware processors and any distributed telecommunications node equipment to determine that these appear adequate and consistent.

> **Procedure 8.9.1.2.**   Review procedures for sending documentation updates to remote distributed processing node locations and determine whether control logs are maintained.

> **Procedure 8.9.1.3.**   Visit a selected distributed processing site to determine whether users understand and comply with the physical security related procedures of the distributed processing system.

> **Procedure 8.9.1.4.**   Determine whether a centralized procedure exists to record any suspected physical security problems and to answer any remote site security questions.

Objective 8.9.2.   There should be adequate backup and recovery procedures throughout the distributed system to assure ongoing system integrity.

> **Procedure 8.9.2.1.**   Review published procedures, including schedules, for backing up data from any separate distributed processing nodes and determine whether they appear to be consistent.

> **Procedure 8.9.2.2.**   Examine records covering any recent distributed system recovery covering two or more nodes and assess whether published procedures appeared to be effective.

> **Procedure 8.9.2.3.**   Visit a selected distributed processing remote processing node site and review the status of file backups and off site storage of those files.

Figure 8.9. Control objectives and audit procedures for distributed processing physical security reviews

Procedure 8.9.2.4. Examine the information systems organization's overall disaster recovery plan (see Chapter 10) and determine whether appropriate consideration has been given to the recovery of multi-node distributed processing systems.

Objective 8.9.3. Access to distributed program libraries and data files should be controlled to maintain the integrity of related application programs.

Procedure 8.9.3.1. Determine that only object code versions of programs are distributed to distributed processing sites to prevent local modifications.

Procedure 8.9.3.2. Review program library versions or file sizes at several remote, distributed processing locations to determine that each has the correct version of key distributed application programs.

Procedure 8.9.3.3. Determine that password schemes or other information security procedures exist at all distributed locations to control access to data.

Procedure 8.9.3.4. If the distributed system includes the use of a fourth generation language or similar processor, determine that procedures are in place to prevent the unauthorized updating of data files with that processor.

Objective 8.9.4. Effective data security policies should be implemented at all distributed processing node locations.

Procedure 8.9.4.1. Review physical security controls at selected distributed locations, and if security procedures are restrained by equipment limitations at any site, determine that restrictions have been placed over operator actions at those locations.

Procedure 8.9.4.2. If the distributed application requires users to access other nodes in the distributed network, determine that adequate information security procedures exist at each node.

Procedure 8.9.4.3. Interview selected users at a distributed processing site to assess their understanding of data security concerns regarding the distributed processing system.

Figure 8.9. Control objectives and audit procedures for distributed processing physical security reviews *(continued)*

Objective 8.9.5. There should be overall change procedures for installing new hardware, data files, or program components on the distributed system.

Procedure 8.9.5.1. Review procedures for introducing changes to the distributed processing system and assess whether adequate attention has been given to maintaining overall system integrity.

Procedure 8.9.5.2. Determine that a test system has been established to simulate the distributed processing system and to monitor the effect of changes to that system.

Procedure 8.9.5.3. Review procedures for resolving problems which may have been introduced to the distributed system due to an error in a new system modification and assess the adequacy of those procedures.

Procedure 8.9.5.4. Review change control documentation records at the central control site and assess their correctness through a review of the actual equipment and programs at a selected distributed site.

Figure 8.9. Control objectives and audit procedures for distributed processing physical security reviews *(continued)*

Computer Physical Security Questionnaire

The Internal Audit Department is conducting a survey of departmental computer security practices throughout the organization to develop improved security and protection procedures for all departmental computer systems. Please complete one of these forms for each major type of computer in your department. For example, if you have a minicomputer as well as several similar microcomputers, please complete one form for the minicomputer and one for the group of microcomputers. If you have questions, please contact Internal Audit.

Prepared by: _____ Title:_____

Department:_____ Date:_____

Type of computer system: _____ How Many: _____

System purpose (general business, etc.):_____

Person responsible for operating system: _____

System Configuration:

    Computer model:_____ Date installed_____

    Memory size:_____ Disc capacity: _____

    Number and type of disc/tape drives: _____

    Types of special processor boards/features: _____

    Type of printers:_____ Other output devices:_____

    Number and type of terminals attached to system: _____

    Does the system have a modem?: _____

    Is the system connected to other computers?:_____

System Software:

    Operating system:_____

    Version:_____

    Types of software installed:_____

    Approximate number of applications:_____

    Is system used for programming?:_____

System Environment:

    Is system equipped with:

        Surge protector?_____

        Uninterruptible power supply?_____

        High speed back-up tape/disc drive?_____

        Password and logging software?_____

Figure 8.10. Computer physical security questionnaire

Virus protection software monitor?_____
Power On/Off lock?_____
Chassis lock?_____
Is there a lockable storage area for:
Program diskettes?_____
Data diskettes/tapes?_____
Documentation manuals?_____
Sensitive forms?_____
Can computer room be locked during off shifts?_____
Is there adequate air conditioning and power for the system?_____
Types of fire control devices available:_____
Are programs and files backed up?_____
Who is responsible?_____
How often are backups performed?_____
Where are backups stored? _____
Have backups been used to restore system?_____

Figure 8.10. Computer physical security questionnaire *(continued)*

A checklist for field operations guidance. If it is determined that field data processing operations do not have a proper appreciation of physical security risks, the team can develop a checklist for distribution to these remote operations. This checklist can be informal but should identify potential risks and suggest corrective actions. An example of such a checklist is shown in Figure 8.11.

A Hybrid Approach to Risk Analysis

The auditor involved in a physical security risk analysis should select an approach that appears best for the organization under review. Each of these risk analysis approaches has advantages and disadvantages. Generally, the quantitative approach is best suited to the large, centralized computer environment while a management oriented qualitative approach works best for a distributed or small systems network. This section discusses a hybrid approach to risk analysis or evauation which should work in many organizations. The explanation is based on a hypothetical example.

Assume that the data processing function where the physical security evaluation will be performed has a larger central computer with a network of terminals and remote job entry processors. In addition, there are a variety of microcomputers and several freestanding process control minicomputers. Upper management has requested that the audit department work with data processing management to identify physical security exposures and recommend steps to improve physical security.

Identifying Physical Security Exposures

The first step is to designate a management team to perform an analysis of physical security exposures. The team's purpose is to engage in a detailed, "What happens if it breaks?" analysis covering all aspects of data processing physical security. The group should discuss the organization's exposures if key elements of the data processing system unexpectedly become unavailable or compromised due to some physical security event.

In performing this type of review in the modern data processing center, it is difficult to separate totally physical security issues from the information security concerns discussed in Chapter 9. For example, the team could raise the question of the potential exposure if the on-line order entry system became unavailable. This might occur if there had been a failure in the telecommunications system or if there were an information security breach, such as a

Computer Physical Security Questionnaire for Field Operations

NOTE: This is an example of the type of checklist that can be distributed to help remote operations of an organization better establish security controls over small computer systems.

1. Are full system backups taken at least daily?
2. Are at least three generations of backup files retained?
3. Is at least one generation of backup files stored in an off-site location?
4. Are backup files kept in a locked storage area?
5. Are computer diskettes, documentation manuals, and other materials kept in a locked storage area when not in use?
6. Is the computer system locked when not in use?
7. Are backup and remote storage procedures documented?
8. Is there an inventory listing of all computer system components?
9. Are normal organization asset tagging procedures used for the system and its components?
10. If the system is located in a severe environment, is the system properly shielded from dust and other hazardous conditions?
11. Are electrical power and air conditioning resources adequate?
12. Is the computer system equipped with an electrical surge protector?
13. Is the system equipped with an uninterruptible or standby power system?
14. Are devices such as floor mats used to protect equipment from static electricity?
15. If the system is equipped with a modem, are there procedures to screen incoming transmissions?
16. Have arrangements been made for an alternate processing facility in the event of a long term system failure?
17. Are key applications on the system protected with a password system?
18. Are passwords changed on a regular basis?
19. Are there policies in place to prohibit the introduction of unauthorized software on the system?
20. Is use of the system restricted to authorized individuals?

Figure 8.11. Computer physical security questionnaire for field operations

"hacker" breaking into the system and destroying key files. The evaluation team also should consider information security concerns as part of its analysis; information security presents as great a concern as physical security in the modern data processing center.

The team should consider the various physical security exposures outlined in Figures 8.1, 8.2, and 8.3. Also, they should consider the information security exposures discussed in Chapter 9. It may be necessary to use surveys, such as Figure 8.10, to gather data about remote locations in the organization. Figure 8.12 contains a questionnaire for such a management review of physical security exposures within the organization. The questionnaire is general and should be tailored to the particular organization under review.

The management review should identify three levels of exposures:

1. *Exposures requiring immediate correction.* In some instances, the team may identify security exposures within the data processing function that do not require a detailed risk analysis. For example, they may find that a backup contingency site has not been tested or a remote minicomputer site does not store key files in an off-site location. Such exposures should be corrected at once without extensive detailed risk analysis.

2. *Exposures requiring a detailed review.* In some instances, the review team will conclude that they need more information before identifying all physical security exposures. For example, management at a remote minicomputer site may have provided ambiguous answers to the security survey. The management team may require detailed information and ask the audit function to perform a physical security review of the location in question. Because of limited audit resources and management's need to complete this overall review, these audits should be restricted to areas where the team needs this information to make a proper judgment.

3. *Exposures to consider for evaluation.* Some data processing physical security exposures may not require a detailed audit analysis nor present areas which can be corrected without further cost versus risk considerations. Rather, they will require a more formal analysis.

This three-step process will result in a working list of exposures. Except for those which are identified as needing immediate action, the next step is to evaluate the exposures.

Questionnaire for a Management Review of Physical Security Exposures

NOTE: This set of questions is designed to allow management to evaluate data processing risks in the example organization described in the text. A similar set of questions can be tailored for a risk evaluation exercise in other organizations.

A. Central Computer Operations

1. What would be the cost to the organization if all central computer applications were not available to operations for an extended period of time?
2. Is business interruption insurance coverage adequate?
3. Is there a backup location for processing computer applications in the event the central site became unavailable?
4. Has the processing at the backup site been tested?
5. Is there an overall plan covering these contingency operations?
6. Are files, programs, and documentation kept in an off-site location?
7. Are there adequate fire detection and prevention devices in the computer room?
8. Are there hazards in the building or at nearby locations that could impact central computer operations?
9. What is the cost to the organization if power to the computer system were out for a limited amount of time?
10. What would be the impact of a telecommunications failure on computer systems operations?
11. Could the organization revert to manual procedures for any of its key applications in the event of a computer systems failure?
12. Would computer operations be the potential target of violent labor unrest or a civil disturbance?
13. Could someone break into the computer facility without immediate detection?
14. Could someone break into the computer system files through a remote terminal?
15. Does the password system prevent an authorized user in one application area from getting into another area?
16. Could a competitor gain access to key data files such as cus-

Figure 8.12. Questionnaire for a management review of physical security exposures

tomer lists, product formulae, and financial data?
17. Could an unauthorized person gain access to confidential computer system output reports?
18. Could an unauthorized person gain access to any critical output reports produced for distribution?
19. Are background checks performed on new employees?
20. Could employees easily modify a program or application, without detection, to defraud the organization?

B. Departmental Microcomputer Operations
1. Are there critical applications, such as financial analysis reports, operating on any of the microcomputers?
2. Are there policies and standards requiring that microcomputer systems be backed up?
3. What would be the impact on the organization if a given microcomputer system and its files were not available?
4. Could the microcomputer system be used to access or manipulate central computer files or programs?
5. Could unauthorized outsiders access the microcomputer through a modem connection?
6. Could an undetected virus program be introduced to the microcomputer system?
7. Could an unauthorized person gain access to critical files or programs on the microcomputer system?
8. Would it be possible to steal a valuable software program or circuit card from the system?
9. Could the microcomputer suffer from a file failure or other malfunction through a static electrical charge or an electrical power disturbance?
10. What would be the impact on the organization if key microcomputer specialists were not available?

C. Specialized Departmental Computer System
1. What would be the impact on the local department as well as the total organization if the specialized computer were not available?
2. Does the specialized department understand the need to back-up computer files and data?
3. Does the specialized department follow the overall information

Figure 8.12. Questionnaire for a management review of physical security exposures *(continued)*

systems policies for controlling access to computer systems program and data files?
4. Would it be possible to steal a valuable software program, specialized equipment, or other materials from the computer system site?
5. Could the specialized computer system suffer a failure due to the environment in which the processor has been implemented?

Figure 8.12. Questionnaire for a management review of physical security exposures *(continued)*

A Cost/Risk Security Evaluation

The approach discussed in this section is a simplified version of the quantitative risk analysis evaluation discussed earlier in this chapter. It does not result in a detailed set of probable expected losses but allows management to consider high risk exposures which can be corrected at reasonable cost.

The first step in this approach is to rank the list of exposures identified by the management or audit review by probable risk. However, rather than attempting to calculate probabilities for each, they should be ranked on a scale of 1 to 100 according to their expected occurrence within a one year period. For example, the risk that the mainframe chiller system might have a plumbing failure which would flood electrical equipment probably would be ranked quite low. While there may be a plumbing failure some time in the future, it probably will not happen in a given year. Similarly, the risk that a freestanding microcomputer with key files on hard discs could be subject to unintentional or malicious damage might be relatively high. This first step should be simply a relative ranking with attention given to separating the high and low expectation risks.

The second step is to rank each of the identified exposures according to the expected losses or costs which would result from any single exposure. Following the above examples, the expected loss of the chiller failure would be very high while the cost of restoring microcomputer disc files would be relatively low. Since it may be difficult to derive exact costs, the management evaluation team should use its best estimates in this ranking exercise.

The exposures should then be ranked for a third time, according to the cost to implement corrective procedures for each exposure. This is really a two-step process, as potential corrective procedures should be considered for each exposure followed by an estimate of the relative costs. If there are several potential corrective solutions for one risk, the lowest cost solution should be selected initially.

The final step of this three way ranking of physical security exposures is to plot them on a three dimensional cost/risk evaluation matrix. This three dimensional chart has relative risk on the horizontal axis, relative cost of exposures on the vertical axis, and cost of corrective action on the third dimension axis. Such a cost/risk evaluation matrix is illustrated in Figure 8.13. The various exposures can then be plotted on this three dimensional matrix to determine which are the high risk, high potential loss, and low implementation cost exposures. These should be considered for implementation first.

The auditor performing this analysis probably will not have the ability to easily plot in three dimensions. Several microcomputer spreadsheet packages allow this, but the most popular ones do not. Another approach is to divide the matrix into two two-dimensional plots by dividing the costs to implement into "high cost" and "low cost" implementations. This will result in two plots as shown in Figure 8.14. These plots are each subdivided into four quadrants to separate high and low risks and high and low exposure costs.

The eight quadrants in Figure 8.14 have been numbered. If all the ranked exposures are plotted between these eight quadrants, the auditor will be able to determine which are the more critical exposures with the most cost effective solutions. For example, all risks which are plotted in quadrant one have a high risk of occurrence, a high exposure cost, and a low cost of implementation. After the exposures in quadrant one, the next group to consider for corrective action appear in quadrant five. The auditor should avoid suggesting corrective action for exposures in quadrants with a high cost of implementation, a low exposure cost, and a low risk.

This relative ranking and plotting approach to risk analysis ensures that the most cost effective projects are implemented first. Since most organizations have limited resources which prevent implementing controls to cover all security exposures, only the highest exposure ones can be considered.

The evaluation team should use this initial plot of costs and risks to determine potential candidates for recommended action. It may look also at how the various exposures appear on the plots and decide that a given one is misplaced. It could then be reranked and replotted.

This approach can be an effective method of analyzing data processing security risks. It uses relative values without requiring the detailed computations of the formal quantitative risk analysis. While detailed computations are not necessary, the approach allows relative rankings which often are a better approach than the "best guess" techniques used for many qualitative risk analysis approaches.

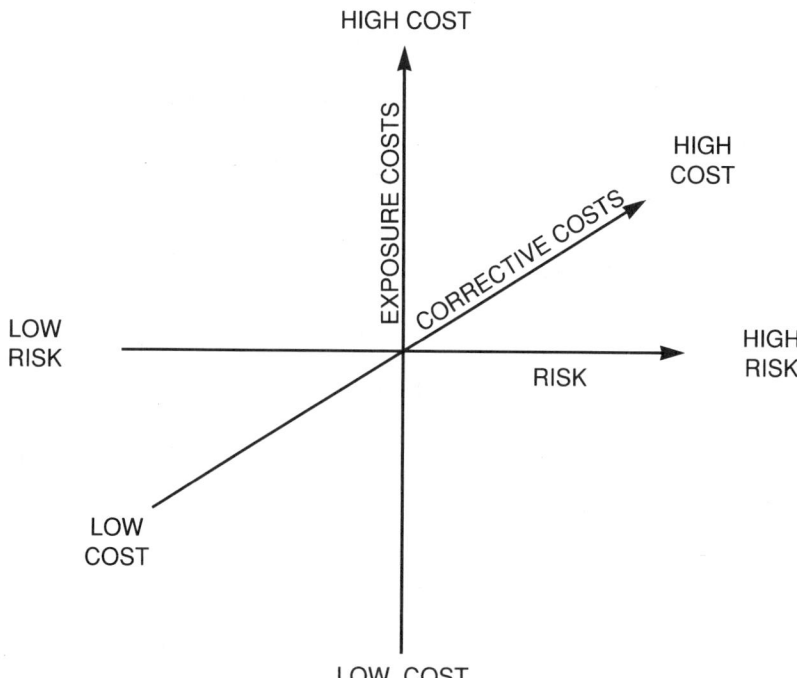

Figure 8.13. The cost/risk evaluation matrix

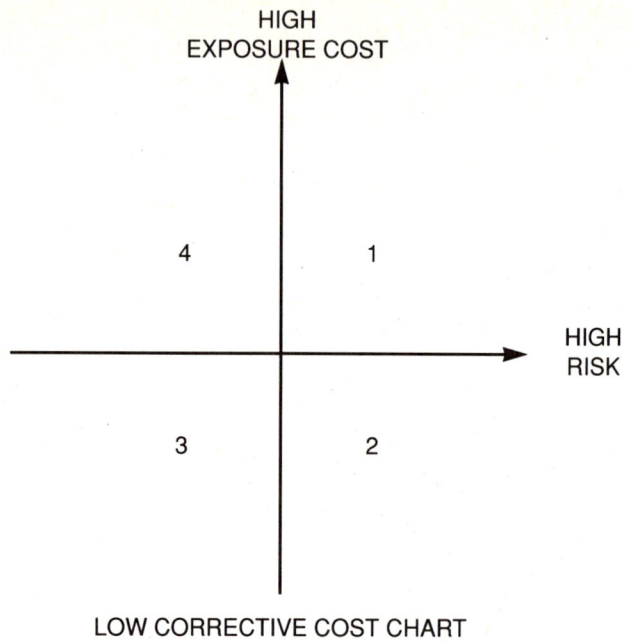

LOW CORRECTIVE COST CHART

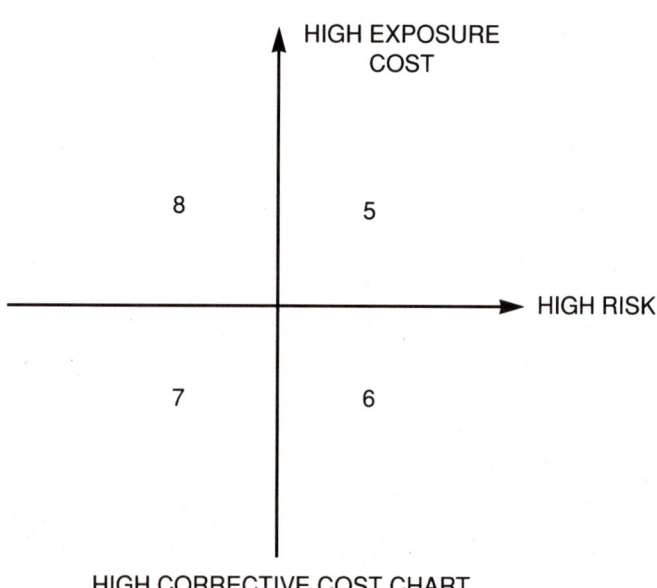

HIGH CORRECTIVE COST CHART

Figure 8.14. The cost/risk analysis quadrant charts

Presenting Risk Analysis Findings to Management

The prior section reduced the process of risk analysis to a workable level that avoids the complex calculations and estimates of probabilities associated with some risk analysis approaches. However, auditors often find it necessary to reduce risks to an even more understandable manner when presenting findings to management. This is not because of the failure of upper management to understand complex computations. Rather, upper levels of management in most organizations frequently are generalists and risk takers. Auditors should keep these two concepts in mind when presenting risk analysis findings.

The typical upper level manager will not be interested in complex charts or tables of subjective probabilities when reviewing the results of a data processing physical security risk analysis review. Rather, the typical manager will recognize that data processing plays a significant role in the ongoing success of the organization and will understand the costs associated with an extended interruption in data processing services. Auditors should consider presenting their risk analysis findings in that context, asking "What would be the costs to the organization if there were major data processing failures and what would it take to prevent them from occurring?"

The auditor first should determine the business costs of an extended interruption in data processing services. These costs should be estimated based on the actual cash flow losses that would result if the system were unavailable for one day, two days, and so forth. These losses can be estimated from statistics such as missed orders or delayed shipments.

When developing these estimates, the auditor also should consider available manual and backup data processing recovery techniques. For example, some orders still can be processed manually if the automated system is down. In addition, limited data processing backup facilities probably can provide some support to help restore data processing.

Care should be taken to avoid arguments over-emphasizing trivial risks when estimating data processing losses. An upper management team accustomed to accepting a level of risk in many of their decisions often will have trouble accepting arguments that trivial data processing risks can lead to a disaster. Often it is easy for an auditor to lose perspective and develop such a scenario regarding data processing risks.

After developing a set of estimates of probable losses associated with a full or partial loss of data processing services, the next step is to develop a list of the high probability exposures which could cause such losses as well as their probability of occurrence. The estimated costs of alternative strategies for corrective actions also should be presented. When there have been failures or partial failures in the past, these should be illustrated as part of the presentation.

The auditor should be able to convince management of the exposures associated with data processing physical security and the costs associated with them. Of course, the auditor should make sure data processing management is in agreement with these estimates and exposures. If audit findings are presented in terms of relative risk, management should be able to relate to the data presented and make a sound decision for corrective actions.

IMPLEMENTING PHYSICAL SECURITY CONTROLS

Much of this chapter has discussed identifying physical security exposures and evaluating the associated risks. Frequently, the auditor also plays an important role in helping implement effective physical security controls. While it is data processing's responsibility to actually install such controls, the auditor should review existing procedures and recommend corrective actions where appropriate.

Auditors often are in a unique position to recommend physical security improvements. As discussed previously, management looks upon auditors as the data processing security "experts," because auditors often are the first persons to suggest data processing physical security improvements.

Today, data processing management usually has a good understanding of physical security procedures. In addition, many larger data processing functions have data security professionals on staff; however, these persons sometimes are more involved in information security (such as the access control software) than physical security. In addition, their involvement usually does not extend beyond the formal business data processing function, even though there should be security concerns in such areas as office automation and departmental computers.

Improving Security in the Computer Room Facility

The central data processing operations facility in the larger organization often is vital to the overall operations of that organization. The equipment is relatively fragile and quite complex. It can be damaged easily through either unintentional or malicious acts. Depending upon the relative risks and exposures, it is important to have strong computer room physical security controls. These controls cover the following broad areas.

Computer Room Access Controls

It is important to limit access to computer operations to authorized persons. In a smaller organization, this can be accomplished through informal procedures such as placing the equipment in a lockable room and restricting access to those who can be recognized and have a right to enter the facility. Controls must be much stricter in a larger organization and larger computer facility.

There are a variety of access control tools and techniques. In earlier days, access was limited by such ineffective methods as an "Authorized Person Only" sign coupled with a buzzer lock which often could be activated by reaching through an open window. Some rather sophisticated access control devices are now available. These include:

Cipher locks which require a combination or code to gain entry

Key cards which can be matched against computer records

Devices which match such things as voice patterns, fingerprints, or signatures

It is essential that each of the above devices be monitored and updated when there are personnel changes. All too often, an organization will install a sophisticated access control device but fail to update it for such events as employee terminations or transfers.

Computer Room Physical Facility Controls

During the period of political and social unrest in the late 1960s and early 1970s, computer centers often were the targets of terrorist actions in the United States and Western Europe. Terrorists recognized that the computer centers of prominent organizations were vulnerable and were symbolic targets for destruction.

Organizations always should assess the relative vulnerabilities of their computer centers to acts of terrorism, riots, or other malicious acts. If there is any potential concern about such actions, steps should be taken to limit exposures. This might require management to move computer operations to a separate "block house" facility. However, many organizations can limit this exposure by placing computer rooms in an unobtrusive area within a normal operations building without windows and with secure walls.

Computer Room Environmental Controls

Larger, mainframe computer systems require precise environmental controls over such services as air conditioning and electrical power. Large transformers must be mounted outside of the computer facility to supply proper power while large compressors are required for cooling. These devices are vulnerable to failure and malicious damage. In addition, a larger computer installation requires a system of detection devices to report fires or failures of the environmental system and requires preventive devices such as fire control systems.

The auditor can assist data processing management by surveying the environmental vulnerabilities of a larger data processing center. Many of these issues are highlighted in the checklist in Figure 8.15. While data processing management often recognizes that the lack of a backup power supply or the outside location of the air compressors represents a security exposure, the auditor can be an effective aid in reporting these less obvious but equally important vulnerabilities to upper management.

While auditors can be effective in pointing out environmental control vulnerabilities in larger computer centers, they should approach these same issues with a degree of reasonableness at smaller computer sites. For example, it may not be necessary to have a full Halon fire control system protecting a smaller mini or microcomputer system. The risk of loss of data or processing capability may not be that great. In addition, the cost of the fire control equipment may be high relative to the cost of the computer equipment. Auditors always should consider relative risks and costs when making such evaluations.

Physical Security Over Computer Media

In addition to protecting the actual computer equipment, it is critical to have proper physical security controls over computer media such as tape files or special forms. Some of these matters are addressed in the data center's disaster recovery plan discussed in Chapter 10, which considers arrangements to store key tapes and other media in an off-site location. In a larger data center, some physical security is provided by access controls which prevent tapes from being transported freely in and out of the computer room. However, the auditor should look for a system of physical security controls to ensure that tapes, special forms, and other media are properly protected.

While the auditor often should temper physical access control comments when dealing with a smaller computer center, media backup procedures are very important in the smaller system environment. A formal, limited access mainframe computer room will usually be structured so that it is difficult to access key tapes and disc files. This is not true for many mini or microcomputer systems. The auditor should monitor aggressively such controls to determine that media and forms are locked up properly when not in use and are controlled.

Figure 8.15 is a checklist to assist the auditor in reviewing physical security controls in the computer center and recommending improvements. This checklist is particularly valuable for departmental computer centers. These department machines may be classic mainframes or minicomputers. Often, they are used for such things as engineering automated design processing. Although management recognizes the importance of physical security controls for business data processing controls, often it ignores these smaller, departmental or specialized computer centers.

The checklist can be used to review physical security controls and suggest improvements for all computer sites within the organization. The effective auditor always keeps relative risks in mind when reviewing such controls and suggesting improvements. It can be argued, for example, that a computer center used for engineering product design is at least as critical to the organization as a small divisional business process system. Although a computer center may be outside the traditional area of accounting and business data processing systems, it still requires some strong physical security controls.

Improving Overall Data Processing Physical Security

All too often, auditors and management view data processing physical security as an issue that belongs within the computer center only. However, in the modern organization, data processing physical security often extends into the office. Auditors should be aware of physical security concerns in such areas as office automation, the information center, and office microcomputers. All of these are used for some levels of data processing. There are physical security exposures, risks, and areas for improved controls associated with each.

Information security on office automation equipment will be discussed in greater detail in Chapter 9. However, physical access controls often are important here as well. While not all office word processors should be kept in a locked, limited access facility, they should be protected as much as possible from tampering, snooping, and potential theft. It is important that media also be secured. Often this can be accomplished by keeping it locked during non-business hours.

Exposure Checklist for Evaluating Computer Center Physical Security Risks

NOTE: This checklist is designed for physical security reviews of departmental computer systems found, for example, in engineering laboratories or manufacturing process control areas. These may be large or small computer systems but often lack the formal procedures found typically in an information systems organization.

PHYSICAL SECURITY CONTROL	YES	NO	N/A
1. Is the computer facility located away from areas where it might be exposed to hazardous materials?			
2. Is entry to the computer facility limited to individuals with a need for access to the equipment?			
3. Have specialized output devices, such as plotters, been moved outside of the main machine area to allow needed access?			
4. Is the machine room facility locked during non-prime shift hours?			
5. Is the machine room equipped with adequate fire detection and prevention devices?			
6. Are other environmental controls, such as air conditioning and humidity monitors, adequate for the specialized computer equipment?			
7. Are there adequate, lockable storage areas in the computer room for documentation, files, and other media?			
8. Are there procedures to back up all system files on a regular basis?			
9. Are backup copies of key files and programs kept in a secure, off-site storage area?			
10. Are there logging and authorization procedures to control the movement of tapes and other media in and out of the computer room facility?			
11. If the departmental computer system serves criti-			

Figure 8.15. Exposure checklist for evaluating computer center physical security risks

	YES	NO	N/A
cal organization functions, have provisions been made for an alternate processing site in the event of a disaster?			
12. If the computer system is equipped with a modem, are there procedures to monitor and accept only authorized inputs?			
13. Is there an information security software system in place to control access to programs and files?			
14. Are there procedures in place to limit the distribution of key output reports to authorized individuals?			
15. Are there adequate physical asset controls over the computer, its peripherals, and any related specialized equipment?			

Figure 8.15. Exposure checklist for evaluating computer center physical security risks *(continued)*

There should be some of the same physical security concerns over office microcomputers as there are over other office automation equipment. While typically it is not appropriate to keep an office microcomputer in a locked, limited access facility, steps should be taken to keep it secure during non-business hours. Though mainframe machines can be subject to malicious damage, microcomputers can be vulnerable also to dishonest employees who potentially can steal such things as component boards, microprocessor chips, or even entire units.

There are simple controls that can be installed to enhance office microcomputer physical security. These include locks that prevent equipment from being moved or opened as well as asset tags consistent with tags used throughout the rest of the organization.

Figure 8.16 is a checklist to help the auditor in reviewing data processing physical security controls within the office. The auditor should be aware of these security vulnerabilities when performing other operational or financial reviews in the office environment. While the typical office business machine does not require the formal, physical security controls that would be found in a large mainframe environment, the checklist points out some areas to improve physical security. However, the most important thing that the auditor often can accomplish in this area is to educate office personnel in the importance of data processing physical security controls within the office.

Exposure Checklist for Evaluating Office Automation Physical Security

PHYSICAL SECURITY CONTROL	YES	NO	N/A
1. Are word processors and office microcomputers locked during non-business hours?			
2. Are desk anchors or other locking devices used to prevent theft of processors, component boards, and peripheral devices?			
3. Are machines marked with asset tags and component boards with identification marks?			
4. Are diskettes, documentation manuals, and related materials stored in locked cabinets when not in use?			
5. Are electrical surge protectors installed on all word processors and office microcomputers?			
6. Are mats installed in the work area to protect equipment from static electricity?			
7. Are there an adequate number of fire extinguishers in the office automation work area?			
8. Are there procedures in place requiring that backup copies be made regularly for all data and program files?			
9. Are logs maintained to control file backup activity?			
10. Is at least one generation of backup files maintained in a secure, off-site location?			
11. Are passwords used on all devices for access control?			
12. Are device passwords changed on a regular basis?			
13. If office automation equipment is on a local area network, are there information security controls to prevent one device from accessing files on other devices?			

Figure 8.16. Exposure checklist for evaluating office automation physical security

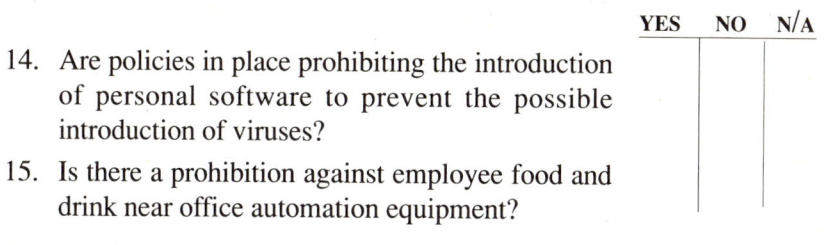

	YES	NO	N/A
14. Are policies in place prohibiting the introduction of personal software to prevent the possible introduction of viruses?			
15. Is there a prohibition against employee food and drink near office automation equipment?			

Figure 8.16. Exposure checklist for evaluating office automation physical security *(continued)*

Implementing a Physical Security Policy

The preceding paragraphs and checklists highlighted a variety of areas where auditors can suggest data processing physical security improvements encompassing the mainframe computer center and other areas of the organization. In general, many of these point to an overall need for physical security policy inside the organization. The auditor who recommends physical security improvements in and outside the computer center often acts as a catalyst for initiating a total security policy within the organization.

Frequently, large organizations have a plant security function that does little more than check for illegally parked cars in the organization's lot. As part of the auditor's findings and recommendations regarding data processing physical security vulnerabilities, it may be appropriate to suggest that plant security become more involved in these issues. They could report violations such as unlocked office computers found during their evening rounds or question unfamiliar employees working in the information center during weekend hours. In other words, they could become an effective force to monitor physical security within the organization.

The auditor reviewing data processing physical security should consider how these security practices fit overall organizational security policies. Often, the auditor finds that the latter have not been updated to reflect modern office automation and departmental microcomputers. If that is the case, the auditor should recommend that overall security policies be established within the organization. These should reflect the data processing physical security risks and exposures in the modern organization as well as good security practices.

CHAPTER 9

Ensuring Information Security and Integrity

INTRODUCTION

In earlier, batch-oriented computer systems, information security was achieved with proper physical security controls over the data center. Today, a physically locked computer room is not enough to protect data and programs. Security violations occur through improper access to user terminals or even through intrusion by outside "hackers" using personal microcomputers. Overall information security controls and procedures are necessary to protect data processing information systems from deliberate or accidental access.

Controls over access to computer data and programs is an important general control that should be considered when the auditor is reviewing a data processing system. However, a good system of information security goes beyond simply establishing information access controls. Information security involves establishing overall policies and controls that prevent or detect any unauthorized computer system access attempts.

Auditors often become involved in establishing information security policies and procedures. Even though an organization's management may have a formal information security function, auditors frequently are sought as advisors and reviewers of that function. Auditors should be aware of the exposures and the controls for establishing effective information security within an organization.

The controls necessary to establish effective physical security and information security often are closely related. However, *physical security controls* usually refer to controls over access to computer equipment, media, and physical facilities as discussed in Chapter 8. *Information security controls* refer to controls over access to data and programs, including access to telecommunication lines. While it is easy to separate physical and information controls in a larger data processing center, this separation often becomes fuzzy when reviewing security controls in a smaller mini or microcomputer system.

This chapter introduces the auditor to the types of information security exposures encountered in a modern data processing environment. The chapter also introduces tools and techniques which the auditor can recommend to establish effective information security and the policies necessary to implement them. Many new tools and techniques are becoming available as concern increases over computer crime, computer viruses, and personal privacy.

Some larger data processing organizations have a formal position of data security officer. The auditor may be asked only to review the practices and procedures of that function or be expected to simply coordinate review activities with the data security function. Smaller organizations tend to look to the auditor as the overall advisor on information security. An auditor knowledgeable in information security practices can fill an important void in such an organization.

INFORMATION SECURITY EXPOSURES

All too often, auditors focus on physical security when performing reviews, ignoring information security exposures. Auditors have been known to place considerable emphasis on physical access controls for a computer room while paying little attention to access controls over terminals in user areas. The typical modern organization faces a much greater security exposure today because of uncontrolled access to data files through user terminals.

Management and society in general are becoming increasingly aware of information security exposures. From time to time, there are reports of someone using a computer to embezzle money from a bank or change grades on a college transcript. There was a popular motion picture several years ago, "War Games," which described how teenage children supposedly used a personal computer to violate strategic defense computer systems. There also have been reported instances of "virus attacks" which have disabled computer systems. While auditors frequently read of security-related criminal acts, they frequently fail to identify potential exposures within their own organization.

Exposures Due to User-Friendly Access Features

Microcomputers and easy to use, on-line software packages have introduced data processing skills to many individuals with no formal training in the area. Users discovered they can use fourth generation language tools, for example, to access or update data files. In addition to the specialized retrieval languages, many applications are designed to allow end users to update, access, and otherwise control the application. While authorized persons input transactions to add, change, or delete records in a user controlled system, there is always a risk that unauthorized persons will use the same systems improperly.

What is the risk or exposure here? Records can be altered to transfer assets or modify accounting records improperly, or sensitive information can fall into unauthorized hands. For example, a salesperson could modify sales performance statistics to earn a larger commission check, or a clerk could access payroll records to determine the compensation of fellow workers.

Customer enacted devices such as automatic teller machines (ATMs) are a type of user-friendly computer that presents an information security exposure. These are computer terminals which permit outsiders to initiate transactions through a computer system. Some examples of other types of user-friendly systems that create security concerns are:

Consumer shopping systems. Consumers can access a computer system through a modem attached to a home computer to place orders for personal merchandise. In some instances, only a touch tone telephone is needed for access. Some systems allow users to order such things as home electronic components, airline tickets, and even groceries.

Industrial ordering and consignment shipping systems. A growing number of industrial organizations are developing Electronic Data Interchange (EDI) systems. Customers use computer terminals to place orders with vendors without going through a formal, paper oriented order entry function. There are similar systems allowing the customer to order a large quantity of specialized goods to be stored at the vendor's warehouse. The customer then uses a computer terminal to initiate shipments. These systems were discussed in Chapter 4.

Manufacturing inquiry systems. With the introduction of "just in time" approaches to manufacturing, increasingly customers are asking for the right to access vendor manufacturing files to determine the status of ordered items. While such systems tend to be of an inquiry nature only, outsiders are given the right to penetrate a computer system.

Cash transfer and stock trading systems. Treasury management and stock trading systems have been used for transactions between larger organizations for some time. However, frequently such systems are now designed to support smaller, less sophisticated users.

All of the above types of systems are designed to be easy to use, and to increase productivity and efficiency for an organization through flexible systems with lower personnel costs. Auditors can expect to see more and more such applications in the future, in areas which will be limited only by the creativity of systems designers.

Frequently, applications designers do not recognize exposures associated with such user-friendly systems. While considerable emphasis has been placed on controls for cash applications, such as ATMs, the same controls often are not considered for other types of applications. In addition, in their attempts to build such systems as convenient and user-friendly as possible, designers intentionally may reduce systems controls.

Auditors should be aware of the information security exposures associated with applications that are, perhaps, too user-friendly with reduced controls. Some of the areas where controls can be reduced to achieve increased productivity include:

Poor password system controls. Data processing applications normally have some type of log-on or password requirement for systems users before they can access a given application or function. In the interest of making such password systems easy to use, systems often are designed with short, easy to remember passwords which are not changed on a regular basis. Unauthorized persons, both inside and outside the organization, potentially can gain access to the computer application because of these poor controls. The exposure to the organization depends upon the type of application being accessed. For example, it may allow an unauthorized person to access a file to review restricted data such as a personnel record. However such an access also might allow a perpetrator to set up a voucher for check payment, or even destroy files and data.

Weak controls over system access. Many user-friendly systems are set up so that users, either inside or outside the organization, can dial up the system to begin processing with a home microcomputer and a modem. Systems access is an area where controls can be enhanced without a significant decrease in user-friendliness. For example, there are call back devices available which will dial back the terminal attempting to access a system on the terminal's assigned telephone number. The connection to the main computer application will be made only after the call back is made. Other techniques require that the dial-up user input a separate personal identification number. Some of these tools and techniques will be discussed later in this chapter.

Poorly designed application controls. In the interest of being user-friendly, some applications are designed with relatively poor controls. For example, applications increasingly are designed to operate with a natural language rather than a strict coding structure. Because of the ambiguities which can be built into such systems, errors can result or improper actions can be taken. If the poorly designed application controls are coupled with an inadequate audit trail, the risks of information security exposure increase. Such exposure is not unique to user-friendly applications. However, because the system designer attempts to make these systems as easy to use as possible, controls are compromised more often than in a formal, controlled application. This is an important area for auditors to be aware of the need for controls.

Auditors should also be concerned with computer crime or fraud and with the possibility of invasion of privacy. However, the major exposure to most organizations is in the area of normal, user accessed business applications. Many security violations initially occur because an employee tries to find out if something works.

The exposure to the organization through information security violations depends upon the type of application. System information violations can result in improper financial data being generated, in users making improper decisions based on incorrect data, or in unauthorized persons gaining access to restricted information.

There is a rather fine line between a criminal act and the types of exposures discussed here. However, whether considered criminal or not, security violations expose the organization to poor decisions or incorrect actions. Following are some examples of these types of exposures:

1. A manufacturing foreman might use a weak password system to gain access to a labor reporting system to alter performance statistics within his area. The foreman will "justify" this by feeling that he can best

correct personnel performance problems internally rather than exposing them to outside management scrutiny.

2. A remote sales branch might use poor access control procedures over a
   sales system to determine which leads have been developed by a
   competing branch. In the interest of "friendly" rivalry, the one branch
   might be able to do better in sales than its related branches by attempting to close sales with the same customers.

3. An accounting supervisor might use the poor systems controls in an
   application to bypass certain required backups. The accounting supervisor will "justify" this since it will speed up the period closing process.

Each of the above examples represents an information security exposure.
The organization would have trouble attempting to prosecute any of them as a
computer crime. Nevertheless, they are information security exposures which
point out system weaknesses and which may lead to even more significant
exposures. Often, they can be corrected through good applications controls.

Auditors should work closely with systems designers in constructing the
appropriate controls when such applications are in their pre-implementation
stages. Procedures for reviewing such controls are discussed in Chapter 7.

Computer Crime and Espionage

From time to time, the press has articles about computer-related crime. These
articles often describe a major embezzlement where a computer system was
used to transfer or divert funds improperly. Other times, they describe a non-
cash crime where, for example, a student accesses a university computer
system to change grades on transcript records. However, because computer-
related crimes are not as dramatic as armed robbery, such events get little
attention. Table 9.1 summarizes some of the major computer-related crimes
discovered in recent years.

Although dramatic computer-related crimes sometimes receive attention
when they occur, other little-noted computer-related criminal acts take place
regularly. In 1984 the American Bar Association surveyed 1,000 public and
private organizations on their experiences with computer-related crime over a
twelve month period. More than 25 percent of those responding reported
having sustained " . . . known and verified losses due to computer crime," with
losses ranging up to $10 million per year.

Organizations sustaining losses due to computer-related crime often do not
report them because of embarrassment or because the organization has secured

DATE	CRIME DESCRIPTION & LOSS	NATURE OF CRIME
1978	Funds transfer theft —$10,300,000	Funds transferred from bank to overseas accounts by outside consultant.
1980	Data systems violations —Unknown	High school students penetrated various company and government systems and damaged files.
1981	Check "kiting" scheme —$21,000,000	Bank manager used terminal to embezzle funds by generating illegal deposits and credits.
1982	Unauthorized computer use —$200,000	Computer time used by a college data processing director to run own private service bureau.
1984	Illegal welfare payments —$300,000	Employee of a county agency entered unauthorized transactions to set up payments to self.
1985	Destruction of records —Unknown	Terminated employee of an insurance company introduced a "Trojan horse" to destroy 168,000 commission records.
1988	Computer systems sabotage —Unknown	Student introduced worm virus into computer networks to disable up to 250,000 systems.

Table 9.1. Major reported computer crimes

restitution from the offenders. Even when reported, prosecution authorities tend to view computer-related crime with less concern than violent crime. In addition, it is difficult to secure convictions due to the technical nature of such criminal acts.

Auditors should be aware of the methods commonly used to commit a computer-related crime and the legal remedies available to punish offenders. Where there are suspicions, the auditor can perform procedures to detect such activities. Of interest here are activities involving the use of a computer to perpetrate a fraud or deception, as opposed to the theft of a computer. Virtually all known crimes where a computer was used as an instrument of the crime have followed variations of one or several of the methods discussed below. While some of these are quite technical and require specialized knowledge to fully understand them, the auditor should have a general understanding of these typical methods of computer-related crime.

Data Diddling

This is the most simple and common method of computer-related crime. The expression simply refers to the modification of data before or during its input to a computer system. Examples include forging input documents, inputting improper transactions, or otherwise violating applications controls. While this is a common method, it is also one of the easiest to prevent and control. A strong system of general applications controls is needed.

Salami Techniques

A programmer in a financial institution could calculate interest for customer accounts and round down each of these calculations. The balances of less than a penny could be placed then in the programmer's personal account. This is called a "salami" technique. By taking a very small slice, no one realizes that anything is missing. However, these small balances can aggregate to large amounts.

Salami techniques date back to before the days of computers when individuals rounded down pennies in manual systems. This is probably the most common example given when a layman describes a computer-related crime. Salami techniques are difficult for auditors to detect. Sophisticated programmers can make the program code for the improper distributions very difficult to detect and also can place the salami shavings into the account of an unknown accomplice. Auditors can detect such activities only through detailed reviews of program code, a normally difficult task, or through fairly sophisticated computer assisted audit techniques.

Trojan Horses

This method consists of the placement of covert computer instructions into a program that sometimes will perform unauthorized functions but usually will allow the program to perform its intended purpose. For example, machine language level code could be placed into a COBOL program which would allow a perpetrator access to operating system restricted files while that program is executing. Because this access takes place during a normal program execution, it probably would not appear on normal log files.

Trojan horse methods have been used for both large scale computer frauds and for sabotage. Only a very skilled technical programmer can construct a trojan horse, but once constructed it is usually difficult to detect. Strong controls over program library updates as well as active management of programming activities are the best preventive methods.

Logic Bombs

There are numerous recorded instances where a programmer built a logic bomb into an application. For example, programmers have inserted code into payroll systems that cause the application to "self destruct" if the programmer's employee number does not appear on the input employee master. Thus if the programmer were ever terminated, files would be destroyed in the next cycle.

Logic bombs usually are constructed through the use of trojan horse methods. They are equally difficult to identify. The auditor should look again for strong controls over access to computer program libraries.

Computer Viruses

Users of personal microcomputers have created an atmosphere where programs often are copied and shared. A computer virus is a malicious program introduced from a shared disc that can damage all programs on the computer to which it is introduced. Similar to a pathological virus, the virus program then may replicate itself and move elsewhere. For example, a virus program was introduced to a major university in 1987. Once loaded, the virus copied itself to other diskettes through an altered "Format" command and then deleted all files on the machines on which it was copied.

Computer viruses are a new threat to the computer security professional. Now they are found primarily on microcomputers, but they probably will soon appear on other classes of computers. Although some software tools are being developed to detect viruses, developers of virus programs tend to stay a step ahead of the protection tools. The best protection against viruses is to install very strong policies prohibiting the introduction of any unauthorized programs to a computer system.

Asynchronous Attacks

A multiprocessor computer system may have a variety of programs simultaneously using various shared peripheral devices. Requests for services and data cannot move back and forth at once across the channels connecting these devices. They must move in an asynchronous manner, or in one direction at a time. The commands to go the other direction must wait for oncoming traffic. Asynchronous attacks represent a sophisticated technique of accessing the data that is waiting to be sent across high speed channels to gain access to unauthorized data, other programs, or privileged areas of the computer's operating system.

These types of attacks require highly sophisticated procedures, and might be used for espionage in certain critical environments. Because of their complexity, the typical auditor needs strong technical assistance to detect them. In a highly sophisticated or secure computer environment, evidence of asynchronous attacks might be revealed through unexplainable deviations in the performance of the operating system.

Electronic Piggybacking

This is a method for gaining access to a controlled computer system by tagging along with a valid user. A hidden computer terminal can be connected to the lines of a valid terminal through wiretapping or the use of other types of switching equipment. When the legitimate computer terminal is signed on to the system but is not using it, the unauthorized terminal can access the main computer system undetected.

There are simple controls to protect against electronic piggybacking. For example, all terminals could automatically log off the system after so many minutes of nonusage.

Wire Tapping

As computer systems become more interconnected through telecommunications networks, the risk of criminal activity through wire tapping to gain information or change data becomes greater. It is easy to establish a wire tapping scheme. The required equipment is available at local radio and electronics supply stores for minimal investment.

The best method of protecting transmissions from wire tapping is the use of encryption techniques. Encryption is the translation of data transmissions into a private code which must be converted back by the receiver. If data transmission security is a concern, these devices can be installed easily on both ends of the transmission.

High-tech Crimes

There are a variety of other computer-related crime methods which are normally a concern only in highly sophisticated environments such as in a government or military computer installation. For example, unless it is shielded properly, a microcomputer gives off electronic inferences which could be picked up by a remote listening device. The auditor operating in the typical business setting or even in state or local government environments, generally will have no reason to worry about the risk of such exposures. For the auditor interested in more information in these areas, the National Institute of Standards and Technology has several publications.

Computer Crime Legislation

Initially, prosecutors tried to use existing statutes in seeking convictions for computer-related crimes. These included various fraud and theft laws at a state level as well as the federal wire fraud law for acts involving interstate criminal activities. In many instances, legal authorities found that these laws did not apply to acts of computer-related crime.

Computer-related crime is difficult to translate into dollar losses, as is done with normal theft or destruction of property. If a malicious intruder accesses a computer system and erases a programmer test library, what is the loss? The programs were under development and cannot be translated easily into a monetary value. An intruder using an existing dial-in telephone line could not be accused of physically breaking into the data center. In many instances, existing laws just did not easily fit this new type of criminal activity.

In recent years, there has been much activity revising existing statutes and drafting new ones to identify properly and specify penalties for computer-related crimes. The auditor should be aware of changing statutory legislation in this area.

Most computer-related crimes in the United States are prosecuted under state rather than federal statutes. The statutes are quite different from state to state in how they define computer-related crime and the penalties for such acts. Unfortunately, there presently exists no "model statute" which generally has been adopted by most states.

Another related development in the area of computer-related crime is the establishment of computer crime units within city or state police forces. For example, Illinois has established a computer crime investigative unit within its state police department, responsible for assisting local police units in investigating and prosecuting computer-related crimes.

The effective auditor should become aware of existing computer-related crime laws and available investigative help. While it is not the auditor's role to be the internal policeman for computer-related crimes, the auditor should be aware of existing remedies. In a larger organization, the auditor should discuss computer-related crime concerns with the organization's legal or security staff to determine that they have an understanding of the problem and of existing legal remedies. In a smaller organization, the auditor should make management aware of the potential exposures due to computer-related crime.

The auditor reviewing data processing controls and applications always should be aware of potential exposures which could lead to criminal related activities. In addition, when such activities are encountered, the auditor should attempt to influence management to press for prosecution. Perpetrators should see that there are penalties for such acts.

Personal Privacy Concerns

Data processing systems can store and catalog vast amounts of information about individuals. Information can be gathered from several different sources and then compared. Some data comparison activity is relatively harmless. For example, advertisers or direct mail marketers will match an individual's ZIP code with auto registration lists, available for sale by many state governments, and with magazine subscriber lists to target advertising in direct mail campaigns. Other types of data comparisons can be more malicious. For example, several years ago Minnesota proposed matching boat registration records with tax records to identify candidates for tax audits on the theory that owners of larger boats may not be paying their full share of taxes.

Several countries have passed strict laws protecting the personal privacy of data on computer files. In the early 1970s, for example, Sweden passed a computer privacy law that allows individuals to examine their personal data kept on government computer files.

What constitutes a violation of personal privacy in a normal business environment? There is probably no single satisfactory answer. However, any time that information gathered for one purpose is revealed to outsiders for other purposes, privacy may be violated. If a financial institution were to match loan application files, which contain much personal information, with demand deposit and savings balances to publish a list for sale to outsiders, personal privacy certainly would be violated. If information security at that same institution was so weak that an outsider could penetrate the system, the institution would be at fault for not protecting the privacy of its customers.

The auditor should be aware of personal privacy issues and concerns when reviewing information security controls and procedures. Poor information security can allow improper access and even computer-related crime. In addition, potentially it can violate the personal privacy of employees, clients, or customers of the organization. If violated in a way that causes harm, individuals could seek a remedy under existing privacy laws or a suit for damages.

Software Piracy

In recent years, software piracy, the theft of copies of purchased computer software, has become a new security exposure. When organizations primarily used large, mainframe based systems, the theft of a copy of a purchased software package was of limited concern because the thief or accomplice had to have the same type of mainframe computer to run the package. All this has changed with microcomputers.

Many individuals have microcomputers in their home or have colleagues inside or outside the workplace with them. A microcomputer software package which will run on one machine will generally run easily on another. Although a software vendor may sell one of its microcomputer software products to a given organization for operation on a single machine, it is very easy for an unscrupulous employee to make copies of the software diskettes for use elsewhere.

In the earlier days of microcomputers, software vendors put controls on their diskettes that prevented them from being copied. However, other vendors readily sold software products which allowed the copying of "uncopyable" software. Because of user complaints over problems in making legitimate backup copies or copying programs to hard discs, there is a trend away from software copy protection schemes. However, most software product packages now come with a set of terms and conditions which require the purchaser to make copies only for legitimate backups. These terms are printed on the package along with a warning that if the purchaser breaks the wrapper, acceptance of the terms is indicated. "Shrink wrap" laws validating these procedures exist in several states.

There have been several lawsuits against organizations who openly pirated software. For example, an organization was sued several years ago for allegedly distributing illegal copies of a microcomputer spreadsheet product to remote sales offices. However, even if not subject to a legal suit, an organization making illegal copies for itself or "winking" at employees who make their own copies is exercising poor ethics and could be subject to public embarrassment.

The auditor should be aware of the potential exposure or embarrassment to the organization through acts of software piracy. This may require reminding management that multiple use copies of software products are illegal. It also should require a formal organization policy statement against the making of such copies. It might be appropriate for employees to sign a statement, placed in their personnel file, acknowledging this policy.

INFORMATION SECURITY TOOLS AND TECHNIQUES

As stated earlier in the chapter, the auditor often is expected to be the computer security "expert" within the organization. This role is easier when reviewing physical security controls because the requirements of a secure physical environment are not too different whether the organization is using a large IBM mainframe machine, a DEC supermini machine, or another vendor's minicomputer. It takes more, however, for the auditor to become an "expert" in the area of information security, because the tools and techniques differ depending upon the types of hardware used, the terminal network surrounding that hardware, and the operating system used.

Discussed here are some of the information security tools and techniques which may be available for a given computer hardware environment. The effective auditor will become aware of the specific products and procedures which are practicable in the auditor's own data processing environment.

Access Control Software Systems

As previously discussed, information security exposures include computer criminal acts, violations of personal privacy, and other unauthorized attempts to access computer data files. In the days of batch oriented systems, strong physical controls often were all that was necessary to establish appropriate controls. However, the modern data center with a large network of terminals requires strong information access controls.

There are typically three levels of access controls which should exist in any computer information security system as follows:

User identification. This is the most basic level of information that should be required when gaining access to a computer system. Usually a log-on ID code provides this by identifying the physical security unit and the user or a group of users. This is not particularly confidential information as log-on IDs are recorded on computer system log files.

Password or personal identification number (PIN). This is a private number or code which is used to authenticate the person using a log-on ID. For example, in a banking ATM system, the user inserts the bank card, which is the log-on identification, and then enters a PIN, which is the authenticating password.

Authorization mechanism. A software procedure is necessary within the computer system to verify that a log-on and password combination is correct and to grant the rights to access various computer system resources. These rights are specified by authorization rules. For example, one password user can only read but not update inventory status records while another can update only certain sales records.

The authorization mechanism is the most important component in the information security system. The auditor should become aware of the authorization software package installed on a computer system as well as the extent to which that package is used. Sometimes an authorization mechanism, or access control software package, will be implemented in such a limited manner that the package accomplishes little.

For larger computer systems, authorization mechanisms or access control software packages are added as an addition to the operating system, purchased from either the computer hardware manufacturer or an outside software vendor. A limited number of organizations also have built their own systems, but generally these are rudimentary at best. On minicomputers, the hardware vendor usually has some type of access control software. Microcomputer systems, if they use an authorization mechanism at all, generally use packages supplied by an outside vendor.

Authorization mechanisms or access control software packages usually contain the following features:

Log-on ID and user password authentications

Log-off facilities to disable a device if a user attempts an invalid password more than a fixed number of times

Authorization controls which allow for specific facilities and data file access rules

Security reports which flag and report unauthorized access attempts

The auditor reviewing an access control software package should look for each of these system elements. The auditor also should become familiar with how the system establishes its rules and how those rules are maintained.

Most larger computer access control systems follow either a *default principle* or an *active request principle*. For a large scale IBM system, an example of a "default principle" system is Computer Associates' product ACF-2. No file or data set can be accessed by a user unless a rule is established within ACF-2 allowing that access. First time users of the product sometimes have installed it on the system without first establishing a rule for the installer to get back and set rules for others. Without a rule, the installer could not get back into the system!

An example of an "active request" product is IBM's own access control package called RACF. Rules are established to restrict access. If there is no rule, there is no access restriction. Other software products, of course, follow combinations of the above two principles.

Which approach is best? This depends upon the data processing organization, the types of information carried on computer files, and the organization's commitment to information security. Establishing "default principle" rules requires a major effort in a larger computer system. Because of this, it often is easy for a computer security function to establish only broad rules which open access to more users than necessary. Conversely, in an "active request" environment, security often will be limited to only those files or data sets where there is a particular management concern.

Auditors frequently are asked to review the selection procedures of a new access control software package. This is a major system installation which impacts the security and integrity of all applications installed within the computer data center. Figure 9.1 provides a checklist for reviewing the purchase of an access control software package.

Of course, the auditor sometimes will encounter a computer operations organization where there is no access control software package or where the package installed is ineffectual. In this situation, the auditor should emphasize the potential information security exposures to management and strongly recommend the installation of an access control information security package.

Establishing Effective Password Systems

An effective system of passwords is the key element in any system of data processing information security. Unfortunately, these same passwords often are used in only a cursory manner. For example, auditors routinely encounter computer systems where passwords have seldom been changed or where all passwords happen to be the first names of the office employees using them. In

Checklist for the Review of Access Control Software

ACCESS CONTROL SOFTWARE FEATURE	YES	NO	N/A
1. Is the information security access control software supported by an established vendor for the hardware and operating system?			
2. Is there a software vendor user group which meets periodically and requests enhancements to the security software package?			
3. Does the software vendor provide training on the use of the software package and on general information security concepts to the security administrators who will use the package?			
4. Can the security software be used to control all types of file structures and programs operating on information systems processors?			
5. Does the software provide a default level of control so that access is denied if a rule has not been established for allowing such access?			
6. Does the software provide flexibility in constructing access rules so that they will be consistent with the overall requirements of the organization?			
7. Does the software allow the security administrator the option to establish passwords of at least eight characters to allow more secure codes?			
8. Are passwords and other key elements of the security software encrypted so that they cannot be retrieved by a systems programmer or other knowledgeable persons?			
9. Does the software provide log-on protection by withholding any information until a password has been accepted?			
10. Does the software provide automatic lockout facilities so that a given number of improper password attempts will lock a perpetrator from a terminal device?			

Figure 9.1. Checklist for the review of access control software

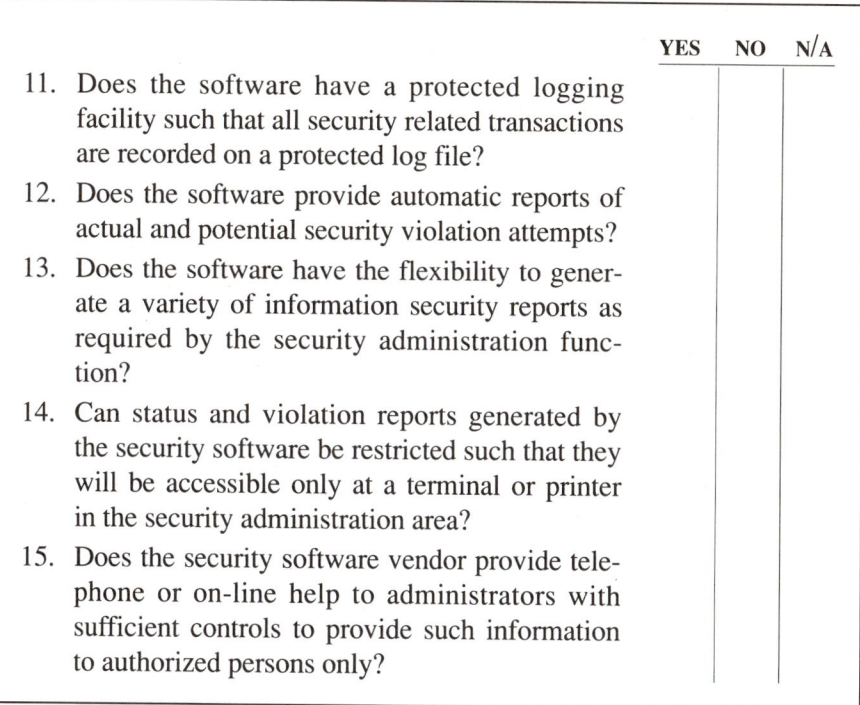

	YES	NO	N/A
11. Does the software have a protected logging facility such that all security related transactions are recorded on a protected log file?			
12. Does the software provide automatic reports of actual and potential security violation attempts?			
13. Does the software have the flexibility to generate a variety of information security reports as required by the security administration function?			
14. Can status and violation reports generated by the security software be restricted such that they will be accessible only at a terminal or printer in the security administration area?			
15. Does the security software vendor provide telephone or on-line help to administrators with sufficient controls to provide such information to authorized persons only?			

Figure 9.1. Checklist for the review of access control software *(continued)*

other instances, these passwords are written down and taped to computer terminals for all to see. These are examples of instances where users of the computer are not aware of the importance of information security.

The use of a proper password allows the access control mechanism of the security software to authenticate the terminal user. Passwords also can be the weak link in an information security system. If they are not changed periodically or if freely left for others to view, the most sophisticated information security system can be compromised easily.

The computer's access control software sometimes will have the ability to assign passwords while in other instances a computer security function will take that responsibility. There are numerous theories and approaches as to what constitutes the best type of password system. Some of the considerations for an effective password system are as follows:

Password generation. Some computer security systems require that users generate their own passwords. They may be asked to submit a new password to the system or to the security administrator on a periodic basis.

Some security systems have controls to prevent users from reusing the same password after every one or two cycles. Other systems generate new passwords for users on a periodic basis. More sophisticated systems generate sets of real words while others produce random streams of characters. The latter tend to be forgotten easily and thus frequently are posted on terminals by users who cannot remember the random sequences of alphanumeric password characters. Whatever method is used, it is a good idea to change passwords at least every six months. In addition, they should be changed when a key employee leaves the department or the organization.

Password ownership. Passwords tend to work best when assigned to or owned by one individual only. There are instances, however, where a single password is assigned to a group or department having access to one file. For example, a data entry function with access to a specific data collection file could be assigned just one password for the department.

Length and composition. There are numerous studies on what is the best length for a password and whether it should contain just alphabetic characters, alphanumerics, or the entire ASCII character set. Many have recommended that a password of four to six upper case alphanumeric characters works well in most non-government security environments.

Password storage. It is easy to compromise a password system if the passwords are stored in full text in a computer file. Ideally, the master file of passwords should be stored in an encrypted manner so that only a data security administrator with the encryption key can read them.

Password entry. Passwords can be read freely and copied if they are displayed on the terminal as they are entered or if they are printed on control reports. Ideally they should be neither displayed nor printed but transmitted directly for verification as they are entered on the keyboard.

Authentication. When a user logs onto a terminal, the password is used to authenticate that user. In some systems, that terminal stays open and active even if the user leaves it without signing off. A better system is to require that a user reenter passwords after five or ten minutes of terminal inactivity.

The password system in use is an area that the auditor should consider when reviewing both controls over an application and general controls. While it is difficult to suggest one good approach to password authentication, the auditor should look for a system that does an adequate job of authentication and provides sufficient security for the given application under review. The auditor should look for a much tighter password system in a treasury management application where fund transfers are authorized than in an application covering inventory movements of bulk goods in a small manufacturing organization. Figure 9.2 provides a table of control objectives and audit procedures for reviewing password controls and procedures.

Some organizations are beginning to implement an improvement to the classic, password-based security system through what is called "see through" security. With this type of system, the user possesses a handheld device which contains encryption code and a key unique to that individual. The host computer has the same algorithm as well as tables of user IDs. Host computers sometimes mount these devices on the front end of the system giving the name "see through." This type of security works as follows:

A user requests access from the host by keying in the user ID through the separate security device.

The host device generates a random number and uses its encryption algorithm to encrypt that number. The host also uses its algorithm key to generate a password.

The host then sends this encrypted random number to the user.

The user identifies himself to his device with his password and then enters the random number received from the host.

The user's device generates the password and sends it to the host. If this password compares to the host's generated password, access is allowed.

Although it sounds complex, "see through" security represents a strong concept in user authentication. Its main feature is that it generates a one time password and, thus, eliminates many of the password problems discussed in this chapter. Figure 9.3 is a conceptual illustration of "see through" security. Many commercial devices to support such systems are currently available, and the auditor can expect to see more in the future.

In addition to password-based authentication systems, there are a growing number of authentication devices based on the physical characteristics of an individual. These are called *biometric* devices. They are used currently for high security applications. However, as their cost decreases auditors will find them in increasing use for other information security applications.

Some of the common types of biometric authentication devices are as follows:

Signature and handwriting verification. Individuals sign their names using unique patterns of pressure and velocity. Systems match these patterns

Control Objectives and Audit Procedures for Reviewing Password Control Systems

Objective 9.2.1.   Passwords should be changed periodically for all users and also when a key employee leaves the organization.

> **Procedure 9.2.1.1.**   Interview the data security administrator to determine the frequency and procedures for changing passwords.
>
> **Procedure 9.2.1.2.**   Interview several systems users to determine their understanding of how often passwords are changed.
>
> **Procedure 9.2.1.3.**   Observe the work areas around terminals to determine that user passwords are not taped to terminals or otherwise displayed.
>
> **Procedure 9.2.1.4.**   Review procedures for changing passwords when an employee resigns or is terminated and determine that these procedures are adequate and are being followed.
>
> **Procedure 9.2.1.5.**   Review procedures for authenticating users and issuing new passwords when a user reports that a password has been forgotten.
>
> **Procedure 9.2.1.6.**   As a test of "forgotten password" procedures, request a user to claim to have forgotten the password and observe procedures for reestablishing the password.

Objective 9.2.2.   Passwords should be based on a minimum number of alphanumeric characters and should not consist of easily recognized identifiers.

> **Procedure 9.2.2.1.**   Determine that the information security software requires passwords with a minimum number of alphanumeric characters.
>
> **Procedure 9.2.2.2.**   Determine that policies and other measures exist to prevent users from using easily guessed passwords, such as a spouse's name.
>
> **Procedure 9.2.2.3.**   Make arrangements to review a selected sample of passwords to determine that overall password

Figure 9.2. Control objectives and audit procedures for reviewing password control systems

policies are being followed; destroy all workpaper records of the passwords used.

Procedure 9.2.2.4. Determine that there are controls such that users can not request their same, duplicate password when asked to change passwords.

Objective 9.2.3. Passwords should have an expiration date such that they will expire if not used over a specified period of time.

Procedure 9.2.3.1. Determine that passwords automatically expire if not used within a designated period of time.

Procedure 9.2.3.2. Determine that the security software automatically signs a user off the system if not used within a designated period of time; also determine that the password must be reentered to reinitiate processing.

Objective 9.2.4. Passwords should not be displayed on terminals or output reports nor should they be stored in text form on computer system files.

Procedure 9.2.4.1. Review password sign-on procedures to verify that passwords are not displayed on user terminals.

Procedure 9.2.4.2. Review security violation and other related reports in systems security and user areas to determine that actual passwords are not printed on any reports.

Procedure 9.2.4.3. Request an edit of the password control file to determine that passwords are stored in an encrypted manner.

Procedure 9.2.4.4. Discuss password encryption with systems security personnel to determine that encryption procedures operate in one direction only, so that encrypted passwords can not be returned to a readable form.

Objective 9.2.5. In high security applications, supplementary controls such as second passwords or other procedures should be used along with conventional passwords.

Procedure 9.2.5.1. Determine that second passwords are required for high security and other sensitive applications such as for making overall changes to the security software.

Figure 9.2. Control objectives and audit procedures for reviewing password control systems *(continued)*

433

> **Procedure 9.2.5.2.** Based on an overall understanding of the types of high security applications being processed, assess the need to consider other password procedures or security devices including:
>
> > The use of a security office to assign and monitor user passwords rather than allowing users to assign their own
> >
> > The use of separate personal identification device readers
> >
> > The use of bionic devices, based on hand prints, signatures, or voice recognition

Figure 9.2. Control objectives and audit procedures for reviewing password control systems *(continued)*

to a computer verification file. The systems tend to fail, however, if the person signing is under stress or is suffering from an injury.

Voiceprint systems. These systems compare a person's voice pattern with a digitized recording. The most sophisticated compare voice patterns to a smoothed set of previous patterns to allow, for example, the voice verification of a Boston-born professional who moves to Texas and acquires a Southern accent. Such systems are best suited for a physical security system rather than terminal information security.

Fingerprint systems. It is well known that every person has a unique set of fingerprint patterns. A variety of devices are available to match a fingerprint to a computer master file. Unfortunately, the error rate can be high due to cuts, blisters, and dirt.

Eye retinal patterns. Just as humans have unique fingerprints, they also have a unique pattern of blood vessels in their eye retinas. Devices which scan the eye retina with low levels of infrared light have been developed. Their error rates are low, but the cost of such devices is high.

Palm geometry. Devices have been developed which compare the finger length, curvature, and webbing between fingers to establish a unique verification. Error rates for such systems are low, and they are being tried in banking ATM devices.

Many of the above biometric devices are still in the early development stage as information security authentication devices. However, these technologies can mature quickly. The effective auditor will follow trade journals covering such devices and make appropriate recommendations for their use when they appear to be cost effective.

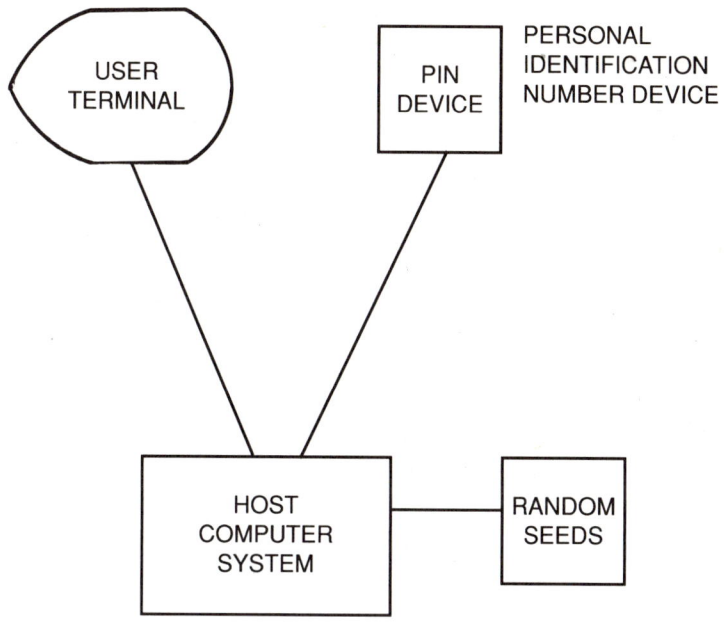

LOG-ON PROCESS

1. USER SENDS ID TO HOST SYSTEM
2. USER IDENTIFIES SELF THROUGH PIN DEVICE
3. HOST GENERATES RANDOM CHALLENGE FOR USER
4. USER ENTERS CHALLENGE INTO PIN DEVICE
5. CHALLENGE AND RANDOM SEEDS ARE CALCULATED AT BOTH PIN AND HOST COMPUTER
6. HOST COMPUTER COMPARES RESPONSES AND EITHER GRANTS OR DENIES ACCESS

Figure 9.3. "See Through" security

Operating System Security

A computer operating system is a complex master program which controls all computer peripheral devices as well as all application programs. In addition, the operating system processes the access control software which provides information security to normal applications. While a user with only a knowledge of the application and the access control system can improperly modify an application data file, it requires a highly skilled technician to modify a computer's operating system.

Secure operating systems have been a subject of major concern among government defense and intelligence computer security experts. Considerable efforts have been expended in defining and developing what has been called a "trusted computer system." The concepts incorporated in such a system are beyond the needs of the typical commercial computer system and the typical auditor. However, this does not mean that the auditor should not be concerned with operating system security. Auditors need to assess the particular vulnerabilities within their own organizations and identify individuals who potentially could compromise operating system security. In a typical data processing organization, the systems programmers are the only persons with sufficient knowledge to modify the operating system. The auditor should determine that there are sufficient controls over that function.

Chapter 2 discussed control objectives and audit procedures for reviewing the systems programming function. If there are good controls over systems programmers, if they consistently report their activities to management, and if data processing management has a basic understanding of the systems programming process, auditors can achieve a preliminary level of assurance that systems programmers are not making unauthorized changes to the operating system.

As a second level of control, the auditor should verify that the access control software is designed so that terminal users cannot gain direct access to operating systems functions. These functions include such things as access to program library files or the various systems directory tables. In addition, the auditor should ascertain that there are no other personnel within the organization with the technical skills to access and modify the operating systems. The main areas of concern might be the applications programming staff and such specialists as database administrators. This sometimes can be determined through inquiry or a review of personnel records.

As yet another level of control, the auditor should determine that there is a good mechanism for the logging and management review of any unusual operations occurrences. These might include situations where a program or peripheral device suddenly does not perform in the usual manner. While it is difficult to attribute such occurrences to operating system security violations, the logging at least will provide a documentation trail for tracing the potential source of any security violation.

None of the steps outlined here normally will provide the auditor with sufficient tools to prevent and control operating systems security violations. However, they will create an environment where violations will be more difficult to perpetrate. If the auditor has strong suspicions of a security violation, a trusted member of systems programming management or an outside specialist with the necessary skills should be used to help review the situation.

Network Security and Data Encryption

A typical data processing system is much larger than just the computer facility, the operating system, and the applications. Many systems today have networks of remote terminals or processors extending beyond the computer's immediate physical location. Communications to these remote facilities normally travel over public or private telephone lines. (Telecommunications networks are discussed in Chapter 4.) Networks can raise a security concern to the auditor.

Although there are many aspects to telecommunications network security, the auditor normally will be concerned with two broad areas:

1. Controls over access to the computer system through improper telecommunications access attempts.
2. Controls to protect data transmitted over telecommunications lines.

There are, of course, other telecommunications security concerns. However, most involve highly secure computer systems as found in government intelligence or defense facilities. Most auditors will not become involved in these latter control issues. Where they are necessary, the organization will certainly have network security specialists to implement them.

Network Access Controls

Many data processing systems are designed so that other computer systems or terminals can dial into the target system to inquire or to update application data. For example, some organizations equip field salespersons or representatives with portable microcomputers for recording activities. These field representatives dial into the central computer over public telephones to transmit data or update files.

Increasing numbers of these network access systems are being implemented in the modern data processing environment. Many industry groups are establishing standards for electronic data interchange (EDI) whereby customers can transmit orders directly to vendor computer systems. Once these orders are processed, the vendors transmit shipping data back to the customer. The auditor can expect to see an increasing use of EDI in many industries due to desires to increase efficiency and become more paperless. The concept of "just in time" manufacturing, pioneered in Japan and now being used extensively in the United States and elsewhere, usually requires EDI.

These network access systems introduce vulnerabilities. If an organization's data processing system is designed so that it can be accessed through a public telephone number, any operator at a terminal or microcomputer with knowledge of that telephone number potentially can access the computer system. For example, in the mid-1980s a group of high school age youths, called "440s" for their local area code, used home microcomputers to penetrate a series of major medical, research, and other computer systems. Although their objective primarily was to see if they could penetrate these systems, they illustrated the vulnerability of larger computer systems to dial-in accessibility.

Even if central computer access numbers are kept private, it can be relatively easy for a perpetrator equipped with a microcomputer to locate access numbers. It is possible, for example, to write a program on a home microcomputer equipped with a modem which will dial sequentially all telephone numbers in an area code. If the answer is a computer data tone, the microcomputer will record that number and go on to the next. The perpetrator can then take the data numbers and later attempt to penetrate the systems.

Microcomputers which serve as both freestanding processors and as terminals to a central computer can create dialup security concerns. The microcomputer could be equipped with a modem which could be accessed from the outside. That outside connection could then use the terminal features of the microcomputer to access the mainframe. The increased use of local area networks (LANs) tied into mainframe processors makes dialup access even more difficult to control. For example, a microcomputer with a modem somewhere on the LAN could provide an entry point to the system. Such a system is illustrated in Figure 9.4.

Even though password systems and other application controls limit access to programs and data once a perpetrator accesses a dial-in number, the auditor should look for controls over actual dial-in access to computer systems.

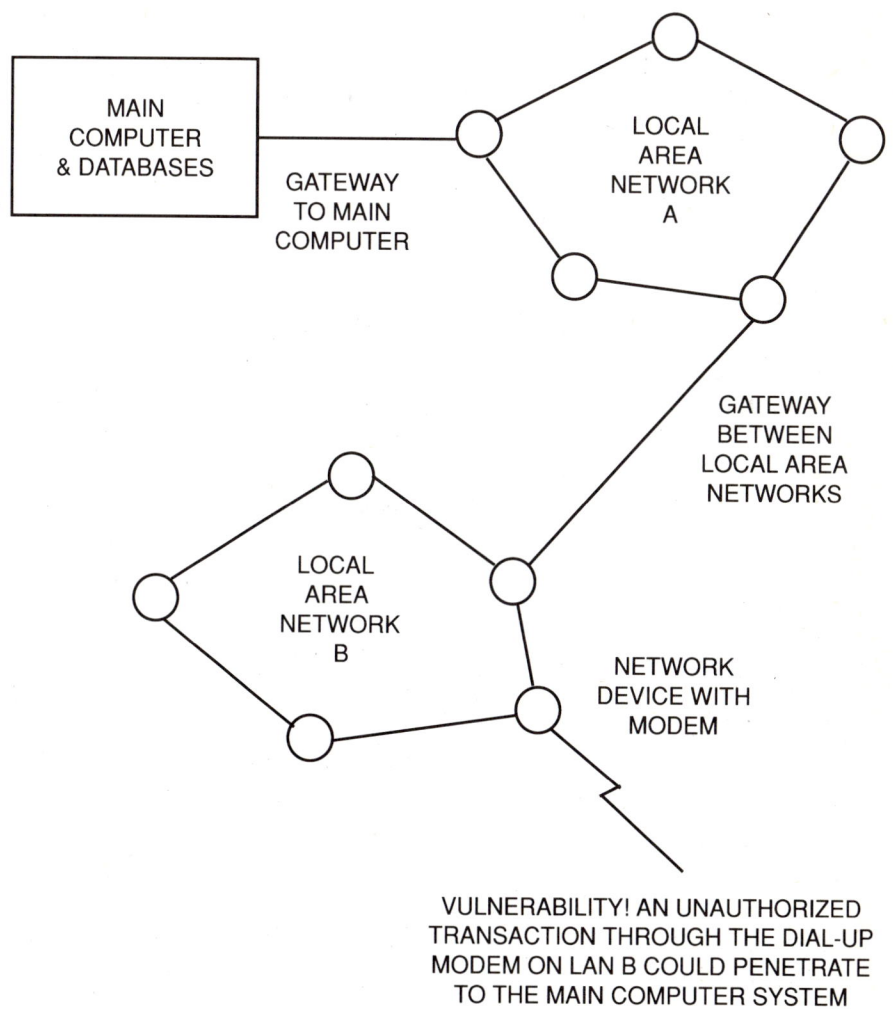

Figure 9.4. LAN access vulnerability

Although a perpetrator might not be able to access a given file or program, that same individual can cause other damage to the overall system.

As a first step, the auditor should inquire into the number of types of dial-in lines into the data processing system under review. In some instances, dedicated lines could be used as an alternative. However, if the nature of the data processing system and its applications seem to require such lines, the auditor should look for some of the following control mechanisms:

Call back systems. Many applications are designed such that dial-in transmissions will come only from terminals at pre-established telephone numbers. A variety of devices are available which will receive incoming calls. They will ask for a log-on ID, match it to an authorized number, disconnect the call, and call back to that authorized number. This type of system is particularly good where incoming calls come from a variety of designated EDI locations. Of course it does not work when incoming calls are from field salesmen who may be using a variety of telephones.

Out only call systems. In many large, multi-location retail operations, it is necessary for the central computer to poll all locations for daily sales data. Rather than allowing incoming calls, systems should be designed such that the central computer initiates all calls to remote locations. Incoming calls would not be recognized by the system.

Tailored log-on IDs. Computer operating software often comes with standard log-on IDs which allow a user to be recognized and asked to enter a password. While it is generally easy to modify these log-on codes, they often remain unchanged allowing a perpetrator to be recognized and invited to start attempting passwords. The auditor should inquire as to how frequently IDs are changed.

Three times and out rules. Once a perpetrator has accessed a computer system, often it is easy to find a valid password by attempting all possible variations. Again, a microcomputer system can keep on trying variations automatically until a valid combination is located. Security software should have a feature that causes the system to disconnect the incoming caller after a fixed number of password attempts.

Limits on demonstration programs. Many of the popular application software packages come with demonstration systems to allow new users to learn the application. Passwords and access rules for these demonstration systems often are published in vendor manuals. Perpetrators such as the "440s" mentioned previously took advantage of this information to access computer systems. Demonstration systems often remain on cataloged public files for extended periods. The auditor should determine that any demonstration programs are used only as long as required for user training. In addition, the data processing function should change the access rules from those published in vendor manuals.

Manual system disconnects. Improper dial-in access attempts are as much a threat on small microcomputers equipped with modems as on larger, mainframe processors. Although the small system may not pose as great a risk due to the limited data and information located on it, such systems do not have the controls to dial back or monitor passwords closely. However, technological vandals have accessed microcomputer systems and damaged key files or programs. When the auditor locates a microcomputer with a modem, it should be suggested that the telephone line be disconnected when not required.

Modem restrictions on LANs. Because a modem attached to one node of a LAN can be an access point to penetrate the entire network, the auditor should recommend that there be no microcomputers with modems on the LAN. If there must be some, they should be equipped with security software to protect the overall network.

The auditor's greatest challenge in reviewing network access security controls often is establishing an awareness of the risks. The auditor often finds that an attitude of "it couldn't happen here" exists among both users and data processing personnel.

Telecommunications Transmission Controls

Once data is transmitted outside of the computer center, it is even less secure. A simple, inexpensive device which can be purchased at a local electronics supply store, for example, can be clipped onto a telephone line to read the stream of data characters. The auditor should be aware of this potential wiretapping vulnerability as well as the risks associated with the data transmitted.

While such devices can record the data transmitted over telecommunications lines, the data is of use only if it can be read and understood. Data encryption is a method to encode the transmitted data according to some algorithm and key so that it can be read only by a receiver who also has the key to decode the message. Many think of data encryption in terms of wartime military actions or espionage activities. However, it is a viable control for many business and other data processing applications.

Codes and ciphers have been in use since the military campaigns of Julius Caesar, and perhaps before. However, computer systems make data encryption and decryption much easier than in the days of manual code books and mechanical coding machines. The basic idea behind data encryption is to use some form of algorithm to code data before transmitting it so that the receiver can apply the same algorithm in reverse to decode it. Someone attempting to intercept the message being transmitted would receive only an unintelligible stream of characters. Figure 9.5 illustrates how this process works.

There are numerous tools and techniques for implementing data encryption. They range from complex, mainframe based systems to microcomputer cards which cost only several hundred dollars per machine. Many modern encryption schemes are based on the Data Encryption Standard (DES) developed by the National Institute of Standards and Technology. Its algorithm transforms data, which is called plaintext, in a manner so complicated that it is extremely difficult to find any correlation between the plaintext and the cyphertext, or encrypted data.

When should the auditor recommend data encryption? Any transmission of highly sensitive data which could be used by others to their advantage is a potential candidate for encryption. For example, cash transfer transmissions which include authorizing passwords could be intercepted to another's advantage and should probably be encrypted.

In many other applications, it is probably not necessary to encrypt data transmissions. For example, a fast food restaurant chain which polls its locations on unit sales statistics probably has no need to encrypt transmissions. The auditor working with management to identify candidates for encryption should use the risk analysis techniques discussed in Chapter 8.

ESTABLISHING INFORMATION SECURITY POLICY

Although many information security tools and techniques are available to the data processing department, they accomplish little without a strong level of information security awareness within the total organization. This includes a general awareness of risks and vulnerabilities as well as procedures for controlling passwords, monitoring information security violations, and promoting a general awareness of security among the user community.

Auditors can play an effective role in establishing an information security policy within the organization. Through reviews covering information security controls, the auditor often has a good overall knowledge of areas within the organization where an information security policy is needed.

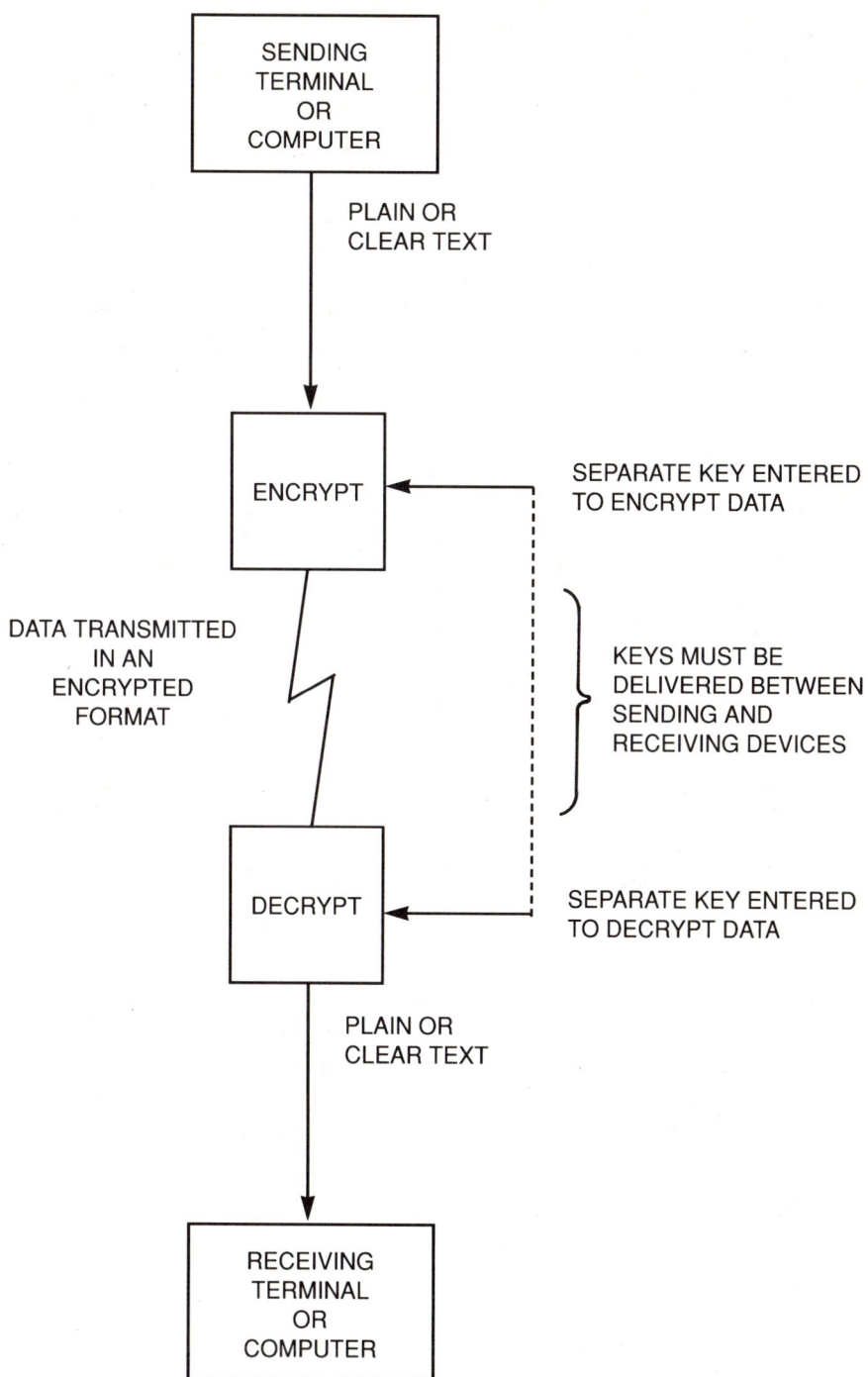

Figure 9.5. The encryption and decryption process

The Information Security Officer

In many organizations, information security has been assigned to individuals as a side activity, and it receives little attention. The systems programming department in a larger organization that has responsibility for installing the data access software often will be given information security responsibility for maintaining that same software. Since this is not a prime function, it often does not receive the attention it should. Similarly, in a smaller organization the data processing manager often takes responsibility for security as an almost casual activity.

Information security is too important to the organization to be a side or casual activity. The auditor should inquire as to who is responsible for information security within the organization under review. If it is a casual, part-time activity, the auditor should point out its importance to management and recommend establishing an information security officer.

In most organizations, the information security function should be responsible for assigning and controlling passwords, monitoring improper access attempts, helping review new applications for security controls, and promoting an overall awareness of information security within the organization. An information security officer or manager can be an effective ally of the auditor in promoting and improving the data processing controls environment.

Nevertheless, auditors should review the information security function for its controls and procedures on a regular basis. If part of the data processing department, this review can be integrated into the overall data processing controls review. Otherwise, it can be considered the subject of a separate examination. Figure 9.6 contains control objectives and audit procedures for a review of the information security function in a typical, larger organization. It assumes that some form of effective access control software has been installed.

The information security officer or manager can be part of the data processing department or, in a larger organization, might be organized as a separate function. If part of the data processing department, the security officer should report at a high level, such as to the data processing director. If not part of data processing, this function often operates effectively as part of the organization's overall security function. In other organizations the information security function reports to the internal audit director as a separate function.

Introducing Information Security to Users

Auditors often have reviewed information security controls within the data processing function, have found them to appear adequate, and then have toured

**Control Objectives and Audit Procedures for
Reviewing the Information Security Function**

*NOTE: These objectives and procedures are designed for a review of
the information security function in a larger data processing organization. Although all of the procedures suggested may not be needed for
the smaller data processing organization, many of these general concepts are still applicable.*

Objective 9.6.1.   There should be an overall management policy
covering information security for the organization which includes
such matters as employee responsibilities and report or document
classifications.

    Procedure 9.6.1.1.   Determine that there is an overall manage-
ment policy for information security which states the importance of information security to the organization and which
defines the responsibilities of the information security function.

    Procedure 9.6.1.2.   Determine that there is an overall classifi-
cation policy for all critical reports or documents, manual
and automated, which classifies documents consistent with
the following general guidelines:

        Company Private—The highest classification for such
things as product related research or financial strategies

        Company Confidential—A middle classification for doc-
uments that could have a detrimental effect if dis-
closed, such as customer lists

        Personal Private—A classification more related to indi-
vidual departments or employees, such as departmental
budget reports

    Procedure 9.6.1.3.   Select a sample of documents, such as the
accounts payable system's vendor lists or engineering
parts lists, and determine that they are classified properly.

Objective 9.6.2.   All employees should be required to acknowledge

Figure 9.6. Control objectives and audit procedures for reviewing the information security function

security policies and should be educated in the importance of information security.

Procedure 9.6.2.1. Determine that there is an overall organization education policy covering information security issues and that employees are responsible for acknowledging their understanding of that policy.

Procedure 9.6.2.2. Review a sample of personnel records to determine that security policy statements have been signed properly.

Objective 9.6.3. The information security function, if part of the overall information systems organization, should report at a high enough level to operate independently.

Procedure 9.6.3.1. If the information security function is part of the information systems organization, determine that it is reporting at a sufficient level to allow it to monitor activities of other groups, such as applications development or systems programming, and to enforce security policies and procedures.

Procedure 9.6.3.2. Interview the management of the information security function to assess whether it is able to act independently.

Procedure 9.6.3.3. If the information security function reports to another function within the organization, such as corporate security or loss prevention, interview members of that reporting function to assess whether they have proper technical knowledge to administer the data processing aspects of that function.

Objective 9.6.4. Information security procedures should give proper consideration to all aspects of security including telecommunications, local microcomputers, and physical security.

Procedure 9.6.4.1. Interview members of the information security function and review activities to determine that it covers all aspects of information security and is not limited to just central computer systems concerns.

Figure 9.6. Control objectives and audit procedures for reviewing the information security function *(continued)*

Procedure 9.6.4.2. Determine that there is an ongoing program of security risk assessment in such areas as:

Local or departmental computers

Local area networks

Telecommunications networks, including dial-up telephone lines

Electronic data interchange networks

Objective 9.6.5. There should be a formal information security procedure for granting access rights to data files and systems.

Procedure 9.6.5.1. Review approval procedures to grant access rights and determine that there is a sufficient level of management approval for granting such rights.

Procedure 9.6.5.2. Select a sample application and determine that all users with a right to access that sample application have been approved properly.

Procedure 9.6.5.3. Review any users who have been granted overall access rights or rights to numerous applications and assess their "need to know" and to access these multiple applications.

Procedure 9.6.5.4. Determine that there is an ongoing policy to review access rights with particular emphasis when employees change job functions.

Objective 9.6.6. Potential violations of security controls should be monitored and investigated.

Procedure 9.6.6.1. Review procedures for monitoring security violations and assess whether these appear adequate.

Procedure 9.6.6.2. Determine that there is a procedure to report repeat access attempt violations to proper levels of management and that corrective action is taken.

Objective 9.6.7. The information security function should be aware of computer crime laws as well as documentation requirements in the event of potential computer crime activity.

Procedure 9.6.7.1. Interview members of the information security function to assess their understanding of computer

Figure 9.6. Control objectives and audit procedures for reviewing the information security function *(continued)*

crime laws as well as potential vulnerabilities within the organization.

Procedure 9.6.7.2. Determine that information security personnel are aware of the need to capture evidence properly, such as files or reports, in the event of a suspected computer crime.

Objective 9.6.8. The information security function should provide guidance on security risks and concerns for new information systems projects.

Procedure 9.6.8.1. Determine that the information security function is informed of all major new systems projects and that the function participates in pre-implementation planning when appropriate.

Procedure 9.6.8.2. Determine that there is a formal information security program to review all proposed new software purchases as well as telecommunications changes for potential security vulnerabilities.

Figure 9.6. Control objectives and audit procedures for reviewing the information security function *(continued)*

user areas to find everyone's password taped to terminals! This illustrates a lack of understanding of information security among users. Security tools and techniques as well as overall policies will have little value to the organization without an overall awareness of information security among the user community.

Chapter 11 discusses the impact of end user computing on the modern auditor. The trend to end user computing increases the importance of having a high level of awareness of information security within the user community. With the growth of information centers and of departmental microcomputers, many security-related decisions are made on a departmental level outside the direct supervision of any information security officer.

As part of any review of the information security administration function, the auditor should look for overall policies and procedures directed at the end user community. In addition, the auditor should monitor security procedures when performing other department level reviews. However, rather than having the computer audit function review this throughout the organization, it is often accomplished by having financial or operational auditors review end user information security procedures as part of their other work.

Figure 9.7 contains control objectives and audit procedures for reviewing information security in end user areas. The audit procedures assume there are adequate overall information security controls within the data processing function and are designed to determine whether end users understand and follow them.

Establishing an Organization Security Policy

An important element in any information security program is a broad statement of security policy. The purpose of the policy is to make all employees aware that information security is a top management concern. If a policy statement does not exist, the auditor certainly should recommend that one be established.

As part of any security policy recommendation, the auditor often will be asked to make suggestions as to the contents of a security policy statement. An important element is a clear statement that *Information is an Organizational Asset*. It should be clearly stated that employees may not take or improperly access data or information of any sort. A violation would make employees subject to penalty.

**Control Objectives and Audit Procedures for
Reviewing Information Security in End User Areas**

Objective 9.7.1. End users should be aware of both overall and departmental information security policies and document classification policies.

> **Procedure 9.7.1.1.** Determine that published copies of organization information security policies are available within the end user department under review and that personnel are aware of those policies.

> **Procedure 9.7.1.2.** Interview selected end users to determine their understanding of security and documentation classification policies.

> **Procedure 9.7.1.3.** Observe procedures for storing computer systems output reports as well as sending them to others to determine that appropriate document classification procedures are being followed.

Objective 9.7.2. Computer systems passwords should be kept confidential by user employees.

> **Procedure 9.7.2.1.** Observe the end user operations area to determine that passwords are not displayed openly on desks or terminals.

> **Procedure 9.7.2.2.** Interview selected members of the end user organization to assess their understanding of the importance of password confidentiality.

> **Procedure 9.7.2.3.** Determine that employee additions, terminations, or changes in responsibility are reported properly to the information security function in order to add, delete, or change systems access rights.

> **Procedure 9.7.2.4.** If a key employee, who has a private password granting access to key applications, is absent from work, assess procedures for gaining access to those information systems to perform that employee's functions.

Objective 9.7.3. End users should be aware of potentially confiden-

Figure 9.7. Control objectives and audit procedures for reviewing information security in end user areas

tial computer system outputs and should take steps to keep them confidential.

Procedure 9.7.3.1.   Interview members of end user departmental management to gain an understanding of which reports and documents should be considered potentially confidential.

Procedure 9.7.3.2.   Observe an end user departmental area to determine that potentially confidential reports and documents are kept in secure areas.

Procedure 9.7.3.3.   Visit the end user area during an off shift period to determine that all appropriate documents are secured and any local departmental computers are locked.

Objective 9.7.4.   End user information security should include local microcomputers and office automation equipment as well as central computer system security.

Procedure 9.7.4.1.   Determine that password security controls are available and are being used for any departmental computer systems.

Procedure 9.7.4.2.   Determine that key programs and files are secured.

Objective 9.7.5.   End users should be aware of procedures for reporting potential information security violations and other unusual items.

Procedure 9.7.5.1.   Determine that clear policies have been established and communicated to end users regarding their responsibility for monitoring and reporting potential security violations.

Procedure 9.7.5.2.   Interview several members of the end user organization to assess their understanding of the procedures for reporting potential information security violations.

Procedure 9.7.5.3.   Review records within the information security function to determine whether the end user department being reviewed has reported any potential security violations; and determine that end users received a proper follow-up to any such reports.

Figure 9.7. Control objectives and audit procedures for reviewing information security in end user areas *(continued)*

INFORMATION SECURITY AND CHANGING TECHNOLOGIES

Technology is making information security more important as well as more complex. We discussed previously some of the security concerns with EDI (or electronic data interchange). While EDI is concentrated currently in a limited number of industries, auditors increasingly find it used. It is not uncommon for a large customer of a small manufacturer using a minicomputer to download orders to the minicomputer using EDI. Both management and auditors must be aware of the control concerns associated with such an arrangement.

The growth of database systems is another area of changing or evolving technologies which affects information security. In older data processing systems, it was relatively easy to develop a relationship or set of rules covering which users could access a given file or set of data. Large, integrated databases often incorporate considerable data sharing. It can be an information security management concern as to who should access the various data elements within such a shared database.

While changing technologies cause information security concerns, neither management nor auditors can "put their head in the sand" and refuse to accept new concepts due to security concerns. In the EDI example, management may have no choice other than to accept the large customer transmission of orders if they want to retain that customer. In such an instance, the auditor can play a key role in working with management to recommend control procedures which will be adequate.

Auditors in today's data processing environment must keep aware of changing technologies and their security and control implications. Such a level of awareness allows the auditor to be an increasingly effective advisor to management on matters of information security and control.

Effective Disaster Recovery Planning

INTRODUCTION

Many modern organizations are dependent upon data processing operations. If the order entry terminal network goes down for only a few minutes in a sales or distribution organization, there can be chaos. Order entry clerks cannot take orders, verify the status of orders, or make credit checks. When such a system is brought back online, there is a feeling of great relief. The problem, however, is much greater when the computer is down for an extended period. While informal procedures often exist for operating during short term periods of system down time, many organizations have no plan for restoring data processing operations in the event of an extended, emergency interruption.

A data processing disaster recovery or contingency plan is a formal set of procedures designed to allow the data processing function to continue to operate at some level in the event of an extended period when normal data processing facilities are unavailable. For example, a fire, flood, or other natural disaster could all but wipe out an organization's central data processing facility. Without an effective plan to restore data processing operations, business units supported by automated systems could be inoperable for an extended period of time.

Auditors frequently are asked to help draft or review a data processing disaster recovery plan. They know the requirements and vulnerabilities of various data processing applications; and they understand the need for backing up key programs and data. As discussed in Chapter 8, auditors are viewed as the organization's data processing security "experts."

This chapter discusses how auditors can help build effective data processing disaster recovery plans. The chapter explains how to assemble a disaster recovery planning team, evaluate data processing vulnerabilities, and build and test an effective disaster recovery plan. The auditor, however, should not be the only person in the organization responsible for such a plan. Rather, an effective data processing disaster recovery plan is the responsibility of top management. The auditor should act only as a catalyst to see that it is developed and then should help test and evaluate the completed plan.

THE NEED FOR A DISASTER RECOVERY PLAN

Virtually all organizations are dependent upon data processing operations and would suffer a severe loss if their computer systems became unavailable for an extended time. In larger organizations, most business systems are highly automated. These systems might include such areas as payroll, accounts payable, manufacturing control, and production process control. Some organizations, of course, are more dependent than others. A hotel chain or airline with an on-line reservation system or a bank or other financial institution with an on-line financial accounting system would have difficulty operating at any level without their computer systems.

Although a modern computer system has become a necessary element for ongoing operations, management often fails to understand the need for effective data processing disaster recovery planning. Data processing services are viewed almost as an outside utility like power, telephones, and water. For true outside utilities, however, extensive backup facilities are generally developed to provide such services continuously. Data processing is like an in-house utility that requires its own backup arrangements.

An electrical utility may be subject to a catastrophic disaster where it could lose an entire generating plant. More commonly, it loses a local transformer because of a storm. Both will shut down power until contingency restoration plans are implemented. That restoration will seem almost transparent to the customer; lights will flicker for a moment while backup power sources take over. Similarly, a data processing operation may be subject to a major or a local disaster. As an example of the latter, a broken water pipe could flood a telecommunications control room and bring the network down. An effective data processing disaster recovery plan needs to provide for all types of contingencies.

Auditors play an important role in calling management's attention to the need for effective data processing disaster recovery planning. If such a plan does not exist, the auditors can emphasize the need for one in a general controls review audit report. Similarly, if such a plan has been drafted but appears ineffective, they can highlight this fact in an audit report. They also can educate management on the importance of such planning.

It must be emphasized that data processing disaster recovery planning is the responsibility of management. Auditors and data processing personnel can help develop the plan and test it for effectiveness. However, if there is an extended interruption in data processing services without an effective recovery plan, management must accept the consequences of such a loss. Some of the significant exposures that management faces include:

Cash flow loss. In many organizations, data processing systems are necessary to record new orders, bill for order shipments, and collect open accounts. Without the availability of these automated applications, an organization's cash flow could be limited. There are numerous other applications in the typical organization which may have a similar impact on cash flow.

Legal liability. A common interpretation of the Foreign Corrupt Practices Act of 1977 is that the officers of an organization are personally responsible if adequate preparations are not made to safeguard an organization's vital documents and other resources. An organization's officers may be exposing themselves to a legal suit if they fail to develop a disaster recovery plan. In addition, management in some regulated industries often have a legal or administrative responsibility to have proper disaster plans in place. For example, if a financial institution is insured by the Federal Deposit Insurance Corporation, it is required to have a disaster recovery plan for data processing operations.

Logistics and business interruption losses. In addition to financial considerations, many organizations are dependent logistically upon a computer system for many key operations. A household moving company may use an on-line system to schedule its trucks and drivers. Without such a system in place, the company might not be able to operate.

Public relations losses. Although it may not result in a direct financial loss, an extended interruption of data processing services could cause major public relations and shareholder relations losses. If an organization is subject to a major catastrophe such as a flood or fire, it will have the short term sympathy of outsiders. However, if vital functions are not restored promptly and orderly, there can be a serious loss of public respect.

Management must consider these issues in determining the need for an effective data processing disaster recovery plan. Even if the organization has only a smaller computer system, there will be a need for some form of disaster recovery plan. While data processing personnel, auditors, and others can help develop the plan, it must have the support of management to be effective.

EVALUATING EXISTING DISASTER RECOVERY PLANS

Many data processing organizations have developed some form of disaster recovery plan as a result of earlier audit report recommendations. In some instances, there was a major effort to develop an effective and operable plan. However, many of those plans end up as thick books that are shelved and never used again. The disaster recovery plan often is neither updated nor tested after its original implementation.

Some existing disaster recovery plans that an auditor may encounter are limited to informal documents. Data processing may have made reciprocal arrangements at a nearby computer site for the right to process there if the organization's own computer were to be unavailable. These plans often are informal and untested. They tend to be ineffective and inoperable.

The auditor should evaluate any existing disaster recovery plans as a first step in helping to build an effective plan for the organization or as part of any data processing general controls review. Auditors find that most existing plans are either based on informal reciprocal arrangements or out of date. The auditor should learn to recognize the types of existing plans to make recommendations for improvement where necessary.

Reciprocal Agreement Plans

In the early days of data processing, backup and recovery plans were relatively simple. Automated applications often were conversions of existing paper oriented systems. It was necessary only to keep the old forms, and the organization could return to manual operation with little problem. In addition, often it was not difficult to informally arrange a backup processing site. In those times, not many organizations had computer systems and the existing machines were not always busy. Because these machines periodically went

down for fairly long periods, it was not difficult for a data processing operations manager to arrange to do backup processing on another organization's
computer on an idle shift.

These informal data processing backup plans are no longer practical in the
modern data processing operation for the following reasons:

Modern applications no longer have manual backups. The era when
most automated applications were merely conversions of older paper oriented systems has long since passed. If the application still replicates an
older paper system, the persons who had knowledge of the older system
often have retired or moved to other areas. Most newer applications, more-
over, have no direct manual counterparts. For example, it would be almost
impossible to replicate a modern manufacturing requirements planning
system using manual procedures.

Computer systems are usually too busy to allow backup use. Although
some organizations still have existing disaster recovery plans based on
reciprocal use agreements with other computer sites, they are probably
inoperable. Most organizations use their computer systems 24 hours a day,
six or seven days a week, and would not be able to allow an outsider to take
over a shift.

Telecommunications and on-line applications limit transport. It is no
longer possible to shift a computer operation to another site without extensive planning. The typical modern, larger computer system has a large
network of terminals and telecommunication links to other sites. These
cannot be moved easily without detailed planning.

Litigation concerns discourage informal arrangements. It was once
common for the data processing operations manager to make informal
arrangements with a data processing manager at a nearby organization.
Today, that same agreement probably would be cleared through the organization's legal staff before approval. Few organizations will promise formally to provide backup services because they may be held legally liable if
they do not.

Despite these shortcomings, disaster plans based on reciprocal agreements
are still encountered. The auditor should review such an existing plan with
attention given to whether the reciprocal agreement appears to be workable
and whether there has been any level of testing. Questions the auditor should
ask when reviewing these reciprocal agreement plans are summarized in Figure
10.1. If the auditor finds the agreement is untested or appears unworkable, an

Checklist for a Review of a Reciprocal Disaster Agreement

NOTE: This checklist is designed to allow the auditor to assess the adequacy of data processing disaster recovery plans which are based on a reciprocal agreement with another organization. The checklist is designed such that "NO" responses represent potential weaknesses in the organization's reciprocal plan.

RECIPROCAL PLAN FEATURES	YES	NO	N/A
1. Is there a formal, signed reciprocal processing agreement?			
2. Has the plan been signed by management levels above the information systems organization?			
3. Has the corporate legal staff or appropriate counsel reviewed the agreement?			
4. Has the plan been updated or reconfirmed within the last 24 months?			
5. Are there specific procedures outlined in the agreement defining what constitutes an emergency condition and requirements for notification?			
6. Is the agreement structured such that neither party can refuse service to the other arbitrarily in the event of a disaster condition?			
7. Are the hours that either party can use the other system defined in the agreement?			
8. Does the agreement contain terms for compensating the other party for data processing machine resources and other services?			
9. Have tests of key systems been run on the other system within the last 12 months?			
10. Does the agreement outline procedures for notifying the other party of planned and actual changes to the operating system or to other operating software or hardware?			
11. Does the agreement cover backup and recovery			

Figure 10.1. Checklist for a review of a reciprocal disaster agreement

	YES	NO	N/A

procedures for the organization's telecommunications network?

12. Does the agreement define only critical applications to be run at the reciprocal site?

13. Are key contact personnel listed within the agreement or in a separate disaster recovery plan document?

14. Are there arrangements for storing key backup files as well as critical documents in a secure location that is accessible to the reciprocal site?

15. Is there a comprehensive information systems disaster recovery plan which incorporates the expectations and responsibilities of the reciprocal agreement?

Figure 10.1. Checklist for a review of a reciprocal disaster agreement
(continued)

appropriate recommendation would be to revise and redraft the disaster recovery plan, as discussed in the paragraphs following.

There may be some instances where a disaster recovery plan based on reciprocal agreements may be the only available alternative. If the organization is located in a smaller city in a relatively remote area, a reciprocal agreement may be one of the few available alternatives. If that is the case, the auditor should look for some evidence of disaster plan testing at the reciprocal site.

Untested or Outdated Plans

The auditor frequently finds that disaster recovery plans remain unchanged and untested once they are completed. Data processing departments and users often devote extensive efforts to develop such plans and then put them on the shelf with little ongoing referral. Such a plan is, perhaps, more dangerous to an organization than no plan at all if management places a false level of reliance on the plan. If there is an extended interruption in data processing services, it suddenly can become apparent that the published plan does not work.

Some of the reasons that cause an untested or outdated plan to be inoperable include:

Key data sets have not been saved. A disaster recovery plan should identify all data sets necessary to recover key systems. If backup records have not been kept up to date, the published plan may not provide the proper guidance to restore key applications.

Recovery site configurations may have changed. There may have been subtle changes in the hardware or operating system at the disaster recovery backup site. These changes should be identified when the plan is updated; they will not be known in untested or outdated plans.

Key personnel designated in the plan may not be available. A properly constructed disaster recovery plan designates various staff members to handle recovery activities. Because staff personnel may change, an outdated plan will not identify recovery responsibilities properly.

Sometimes, the auditor can determine whether a plan is outdated through simple inquiry and discussion with data processing management. However, in many instances, it will be necessary to perform some tests to determine the currency of the published plan. Disaster recovery plan audit tests are summarized in Figure 10.2.

As a result of this investigation, the auditor may find the current plan to be adequate with some elements in need of revision, or may find it to be inadequate and in need of a major revision or rewrite. An adequate plan should contain many of the elements discussed later in this chapter. If the plan is found to be adequate but in need of fairly minor revisions, the auditor should identify those elements and suggest an update.

In many instances, this evaluation of any existing plan will reveal that it is not effective. It may not identify all key systems and data sets necessary for recovery; it may not identify a recovery team to restore data processing operations; or it may not include a tested site for disaster recovery processing. If any or all of these are lacking, the auditor should consider recommending the building of an updated, effective disaster recovery plan.

This recommendation for a new disaster recovery plan may require the auditor to do some "selling" to management. Any audit report with this recommendation should outline shortcomings and deficiencies in the current plan and discuss the potential risks of not having an effective plan. This "selling" is often necessary because management will look at the expense of developing the last plan, realize they have never had to use it, and question whether it is necessary to go through the effort again.

Control Objectives and Audit Procedures for Reviewing a Disaster Recovery Plan

NOTE: The purpose of this table of objectives and procedures is to allow the auditor to review a data processing disaster recovery plan to determine whether it is up to date and adequate.

Objective 10.2.1.   The plan should list key personnel, including users and the actual recovery team, who will be available during the recovery process.

>   **Procedure 10.2.1.1.**   Compare the names of persons identified in the disaster recovery plan with current organization charts to determine whether recovery assignments are current.

>   **Procedure 10.2.1.2.**   Select a sample of home telephone numbers listed in the plan and determine that they are current.

>   **Procedure 10.2.1.3.**   Interview selected personnel designated for tasks in the disaster recovery plan and determine that they understand their responsibilities as outlined in the plan.

>   **Procedure 10.2.1.4.**   Interview selected members of the disaster recovery team to determine that they understand the calling sequence and their responsibilities.

>   **Procedure 10.2.1.5.**   Review procedures for updating personnel assignments in the recovery plan and assess their adequacy.

Objective 10.2.2.   Data sets, file requirements, and key forms for critical applications should be defined in the disaster recovery plan.

>   **Procedure 10.2.2.1.**   Determine that the disaster recovery plan lists critical applications along with the data sets and other media requirements for those applications.

>   **Procedure 10.2.2.2.**   Select several key applications defined in the disaster recovery plan and match file and other requirements to current operations documentation to determine that the disaster recovery plan is current.

>   **Procedure 10.2.2.3.**   Using the file backup requirements of several key applications, visit the file storage backup site to determine that key files are backed up properly.

Figure 10.2. Control objectives and audit procedures for reviewing a disaster recovery plan

461

Procedure 10.2.2.4. On a sample basis, determine that key forms and documentation for critical applications are defined and stored at the backup site.

Objective 10.2.3. The disaster recovery plan should have a current agreement defining an alternative processing site.

Procedure 10.2.3.1. Review procedures for backup processing in the event of an emergency condition and assess whether they appear sufficient.

Procedure 10.2.3.2. If the alternative processing site is defined through a reciprocal agreement, assess the adequacy through use of the checklist in Figure 10.1.

Procedure 10.2.3.3. If the alternative processing site is defined through a "hot site" service agreement, review the agreement to determine that:

The disaster recovery site supplier provides sufficient opportunity to test these processing procedures

The equipment located at the disaster site appears to be compatible with current hardware and operating software configurations

The "hot site" provider has limited the number of subscribers to the service such that contention will not be a major problem

Arrangements have been made to support the telecommunications requirements of key applications

Procedure 10.2.3.4. If the alternative processing site is a "cold" or "shell" location equipped to handle a computer system installation on a short notice, visit the site if practicable to determine that:

The shell site appears to have adequate power, environmental controls, and physical space to support emergency backup processing needs

If other operations are currently located at the shell site, verify that there are procedures in place to move those operations to another site

Commitment letters are on file from computer hardware and equipment vendors to deliver required equipment as soon as possible in the event of an emergency

Figure 10.2. Control objectives and audit procedures for reviewing a disaster recovery plan *(continued)*

Objective 10.2.4. The disaster recovery plan should be tested every twelve months.

Procedure 10.2.4.1. Review disaster recovery plan test results to determine that a test, including a visit to the alternate site, took place within the last twelve months.

Procedure 10.2.4.2. Review documented results from the last plan test to determine that they include an analysis of processing results as well as any recommendations for corrective actions.

Procedure 10.2.4.3. Determine that representative members of a disaster recovery team, rather than just a limited group of software specialists, participated in the disaster recovery plan testing process.

Objective 10.2.5. Disaster recovery plans should be consistent with the overall disaster recovery plans for other organization functions.

Procedure 10.2.5.1. Review disaster recovery and emergency plans for other departments or units within the organization and determine that the data processing plan is consistent with those procedures.

Procedure 10.2.5.2. Determine that procedures are in place to coordinate the data processing disaster recovery plan with other organization emergency plans.

Figure 10.2. Control objectives and audit procedures for reviewing a disaster recovery plan *(continued)*

DEVELOPING A DISASTER RECOVERY PLAN STRATEGY

If an organization has no data processing disaster recovery plan or if the existing plan appears to be outdated or otherwise ineffective, efforts should be initiated to develop such a plan. To develop an effective plan, however, an overall strategy is needed. Strategy development requires the participation of both data processing and upper management. It also requires an analysis of alternative strategies and recovery options.

This chapter discusses some management considerations necessary for effective plan building. It outlines the various team members who should participate in developing the plan and some disaster recovery plan strategies. While auditors play a key role in the plan building process, an effective disaster recovery plan requires much broader participation by all levels of the organization's management.

Management Responsibilities and Needs

Data processing management usually has a strong appreciation for the vulnerabilities of their computer systems and equipment; general management often does not. We discussed how an extended interruption in data processing services might subject the organization and top management to financial and business losses and to legal liability. A first step to building an effective disaster recovery plan is to make upper management more aware of the potential vulnerabilities of the organization's data processing system.

Figure 10.3 is a data processing vulnerabilities checklist which the auditor can use to help develop this awareness. It is quite possible that managers completing such a document will be unable to answer some of the questions. That is acceptable as it points to the need for more data gathering and analysis. The auditor might ask several members of the upper management to complete this checklist. The results can be compiled and distributed informally, without the names of the respondees, to the management team as a stimulus for further action.

Although an organization's data processing systems and applications can be critical, there are other critical areas in many organizations. The engineering drawings kept in a laboratory or drafting department may also be very critical to the organization's ongoing operations. A data processing disaster recovery plan should be tied into a vital records program that is used also for these non-automated items. If such a vital records program does not exist, the data processing disaster recovery plan can serve as a catalyst for developing such a program.

A Checklist for Identifying Data Processing Vulnerabilities

NOTE: This checklist is designed to assist the auditor in reviewing management's awareness of data processing vulnerabilities and in helping management decide that there is a need for a data processing disaster recovery plan when an effective one does not exist.

DATA PROCESSING VULNERABILITY	YES	NO	N/A
1. Does the organization have critical or strategic applications which are necessary for ongoing organization operations?			
2. Is the organization dependent upon centralized or departmental computer systems for financial and government reporting?			
3. Has there ever been a risk analysis performed covering the potential losses associated with key data processing applications?			
4. Do data processing applications, whether purchased software or developed in-house, represent a major investment to the organization?			
5. Even if there are no critical applications, has there ever been an analysis performed estimating the total cost to the organization if the computer systems were not available?			
6. Are all computer systems users aware of the need to back up critical files, and are they performing those backups on a regular basis?			
7. Has consideration been given to the vulnerabilities of the voice or data telecommunications network?			
8. Does the information systems function have a data processing disaster recovery plan?			
9. Has management reviewed and critiqued that disaster recovery plan?			
10. Does the disaster recovery plan include the identification of an alternate processing site?			

Figure 10.3. A checklist for identifying data processing vulnerabilities

	YES	NO	N/A

11. Does management understand the costs and capabilities of the alternate site computer processing procedures outlined in the disaster recovery plan?

12. Is there a formal program for testing the disaster recovery plan using the designated alternative facility?

13. Has information systems ever briefed management on the results of any disaster recovery plan testing?

14. Does the organization have overall security procedures regarding perimeter control and fire alarms?

15. Has the organization's security function given adequate consideration to data processing operations?

16. Has information systems security planning included both departmental microcomputer as well as any other specialized computer systems?

17. Has either the security function or an outside group, such as the local fire inspector, reviewed fire prevention procedures within data processing operations?

18. Has management considered the adequacy of business interruption and loss of records insurance as part of data processing disaster recovery planning?

19. Has management determined that any disaster recovery plan includes adequate backup procedures for other areas such as telecommunications?

20. Has management reviewed disaster recovery planning for other non-data processing areas in the organization's operations?

Figure 10.3. A checklist for identifying data processing vulnerabilities *(continued)*

Management also needs to be aware of the roles of other supporting organizations in disaster recovery planning. Many larger organizations have an in-house security function. While such a department often has limited its responsibilities to normal guard and fire detection duties, it can help build an effective disaster recovery plan for data processing and for the overall organization.

While a smaller organization or one located in a more remote community may not have a security department, other resources often are available. For example, the local fire marshal or fire department can provide advice on fire prevention techniques.

Disaster Recovery Strategy Alternatives

After management needs and requirements have been identified, the next step to building an effective disaster recovery plan is to consider the available strategy alternatives. While some will be modified as the plan is developed, this is a first step to consider potential strategies for data processing recovery. Following is a partial list of strategy alternatives:

Replacement with manual processing procedures. Although many applications cannot revert to manual procedures, some can, at least for a limited period of time. In developing a set of strategy alternatives, it is a good idea to consider which applications can be processed with manual procedures. In a larger computer environment with database files generally this is not a viable alternative. However, in a low transaction volume or a mini or microcomputer setting, it sometimes is a good option.

Identification of alternate processing sites. To perform backup processing, it is necessary to have an alternate site with a comparable computer system. While no decision will be made until the plan is developed and various alternatives explored, it is a good idea to explore the alternatives early. In a larger, multi-location organization, an alternate computer system may be located at another plant or division. In some instances, reciprocal arrangements can be made with another organization. Also, outside vendors are now providing computer centers for backup processing. However, if the computer system is older or an unusual model, such alternate processing sites may be difficult to identify.

Identification of critical data processing functions. Before the identification of specific applications, the general functional areas of critical data processing activities should be identified. If a manufacturing plant is highly automated, manufacturing support systems may be the most critical data processing functions for that organization. Other functions, such as sales and billing, also are critical but might be handled on a manual basis.

Identification of disaster planning resources. When evaluating alternatives, consideration should be given to the available resources for constructing a disaster recovery plan and evaluating it. Key users as well as data processing personnel should be identified preliminarily as possible members of any disaster recovery restoration team. This list of disaster planning resources also should include potential vendor resources.

This identification of alternate strategies is a preliminary exercise at this point in the disaster plan development process. However, this process allows management to begin thinking about the magnitude of the problem as well as potential resources to solve it. At this point, it is not necessary to have answers for all the issues these alternatives raise.

Identification of disaster recovery planning alternatives can be developed by a limited group of personnel prior to the formal plan construction. Many alternatives can be developed by a key member of data processing along with the auditor and a representative of user management. Once developed, these alternatives should provide a basis for securing upper management's approval to develop a formal disaster recovery plan for data processing.

Building the Disaster Recovery Planning Team

Developing and implementing an effective disaster recovery plan is a team effort involving individuals from data processing, user departments, and others. A first step to building such a team is to appoint a chairperson with managerial capabilities and a strong knowledge of the technical issues associated with such a plan.

In a larger data processing organization, the appropriate individual to head up this disaster recovery planning team often is the head of computer security. Because of activities in assigning access rights, this individual often has the necessary strong technical knowledge required to build the overall plan. Depending upon the skills of this individual, it may be more appropriate to have a member of user management head up the planning team with the head of computer security a close technical advisor. The auditor should be a member of and an advisor to this planning team. However, usually it is not appropriate to have the audit function head up the effort as they may be called upon to evaluate the disaster recovery plan at a later date.

In a smaller data processing organization, a key member of user management probably should head up the disaster recovery team with the head of data processing as vice-chairperson. Because of the limited data processing staff in a smaller organization, the head of data processing often does not have time to head such a disaster recovery team project.

In addition to the disaster recovery chairperson, a committee should be assembled to evaluate and build the organization's data processing disaster recovery plan. The size of this team will depend upon the overall size of the organization. For a larger organization, the disaster recovery committee might consist of individuals responsible for the following functions:

Computer operations

Data security administration

Telecommunications and local networks

Systems programming

Database administration

Systems and programming applications development

Information center management

Administrative services, including insurance

Purchasing

Internal audit

Key user application areas

Of course, not all of the above functions exist in every organization. In smaller organizations, individuals often wear multiple hats. For example, a minicomputer-based organization might have one technical support person responsible for data security, telecommunications, database administration, and systems programming. That key individual certainly should be included on the disaster recovery team.

Rather than assembling an in-house team to develop the disaster recovery plan, some organizations elect to use outside consultants to perform or help with this task. There are consulting firms that specialize in developing such disaster recovery plans. In addition, many major public accounting firms have the necessary skills to build a plan for the organization.

The advantage of using an outside resource is the ability to utilize the experience and knowledge gained through building such plans for a variety of organizations. In addition, an outside consultant will construct the plan on a priority basis while insiders may not give it that priority and require a considerable period of time to complete the plan. The disadvantage is that an outside disaster recovery consultant might not know many of the organization's unique characteristics and system details. As a result, plans developed by outside consultants often tend to be generic.

If an organization has a need to construct the framework of a disaster recovery plan in a short time, an outside consultant might be a cost-effective resource to accomplish this. However, members of the organization also should participate in this effort and work to refine the consultant's disaster planning document. Otherwise, frequently, disaster recovery plans developed by outside consultants become the previously discussed ineffective documents that do little more than sit on a shelf.

Responsibilities of the Disaster Recovery Planning Team

The disaster recovery planning team will evaluate data processing risks and then build an effective recovery plan. That process requires considerable analysis and study to determine the applications required for a disaster recovery and effective strategies to make that recovery possible. The auditor often plays an important role in this process by performing some of the survey and analysis work that will be necessary to build the plan.

Much of the work of the disaster recovery planning team will be information gathering. For example, while data processing management and others are aware that an organization may have a complex telecommunications network, it takes the analysis work of the recovery team's telecommunications specialist to define those requirements in a manner that can be documented and reconstructed. Similarly, a complex application requires a series of data sets from feeder applications and, possibly, specialized documents or forms. While some system documentation exists for this, a member of the planning team should review all requirements for the application to determine whether they are documented properly.

After the disaster recovery planning team has analyzed risks and built the organization's disaster recovery plan, those same team members often will have an ongoing responsibility to test and maintain the plan. While certain members of this team have other responsibilities, substitutes should be obtained to allow the members sufficient time to perform their team responsibilities.

EVALUATING COMPUTER DISASTER RISKS

Once the disaster recovery planning team has been established, a good first step for the team is to evaluate potential computer risks. This is a process of identifying and estimating the expected losses from any undesired event. The costs of the safeguards which can be installed to minimize these risks then can be considered. When the cost of installing the safeguard is less than the probable loss that might occur, the safeguard should be considered. Of course, some safeguards should be installed to prevent loss of life or to meet legal requirements, regardless of the cost.

The evaluation of data processing vulnerabilities through a formal quantitative risk evaluation sometimes can be a futile task if the team focuses on numerical probabilities. Some disaster recovery plan teams attempt to evaluate the subjective probabilities of a variety of potential risks. They try to evaluate the probability of a flood, a fire, a hurricane, or any one of a number of disasters that might occur and cause a data processing interruption. Often, considerable effort is put into this type of an evaluation process. The quantitative approach to risk analysis was discussed in Chapter 8.

Numerical probabilities in this setting, however, often are guesses at best. The real concern is that the computer center may be out for an extended period, rather than in identifying the probability that it will be out for a particular reason.

An effective disaster recovery risk analysis should consist of the following six steps:

1. Understand potential data processing risks.
2. Evaluate safeguards currently in place.
3. Consider potential losses due to an interruption of data processing
4. Consider alternatives for backup processing services.
5. Evaluate costs to resume data processing operations.
6. Summarize risks and recovery strategies for management through an effective disaster recovery plan.

Understanding Potential Data Processing Risks

A data processing department faces a variety of risks which could put it out of operation for an extended period. In a single minicomputer operation, such a risk could be a head crash on a hard disc with inadequate backup data. On a more major scale, an earthquake could destroy the physical facility as well as severely limit community support services. The disaster recovery planning team should develop an understanding of what types of risks are possible. These include natural disasters that may affect the entire community, natural or man-made disasters that may impact the facility where data processing is located, and man-made or equipment failures local to data processing.

Natural disasters that may affect the entire community include such things as earthquakes, hurricanes, or floods. Although they occur infrequently, when they do many emergency services for the area become unavailable. Some of these natural disasters reappear with regularity. Some coastal areas can expect a major hurricane or tropical storm every five to ten years. Other major disasters seldom happen. Certain areas lie on geological faults where the last major earthquake was 100 or more years ago. The next earthquake may not occur in the lifetimes of the planning team members.

If the disaster recovery planning team feels there is a reasonable chance of an area-wide disaster, recovery plans should be based on using facilities outside the geographic area. It accomplishes little to store backup tapes a half mile away and have a disaster recovery computer located in the same community if all community facilities become unavailable. However, a pragmatic approach should always be taken in assessing the risk of a major disaster.

Localized man-made or natural disasters will impact only the facility where data processing is located. These include fires, floods, or civil disturbances. Community wide support services such as the fire department remain available. Which of these disasters might occur depends upon the organization being evaluated. For example, a data processing facility located in or near a chemical plant could be disabled or closed due to a chemical plant accident.

The disaster team should best approach this localized type of disaster risk by asking a series of "what if?" questions. What would happen if there were a fire in the building? Are there adequate fire protection facilities? What would happen if a water pipe were to burst and flood the basement? Would the flood impact data processing? This type of analysis should result in an assessment of potential localized hazards and risks.

The last type of risk to consider is localized disasters that impact only the data processing department. These might include the failure of a power transformer, a localized fire in the telecommunications switching room, or the destruction of key data files through employee sabotage. This type of risk may allow the rest of the plant or facility to operate normally while data processing does not. This type of risk, while not as severe, causes considerable local problems for the data processing function. The rest of the organization is operating as normal and cannot understand why data processing is not.

At the end of this risk assessment process, the disaster recovery team should rank each of the identified risks by relative probability of occurrence. It should try to estimate whether the risk may occur within the next five years, whether it may occur beyond five years, or whether it is unlikely to ever happen. The particular risks then can be ranked on a scale from zero to ten for each of these time frames. A zero would mean that it probably would "never" occur in that time period while a rating of ten assumes a fairly high chance. Figure 10.4 is an example of such a risk ranking schedule.

This numerical rating is meant to show a relative probability. There is no need to try to determine whether the probability of a major flood within five years is one in 200 or one in 300. In either event, it is relatively low.

As a last step, the disaster recovery team should review its risk assessment scoring schedules with upper management. This way, it can discuss the rationale assigning relative risks and can consider making any other adjustments based on management considerations. In addition, the team may want to share its assessment rankings with community public safety personnel. Fire department, civil defense, or public safety personnel often are very willing to provide input to this type of assessment exercise.

Evaluating Safeguards Currently in Place

The typical organization has a variety of safeguards or precautionary measures to somewhat protect them from some of the risks identified above. These may include such things as a building fire alarm system, guards and gates to restrict access to the facility, or an emergency electrical power supply. In some instances, these safeguards already may have been factored into the risk assessment relative probabilities, as previously discussed. In other instances, it may be necessary to reconsider the relative probabilities based on a further evaluation of these safeguards.

While risks were discussed as area wide risks, organization wide risks, and data processing only risks, realistically safeguards can be considered only for the organization and the data processing function. Area wide safeguards are the responsibility of community, state, and federal agencies and are assumed to be in operation. The disaster recovery team can do little, in the short run, to improve area wide safeguards. However, they can be improved within the total organization and within the data processing department.

This evaluation of safeguards can be accomplished by a series of interviews and examinations of available resources. This is an area where the auditor on the disaster recovery team can play a significant role. Figure 10.5 contains a questionnaire for evaluating total organizational safeguards while Figure 10.6 contains a similar questionnaire for the data processing function. The audit department may have the answers to some of these questions gathered through earlier operational audits. However, if those audits took place more than a year earlier, it is best to update them.

This safeguard evaluation exercise gives the disaster recovery team an opportunity to understand better the protection measures in place within the

Disaster Recovery Schedule for Ranking Risks

NOTE: This schedule allows the auditor to rank risks to data processing operations on the basis of relative probabilities. The auditor should evaluate each risk in terms of the probability of occurrence within the next five years or in the future following that five year period. For example, the auditor might determine that the probability of a flood impacting the computer center within five years receives a rating of 2, or very low, and a rating of 7 for over five years, indicating that there is a strong probability of a flood over the longer period.

POTENTIAL DISASTER EVENT	RELATIVE PROBABILITY (0–10)	
	UNDER 5 YEARS	OVER 5 YEARS
Area Wide Disasters		
1. Is the area subject to a major earthquake?		
2. Is the area subject to hurricanes or tornadoes?		
3. Is the area subject to major flooding?		
4. Is the area subject to major winter storms?		
5. Is the area subject to other natural disasters, such as forest fires or volcanoes?		
Localized or Man-Made Disasters		
1. Are nearby plants and other facilities subject to fire, explosion, or chemical contamination?		
2. Are there risks of major fires in the building?		
3. Are there risks of major, long term electrical power failures?		
4. Are there risks of major, long term telecommunications failures?		
5. Are there risks of riots or sabotage due to the type of business or of nearby businesses?		
Data Processing Department Disasters		
1. Are there risks of a major equipment failure due to a power or environmental system failure?		
2. Are there risks of an extended equipment outage due to age or obsolescence?		

Figure 10.4. Disaster recovery schedule for ranking risks

3. Are there risks of intentional destruction of equipment or data through sabotage, computer viruses, or the use of destructive devices?
4. Are there risks of intentional modification of key data or programs through deleting data, improper program accesses, and the like?
5. Are there risks of accidental modification or destruction of data or programs through human or equipment errors?

Figure 10.4. Disaster recovery schedule for ranking risks *(continued)*

organization. It also provides an opportunity to make some immediate recommendations for improvement. For example, the questionnaire may reveal that fire alarms have not been tested in over a year. There should be an immediate recommendation for corrective action.

In other areas, this safeguard evaluation process may point to areas for further audit work. It may determine that certain documentation is not stored in a backup site because the documentation does not exist. The auditor may want to review documentation practices for the function that should have created the documentation.

Considering Potential Losses Due to an Interruption of Data Processing Services

The third step in the overall risk evaluation process is to assess, application by application, the potential losses that would occur if data processing services were not available for an extended period. This process again requires that a detailed survey be performed covering all financial and operational applications. In many instances, the audit department may have some of this information from prior application reviews.

The purpose of this exercise is to allow the disaster recovery planning team to rank and prioritize critical applications. Which applications are critical depend upon the organization's size and industry. User areas should be evaluated as to whether they could function without the data processing application for limited periods of time. Also, applications should be evaluated as to whether their unavailability would cause a financial, operational, or legal impact.

Questionnaire for Evaluating Organizational Disaster Safeguards Currently in Place

NOTE: This questionnaire is designed to help evaluate existing organization safeguards to help design an effective disaster recovery plan. In addition to just asking these questions through interviews, the auditor should examine supporting evidence.

POTENTIAL ORGANIZATION SAFEGUARDS	YES	NO	N/A
1. Is access to the facility controlled and limited to authorized individuals?			
2. Is there a formal guard function to screen incoming individuals and, if appropriate, vehicles?			
3. Are employees throughout the organization required to carry identification badges?			
4. Is there a "move ticket" system, complete with required authorizing signatures, to cover employee personal items and equipment which may be moved into or out of the facility?			
5. Are employee briefcases or other bags subject to inspection when employees leave the facility?			
6. Is there a formal sign in and out procedure, along with the presentation of proper identification, for employee visits during non-prime shift hours?			
7. Are sensitive areas within the organization closed or restricted during off hours?			
8. Is there an overall "clean desk" policy requiring all employees to lock sensitive information in their desks when not present?			
9. Is there an overall procedure or plan for facility evacuation in the event of a fire or other catastrophe?			
10. Has the evacuation plan been tested on a periodic basis?			
11. Are there facility wide smoke and fire alarm systems and a fire extinguisher system?			

Figure 10.5. Questionnaire for evaluating organizational disaster safeguards currently in place

	YES	NO	N/A
12. Is the fire alarm system tied into the local fire department?			
13. Is there a formal procedure for handling bomb threats?			
14. In addition to the area wide fire extinguisher system, are properly classified hand held units located throughout the facility at strategic locations?			
15. Are there any emergency or standby power systems in place?			
16. Are facility air conditioning or power transformers protected from improper access?			
17. Are there emergency plans for the entire facility with procedures to be implemented in the event of natural catastrophes?			
18. Have organization emergency plans been formulated into an overall organization disaster recovery plan?			
19. Does the organization have adequate business interruption or vital records insurance?			
20. Are management personnel and other employees trained in evacuation and other disaster recovery procedures?			

Figure 10.5. Questionnaire for evaluating organizational disaster safeguards currently in place *(continued)*

**Questionnaire for Evaluating Data Processing
Disaster Safeguards Currently in Place**

NOTE: This questionnaire is designed to help evaluate existing safeguards within the data processing organization in order to design an effective disaster recovery plan.

POTENTIAL DATA PROCESSING SAFEGUARDS	YES	NO	N/A
1. Are data processing applications backed up as part of normal operations procedures?			
2. Are backup files rotated to a secure, off site location?			
3. Are copies of operating software also rotated to the off site location?			
4. Are copies of application system and other technical documentation also rotated to the off site location?			
5. Have guidelines or standards been developed governing file backups for departmental microcomputers?			
6. Does the central computer site have adequate environmental control facilities?			
7. Is there an emergency or standby power supply system for the computer room?			
8. Is there a zone controlled fire and smoke detection system and a fire control system installed within the computer room?			
9. Is the computer room fire control system linked to alarms at a local fire department?			
10. Has adequate consideration been given to fire controls throughout the computer room raised floor system, including floor to ceiling walls and appropriate hand held fire extinguishers?			
11. Are forms burst and decollated in a manner which minimizes the risk of fire?			
12. Is the computer room located in a secure location with such features as:			

Figure 10.6. Questionnaire for evaluating data processing disaster safeguards currently in place

	YES	NO	N/A

Limited outside windows constructed with breakage resistant materials?

A single controlled entrance with all other doors equipped for emergency exit only?

Limited visibility, including no signs or other unnecessary indicators?

13. Is there a separate, fire controlled facility for the storage of computer tapes?

14. Are telecommunications controls located in a secure location?

15. Is the ceiling to the computer room free of plumbing and other water sources?

16. Have sprinkler systems over sensitive resources been engineered to minimize the risk of water damage, and are plastic sheets available to protect from falling water?

17. Is access to the computer room restricted to authorized individuals through the use of cipher locks, badges, or other controls?

18. Are hiring controls in place within the information systems function to ensure that all employees have the education or experience required for their jobs?

19. Are there procedures to control tapes or related materials which are taken from the computer room?

20. Has information security software been installed to restrict access to programs and data?

21. Are there procedures to control the distribution of output reports to authorized individuals?

22. Is there a formal computer security function to monitor both physical and information security procedures?

Figure 10.6. Questionnaire for evaluating data processing disaster safeguards currently in place *(continued)*

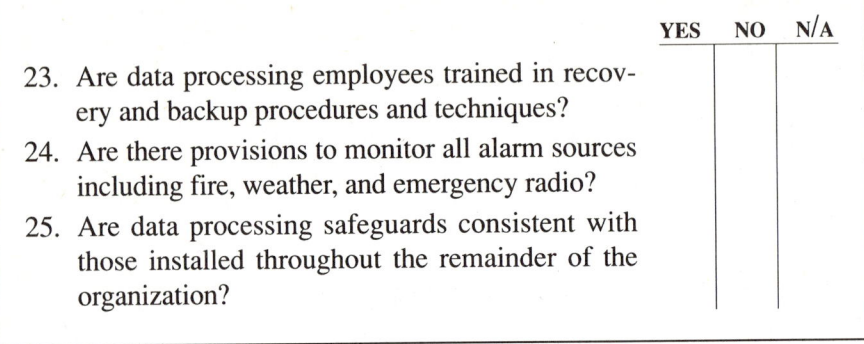

	YES	NO	N/A
23. Are data processing employees trained in recovery and backup procedures and techniques?			
24. Are there provisions to monitor all alarm sources including fire, weather, and emergency radio?			
25. Are data processing safeguards consistent with those installed throughout the remainder of the organization?			

Figure 10.6. Questionnaire for evaluating data processing disaster safeguards currently in place *(continued)*

Figure 10.7 shows a sample form for performing this application analysis. While the prime users in each of the application areas should be interviewed when preparing the form, the disaster recovery team may discuss these survey responses and modify them as appropriate. For example, the personnel manager responsible for a 50-person payroll operating in a minicomputer system might state that data processing services are "essential." The personnel manager responsible for a 5,000-person mainframe computer payroll system would give the same response. The former could operate for at least a limited time using manual methods while the latter could not.

This application loss evaluation helps the disaster recovery planning team in two ways. First, it allows the team to better identify those applications which are most critical. These then can be given a higher priority for disaster recovery planning. Second, the analysis helps to convince management of the importance of effective disaster recovery planning.

Considering Alternatives for Backup Processing Services

The next step in the disaster recovery risk evaluation process is to consider the alternatives for resuming data processing operations. This is an important part of the analysis necessary to build an effective recovery plan. As discussed earlier in this chapter, organizations often spend considerable resources in constructing a "disaster recovery plan" with no arrangements for a backup processing site. If data processing suddenly finds itself without a computer resource, it will be difficult to develop any reasonable strategy to resume processing without some plan for backup computer resources.

There are numerous strategies for emergency backup processing. Each has its benefits, risks, and costs. The disaster recovery team should evaluate these various alternatives, including the relative costs of each option and the

Form for Analysis of Potential Losses Due to an Interruption of Data Processing Services

NOTE: This form is designed to identify critical computer applications and quantify the potential losses that would result if those applications were not available. The auditor should circulate the form to key users and follow up their responses through interviews.

System No.:_____ Name:_____ Date:_____

1. Business function served by the application:

2. Effect of an extended interruption on other business functions:

3. Critical dates or time periods in application processing cycle:

4. Other related or dependent applications:

5. Tolerable duration of a computer service interruption:

   Length of Interruption   *Probable Business Impact or Loss*

   Four hours or less:

   Twelve hours:

   Twenty-four hours:

   One week:

   Over one week:

Figure 10.7. Form for analysis of potential losses due to an interruption of data processing services

6. Remote telecommunications input or output requirements of the application:

7. Special equipment requirements (scanners, special printers, etc.) required by the application:

8. Minimum application requirements for running in a contingency mode:

9. Alternative processing approaches for the application:

10. Other factors and special considerations:

Figure 10.7. Form for analysis of potential losses due to an interruption of data processing services *(continued)*

resources required to make them effective. Based on a recommendation of the disaster recovery planning team, management should decide which alternative to use. They should have data, however, on the benefits, costs, and risks associated with each type of alternative evaluated.

With management's and data processing's growing awareness of the need for backup processing resources, the number of options is increasing steadily. Outside vendors, for example, are now offering backup processing facilities for many different types of equipment. Some of the options for a backup processing facility are discussed in the following paragraphs.

Commercial Recovery Strategies

Outside vendors are offering alternatives for data processing recovery capabilities. Although such services were once available only for large scale IBM and IBM compatible systems, there is an increasing number of services offered for other vendors and equipment types. While a commercial recovery strategy can be the most expensive alternative for many organizations, it also is the most effective and secure one in many instances.

The disaster recovery team should evaluate the existing commercial recovery vendors in its area with the correct equipment type. If the risk of an area wide disaster is relatively high, consideration might also be given to vendors outside the geographic area. There are four principal types of commercial recovery strategies, as follows:

"Hot Site" Recovery Centers

In many larger metropolitan areas, vendors specializing in disaster recovery operations have established "hot site" recovery rooms. These are equipped computer facilities ready for operation in the event of a disaster at a subscriber's own computer site. Such a recovery center accepts only a limited number of subscribers who have the right periodically to test applications and operating software on the disaster recovery machine. Such centers also provide technical support and telecommunications facilities. When a subscriber has an unexpected outage, the disaster center is notified and the subscriber can take over the machine.

The "hot site" disaster recovery approach may be the best short term solution for an organization with a larger computer system. It also can be the most expensive. The recovery centers typically charge a monthly standby fee plus extra fees when used for testing or actual disaster recovery processing. If the site is located in another geographic area, there also may be costs in moving personnel to man the site.

"Hot sites" usually are only available to support the most common vendors of computer hardware. This normally means IBM or IBM compatible equipment. There also are "hot sites" available for other popular computers such as DEC, Honeywell, or Data General; these may be located at a few major urban areas and travel to them would be necessary.

Empty Shell Recovery Facilities

A major problem in restoring data processing operations is having a proper facility to house the machine. Some larger mainframe computers require water cooled chiller devices as well as special electrical power supplies. While it may not be difficult to secure a computer from the vendor, a proper facility to house it may be difficult to locate. To answer this problem, some disaster recovery vendors offer a shell site, an empty computer room equipped with all necessary power and environmental controls. In the event of a disaster, the vendor need only move computer hardware and telecommunications gear into this facility.

Empty shell facilities also will only accept a limited number of subscribers for a facility. They are less expensive than a "hot site." However, it takes time to have the necessary equipment installed, and a shell may not be useful for a short term outage. In the best case, it will require several days to restore data processing services through a shell facility.

Commercial Service Bureaus

Although most are not directly involved in the disaster recovery business, an outside commercial service bureau often offers a possible alternative strategy. They have an operational computer, telecommunications capabilities, and a ready support staff. They also may have facilities to store tapes as well as a service to transport personnel and reports.

The ongoing cost of an arrangement with a service bureau often will be quite low. However, in the event of a disaster, it will be necessary to pay normal machine usage rates which can be expensive. In addition, the service bureau tends to service its regular customers first, so that the recovery operations may not get necessary operating priorities.

Hardware Vendors

Vendors of smaller, minicomputer hardware often have marketing demonstration sites which they make available in the event of a disaster at a customer's site. They typically provide such data processing resources to a customer while a replacement system is being installed. There often is little or no charge for this type of service.

Hardware vendors provide this capability, typically, only for smaller computers and for limited periods of time. They usually will not agree to a contract to provide this type of service. At best, they will provide their customer with a "best efforts" memo of understanding.

Cooperative Disaster Recovery Strategies

Cooperative disaster recovery strategies require two or more organizations to work together to develop a mutual disaster recovery approach. These strategies usually work best for organizations in a geographically remote area or with special equipment requirements. This specialized equipment can include a unique hardware model or a specialized input/output device. These cooperative strategies generally take one of two forms as follows:

Reciprocal Processing Agreements

Many data processing organizations have made arrangements with a similarly configured site at another organization, agreeing to be each other's backup site. Theoretically, this can be a mutually beneficial arrangement since there will be equipment compatibility and few costs to maintain the arrangement. There also are numerous problems associated with this type of agreement as discussed earlier in this chapter.

Reciprocal agreements appear to work best when they cover a specialized item of equipment such as an optical scanner. They do not work when one of the parties to the agreement has a long term interruption and expects to use the reciprocal site over an extended period of time.

Cooperative Recovery Centers

In some geographic areas or industries, data centers with common equipment have joined to form cooperative hot site or shell site recovery centers. These work like the outside vendor hot or shell recovery centers. The advantage is that all users of the site have common interests or requirements. The disadvantage is that the various participating organizations are all part managers of the recovery site.

Internal Systems Recovery Strategies

Rather than dealing with outside providers of recovery services or with cooperative site agreements, many larger organizations decide that it is most cost-effective to develop an internal strategy for recovery processing using multiple computer centers within their organization. Others sometimes decide that the cost of having a backup facility outweighs the relative risks of not having an ongoing data processing capability. If the latter strategy is used, the disaster recovery team should give extra care to documenting the reasons for this decision. Following are internal data processing recovery strategies.

Multiple Internal Processing Sites

Multiple division organizations with a common data processing strategy often make internal arrangements to use the computer centers at other operating divisions or units as backup processing sites. This is true whether the organization has multiple mainframe machines in separate data processing facilities or multiple microcomputers scattered throughout the organization. Provided there is some similarity in equipment and operating software, this often is a relatively easy strategy to implement.

Depending upon the internal culture of the organization, however, the disaster recovery team should take care when attempting to implement such an internal reciprocal agreement. Some organizations have operating units that are so autonomous it may be easier to go to outside recovery services.

Shell or Computer Room Space

Empty shell spaces, discussed above as one of the options available from outside providers, can be maintained by an organization internally. It can construct a full computer room site with proper environmental controls in a separate, remote building. It can also designate bare room space, such as in a warehouse, where a computer could be located in an emergency. Both areas could be used for other things, such as forms storage, on an interim basis. This strategy works well if the other facility is physically remote and not exposed to the same hazards; it also works well if there are long term plans to shift operations to that new facility in the future.

Service Degradation Strategies

Sometimes, an organization makes a formal strategy decision to allow service degradation in the event of a disaster. If a smaller site using a mini or microcomputer, it may decide to revert to manual processing or to cease some functions in the event of an extended interruption. If a larger site, the organization may decide to use the limited capabilities of a smaller branch machine during the interim.

A service degradation strategy brings with it the risk of financial loss, customer dissatisfaction, and operational inefficiencies. However, an organization may feel that the cost of recovery facilities is too great when compared to perceived risks. If the disaster recovery team or top management feels that this is the best strategy, it should be documented carefully. A subsequent loss of services and costs associated with not having a plan may cause others to question that strategy after the fact.

Evaluating the Costs to Resume Data Processing Operations

The next step in the disaster recovery risk evaluation process is to consider the costs and requirements for resuming data processing normal activities, including potential insurance recoveries. This is an information gathering activity which starts by assembling data on the existing operating environment and costs and logistics required to replicate it. In some instances, it may be discovered that certain items of equipment are no longer available. Tentative strategies must be developed to handle this contingency.

When evaluating the costs to recover data processing operations, consideration should be given to the organization's insurance coverage. Organizations should have fire and general liability coverage as well as some form of vital records and business interruption coverage. The disaster recovery team should make a careful evaluation of the types of coverage in place and the extent to which that coverage provides protection and helps the organization resume its data processing capabilities. In a discussion with the individuals responsible for the organization's insurance coverage, this may be the appropriate time to suggest making any necessary changes to that coverage.

Summarizing Risks and Recovery Strategies for Management

As a final step at this point, the disaster recovery team should summarize its findings and report them to upper management. How to summarize potential risks in order to report them to management was discussed earlier in this chapter. While those risks were presented in terms of subjective probabilities, another meaningful approach is to discuss risks in terms of potential economic loss.

Figure 10.8 provides a worksheet for calculating probable economic losses if there were a computer disaster and any of the several disaster recovery methods discussed were used. In evaluating risk and strategy alternatives, the disaster recovery team has been exposed to some of the costs and potential losses. This worksheet allows the disaster recovery team to summarize its findings and recommend a cost effective means of disaster recovery.

BUILDING THE DISASTER RECOVERY PLAN

As emphasized earlier in this chapter, the purpose of an effective data processing recovery plan is to provide guidance for recovering data processing operations. Because of the technical complexity of data processing operations in most organizations, a considerable effort must be devoted to plan development. When completed, the final document should be a working document that is revised and updated constantly.

Typically, there are many other areas of an organization which could benefit from disaster planning. Although it is easy to recognize the vulnerability of data processing, other vulnerabilites often are ignored. Based upon the experiences gained in developing this plan, auditors should consider suggesting similar emergency planning to other areas in the organization.

While developing this plan for situations which should "never happen," the team often gains considerable knowledge about data processing systems and operations. As a result, other recommendations for improving procedures or saving costs may result from the entire disaster recovery planning process. For example, when identifying the necessary data sets for key applications backup and recovery, the team may identify data redundancies which could be corrected through minor system enhancements. Similarly, when looking for compatible equipment for backup processing purposes, the team may find a given item of equipment obsolete and that efficiencies would result in replacing it with a newer unit.

Worksheet for Calculating Probable Economic Loss

NOTE: This worksheet is designed to estimate probable economic losses if a data processing facility were unavailable for an extended period of time, as well as the most cost-effective recovery strategy. This is a three part worksheet. Part I is used to estimate recovery costs for various alternative strategies; Part II is used to estimate business losses per day when key data processing applications are unavailable; and Part III is used to estimate the most cost-effective recovery strategy.

PART I. RECOVERY COSTS

List the various alternative recovery strategies across the columns on top. These may range from a "hot site" facility to taking no action. List the costs associated with each strategy. The cost categories here are only examples. Estimate the number of days to recover for each strategy.

	RECOVERY STRATEGY #1	RECOVERY STRATEGY #2	RECOVERY STRATEGY #3
Annual facility maintenance costs			
Emergency processing costs			
Extra personnel expenses			
Telecommunications costs			
Backup storage costs			
Other out-of-pocket costs			
Total Costs:			
Days to Recover:			

PART II. APPLICATION BUSINESS LOSSES

List the organization's key data processing applications and the estimated losses, or costs to the organization, if each application were unavailable.

KEY DATA PROCESSING APPLICATIONS	ESTIMATED BUSINESS LOSS/DAY
Application #1 _____	_____
Application #2 _____	_____
Application #3 _____	_____
Total Loss per Day:	_____

Figure 10.8. Worksheet for calculating probable economic loss

```
┌─────────────────────────────────────────────────────────────┐
│              PART III. RECOVERY STRATEGY ALTERNATIVES          │
│                                                               │
│  Calculate the total losses, based on days to recover, for   │
│  each strategy, to select the most cost-effective recovery    │
│  method.                                                      │
│                                                               │
│                         ESTIMATED  TOTAL                       │
│                DAYS TO  BUSINESS  BUSINESS  RECOVERY  TOTAL    │
│                RECOVER  LOSS/DAY   LOSS      COSTS    COSTS    │
│                                                               │
│  Recovery Strategy #1  ____  ____  ____  ____  ____           │
│  Recovery Strategy #2  ____  ____  ____  ____  ____           │
│  Recovery Strategy #3  ____  ____  ____  ____  ____           │
└─────────────────────────────────────────────────────────────┘
```

Figure 10.8. Worksheet for calculating probable economic loss *(continued)*

Preparatory Steps

An effective disaster recovery plan requires considerable advance data gathering, documentation, and decision making. Much of this is a one time effort which needs only to be updated on an ongoing basis. The disaster recovery team can gather much of this data and information and, when necessary, call upon systems analysts and others to assist. Following are the key preparation steps to develop a disaster recovery plan:

Backup site selection. Before much other work can proceed on the plan, a decision must be made on a backup processing site. Some strategies, such as the use of alternate site vendors, are expensive and require a major management commitment. Other strategies, such as empty shell facilities, require planning to identify and order equipment for the site as soon after an emergency as possible.

Disaster recovery personnel assignments. The disaster recovery planning team mentioned throughout this chapter functions as a management group for the project. Another group of personnel will be required to actually invoke the plan, including computer operations supervisors, tape librarians, and systems programmers. Steps should be taken to identify these personnel, assign responsibilities within the plan framework, and train when necessary.

Identify critical applications. When performing its risk evaluation process, the planning team should have identified a series of applications more critical than others. There now should be an effort to develop a priority list of the most critical applications. In a disaster situation with limited processing resources, some applications may have to be dropped. The difficulty here is that managers responsible for each application tend to claim that their application is critical. Sometimes, the disaster recovery team can resolve this problem by allocating the costs of the disaster recovery effort to application users.

Inventory critical file and program requirements. Based on the identification of critical applications, all necessary data sets, programs, and database files should be identified and steps taken to assure that all are included in the procedures for off site backup storage. When an otherwise noncritical application provides data files to a critical application, it should be reclassified as critical.

Identify systems software requirements. The operating software necessary to operate key applications should be identified and included in off site backup procedures. If there are any potential differences between this software and that used normally at the backup computer site, steps should be taken to resolve compatibility problems.

Assess telecommunications needs and requirements. Backup processing sites may have the necessary computer power but often will not have the required communications network. Any special equipment needs in the current network should be identified and documented. Based upon the backup processing stategy to be used, preliminary plans should be made for restoration of required telecommunications lines from the backup site.

Identify necessary documentation and supplies. Documentation covering key applications and operating software should be identified, and copies should be stored at the backup site. If any key documentation is missing or out of date, steps should be initiated to bring it current. Any necessary supplies such as printed forms which support key applications also should be identified. A sufficient stock should be retained at the backup site. Vendors should be identified and documented so that items can be reordered with minimal difficulty.

The final item that should be considered is an estimation of the costs of the plan. Because all costs will not be incurred at once, these estimates should be developed on a time-based budget for management review. The disaster recovery planning team also should consider proposing alternatives to management along with the associated costs. For example, one option may be to construct a shell site in another facility within the organization while still another proposed option may be to subscribe to a commercial shell site facility. The costs for each should be outlined in a management proposal along with a recommendation for the alternative preferred by the team.

Completing the Recovery Plan

Much of the time and effort required to develop an effective data processing disaster recovery plan takes place during preliminary phases such as risk evaluation and gathering of preliminary data and information. Documenting all of this in an action-oriented manner will create the disaster recovery plan.

An effective disaster recovery plan should cover two phases—the initial response to the disaster and the restoration of processing at the backup facility. The first phase requires management decisions as to whether to invoke the remainder of the plan into action or attempt to make necessary corrections and continue processing at the existing site. The second phase involves the necessary processing steps at the designated backup sites.

The various sections that would go into a completed disaster recovery plan are discussed in the following paragraphs. This plan defines the responsibilities and actions to be taken. Also, it serves as a training document for all personnel involved in disaster recovery preparation procedures. While the size and content of the plan will vary depending upon the data processing equipment used and the backup strategy selected, many of the items listed below should be included in any plan.

Section 1: Initial Disaster Responses

This section should define the conditions that would constitute a data processing disaster and the individuals responsible for initiating the disaster plan procedures. The section also should list the home and work telephone numbers of all key disaster response personnel and designate a calling sequence. An initial disaster response meeting place that all members can use as a headquarters, such as a nearby hotel facility, should be identified in this portion of the plan.

Figure 10.9 outlines the actions to be taken for this initial response section of the disaster recovery plan. While the operational details vary somewhat from one organization to another, the steps shown in the figure should be described in any plan.

Outline of Initial Disaster Response Steps

NOTE: This outline contains the initial steps that should be built into a data processing disaster recovery plan. A particular plan, of course, would contain more detail and the procedures would vary from one organization to another.

I. **Identification of Disaster Event**
   A. The first person on site should make sure that fire and emergency alarms are sounded.
   B. Senior personnel on site should execute emergency evacuation and response procedures.
   C. Following a call priority list, members of the emergency response team should be notified.
   D. If the loss appears to impact the total data processing facility, arrangements should be made to notify the remote disaster meeting site.

II. **Assessment of Damages**
   A. Key members of the disaster recovery team should assemble at the disaster meeting site.
   B. Senior management should be notified and informed of the status of recovery efforts.
   C. Based on observation or discussion with authorities, the disaster recovery team should assess damages.
   D. Based on this assessment, a decision must be made to implement the full disaster recovery plan, continue working at the present site, or implement a modified plan.

III. **Disaster Recovery Plan Implementation**
   A. Senior management should be informed of any decision to implement the recovery plan.
   B. All members of the disaster recovery team should be contacted.
   C. The existing data processing site should be secured as well as possible.
   D. The backup recovery site should be contacted and notified officially of the disaster situation in accordance with the backup processing agreement.

Figure 10.9. Outline of initial disaster response steps

E. Key management user personnel should be contacted.

IV. Other Initial Steps

A. The nearby hotel site or a location near the backup processing site, as pre-established, should become the Control Center.

B. A log should be established to record all disaster recovery activity and decisions.

C. The local telephone company should be contacted to order additional telephone lines for the Control Center.

D. Hardware and software vendors, as listed in the recovery plan, should be contacted as soon as practicable to define new equipment requirements.

E. Documentation should be retrieved from the backup site to begin to plan processing steps.

F. Organization loss protection or insurance personnel should be informed of the event.

V. Begin Operating at the Backup Site

A. System tapes should be delivered to the backup site to start processing.

B. Key operating software and databases should be loaded as designated.

C. Technical contacts should be established to help with any software problems.

D. Key applications should be loaded for processing following a pre-established priority schedule.

Figure 10.9. Outline of initial disaster response steps *(continued)*

Section 2: Key Application Files and Programs

This section of the disaster recovery plan should formally document the key applications that have been designated for emergency processing as well as the data, program, and other requirements for processing those applications. The locations of any forms or other supplies, backup files, and program documentation should be listed.

Care should be taken to prepare this section of the plan in a format that can be easily updated. Any changes in the procedures for key systems should be reflected in the final plan document. A system of revision numbers or effective dates should be used to assure that the applications documentation is current.

Section 3: Backup Recovery Site Procedures

This section should cover procedures to notify the designated backup site, notify the personnel necessary to begin operations at the site, and begin emergency processing at that site. A major purpose of this section is to document the many procedures necessary for resuming some form of operations at the backup site. Although team members may perform these operations almost without thinking during normal operations, the trauma caused by a disaster often causes otherwise responsible personnel to forget. Figure 10.10 is a checklist outlining in sequence some of the action steps necessary at a disaster recovery site.

The checklist in Figure 10.10 is based on the assumption that the organization will use an existing computer at the designated backup site. This would be the case if a vendor demonstration facility were used, if the organization was a subscriber of a "hot site" recovery vendor, or if a reciprocal arrangement had been made with another organization. If disaster recovery is based on a shell site strategy where new equipment must be installed, additional documentation is required to order and install the necessary equipment configuration.

Section 4: Key Vendor Contacts

The last section of the disaster recovery plan is primarily an appendix that lists the necessary vendor contacts for computer equipment, software, supplies, and other materials. These contacts are necessary to replenish supplies of items needed to restore and continue processing.

Also, this section should list any alternative backup processing sites. Although only one site probably will be designated within the plan, this

A Sequential Checklist of Actions to be Taken at a Disaster Recovery Site

BACKUP RECOVERY SITE ACTIVITY	YES	NO	N/A

Activities Within 6 Hours of Disaster Event

1. Have all members of the recovery team as well as appropriate members of management been notified?

2. Has the alternate processing site been notified that a disaster situation has occurred and that the backup site will be used?

3. Has the Control Center location been notified that the team will be occupying it on an extended basis?

4. Has the existing data processing site been secured with appropriate guards?

5. If key members of the recovery team are unreachable, have arrangements been made to assign alternate personnel to the team?

6. Do all members of the recovery team have a copy of the recovery plan and do they understand their roles?

7. Have procedures been initiated to transport all required backup files and documentation to the alternate site?

8. Have hardware and software vendors been contacted regarding the preliminary disaster assessment?

9. Have telecommunications vendors been contacted to supply voice support for the Control Center as well as data support for the backup site?

10. Have arrangements been made for administrative support such as secretaries and copy machines?

Figure 10.10. A sequential checklist of actions to be taken at a disaster recovery site

	YES	NO	N/A

Activities Within 12 Hours of Disaster Event

1. Have key personnel arrived at the backup processing site?

2. Have backup tapes and documentation arrived at the backup processing site?

3. Do there appear to be any problems with system or application backups which would cause a change in backup processing strategy?

4. Have firm contacts been made with hardware vendors to initiate the ordering of any replacement hardware?

5. Have users of key applications been informed of the disaster situation and of arrangements for continued processing?

Activities Within 24 Hours of Disaster Event

1. Has the operating system been loaded on the backup equipment?

2. Have key databases been loaded on the backup equipment?

3. Has a strategy and processing schedule been established to restore applications from the backed up versions to the most current status?

4. Are all critical applications operational?

5. Are user personnel on site to verify controls in key application?

6. Has a processing schedule been established?

7. Have any necessary job control language changes been made?

8. Are arrangements in place for installing necessary telecommunications linkages?

9. Has a more detailed damage assessment and salvage effort been performed at the original data processing site?

Figure 10.10. A sequential checklist of actions to be taken at a disaster recovery site *(continued)*

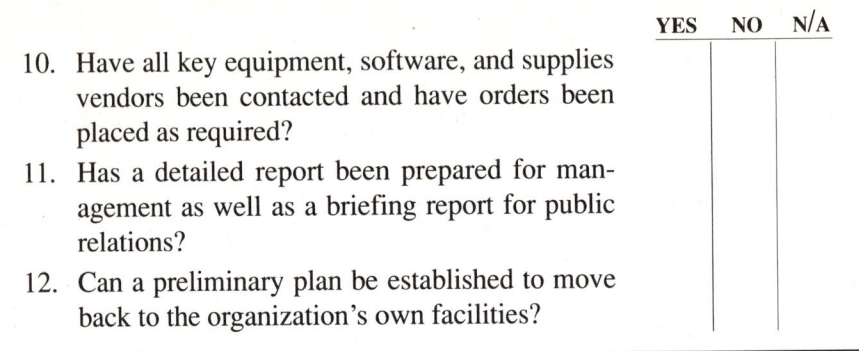

	YES	NO	N/A
10. Have all key equipment, software, and supplies vendors been contacted and have orders been placed as required?			
11. Has a detailed report been prepared for management as well as a briefing report for public relations?			
12. Can a preliminary plan be established to move back to the organization's own facilities?			

Figure 10.10. A sequential checklist of actions to be taken at a disaster recovery site *(continued)*

section should list any others considered as a contingency if the designated backup site does not have the capability to provide the agreed level of services during the emergency period.

MAINTAINING AND TESTING THE DISASTER RECOVERY PLAN

It has been emphasized throughout this chapter that a data processing disaster recovery plan is of little use if placed on the shelf and never updated or tested. Most data processing operations are dynamic, and the applications as well as the procedures to support them change regularly. If the disaster plan and backup data sets are not updated to reflect these systems changes, the plan will be of limited use.

Procedures within the systems development and the operations functions should be established to update the disaster recovery plan on an ongoing basis, detailing any changes. Although the disaster recovery team probably will cease to operate on a regular basis once the plan is published, one person from that group should be assigned the ongoing task of publishing updates to the plan documents, and assuring that they actually have been filed in the plan books. The auditor should monitor this process to assure that it is taking place.

The disaster recovery team that originally assembled the plan should meet no less than annually after the plan is published, to review initial assumptions and suggest other changes to the plan in light of changing factors. For example, a plan may have been based originally on a reciprocal agreement with a nearby firm. However, a subsequent acquisition of a new division may provide another computer center within the organization available for backup processing.

Just as a disaster recovery plan will be of little use if it is allowed to grow out of date, it also will be of little use if it is never tested. Just as an organization should test periodically its fire alarm system, it should test its data processing disaster recovery plan. This is an area where the auditor can play an important ongoing role in reviewing the plan to assure that tests are made, test results are reviewed, and adjustments are made to the plan to reflect the test results.

Disaster recovery plan testing can take two general forms: controlled tests of selected portions of the plan, and surprise, "simulated disaster" type tests. In a controlled test, applications are run in a planned manner at the backup recovery site. With a "simulated disaster" test, these applications are run on a surprise basis.

If the organization has made arrangements with a backup recovery vendor, the contract generally will allow several tests per year as part of the subscription fee. Reciprocal agreements should have some arrangement for reciprocal testing included in the agreement. In any event, the disaster recovery team should designate one or two applications per year to be tested at the backup site. The team as well as the auditor should monitor closely the results of this testing to determine that the designated backups from the published plan allow a correct restoration of that application. If problems are encountered, the published plan should be corrected as necessary.

A "simulated disaster" test does not mean that the head of data processing rushes into the computer room at month end and calls out, "Fire, fire!" to invoke the plan. Rather, a select group of individuals should structure a test of one or several key applications covered by the plan. Confidential arrangements should be made with the backup facility and data processing should be informed on a surprise basis that the given application will be run at the backup site that day.

As much as possible, this type of test should be done on a simulated emergency basis. Data processing personnel would be required to use only media stored at the backup data file storage location to process the designated applications. The results of that test should be monitored closely by the disaster recovery team and the auditor. Such a test will identify areas where files have not been backed up properly or where personnel lack proper training. Corrective action should be taken where appropriate.

Disaster recovery plan testing is an important component of any plan implementation. Although it takes an ongoing commitment of resources to ascertain that the plan is working effectively, this is usually a small price for having an effective plan.

EFFECTIVE DISASTER RECOVERY PLANS AND THE AUDITOR

The implementation of an effective data processing disaster recovery plan is not the responsibility of the auditor, but rather that of general and data processing management. However, the auditor plays an important role in helping develop and implement the plan and in reviewing ongoing maintenance.

The effective auditor should have a good understanding of data processing control and security concepts. In addition, the auditor probably has performed both applications and general controls reviews over many areas in data processing and should understand the strengths and weaknesses of existing backup and recovery procedures. Auditors should review existing disaster recovery plans as part of their ongoing review procedures. Weaknesses and vulnerabilities should be identified and reported to management.

Auditors also should serve as catalysts to improve disaster recovery planning. Just as they will be involved in reviewing new systems under development, they should be involved in the development of planning for data processing disaster recovery. In this way, the auditor can have a strong level of assurance that an important set of data processing controls—backup and recovery controls—is working effectively.

Section IV

AUDIT AND CONTROL OF END USER COMPUTING

CHAPTER 11

Auditing the End User Information Center

INTRODUCTION

Computer applications development was once the responsibility only of the data processing department within an organization. If users required computer generated reports or wanted to automate operations, it was necessary to request help from the data processing function. This is no longer true in the modern data processing organization. Personal computers and powerful end user-oriented mainframe retrieval languages now allow users to create directly their own reports and other applications. This change has had a significant effect on many data processing departments. It also has changed the way auditors must approach reviews of application controls.

End user computing has caused auditors to expand their work scopes. In the past, computer auditors tended to do most of their work within the data processing department while financial/operational auditors stayed in the user areas. Today, this strict separation of functions is no longer efficient. Financial and operational auditors need to understand data processing application controls when they encounter applications developed by end users within the functions under their review. Computer audit specialists also must go beyond the data processing function to understand the controls surrounding these applications.

This chapter discusses end user computing contols whether performed through departmental microcomputers or through information centers tied to mainframe computers. Some of the unique general controls considerations as well as applications controls will be reviewed. However, even if a data processing application is developed using end user computing methods, the control objectives and audit procedures for data processing applications discussed in Chapter 5 still are applicable.

THE MANY FORMS OF END USER COMPUTING

End user computing was started when IBM introduced the concept of the information center in the late 1970s. Although many felt it was a marketing gimmick at the time, IBM suggested that its customers give users terminals, easy to use software tools, and access rights to certain data files from their own dedicated computer system. These terminals all would be located in a central facility, called an Information Center, where users could receive training and technical help and develop their own retrieval applications. Faced with increasing demands for new applications, many data processing functions adopted the information center concept as a way to reduce application development backlogs.

Data processing organizations have taken a variety of approaches to this information center concept. Some felt that "ordinary" users would never be able to develop their own retrieval applications and built information centers for the use of engineers or other specialized technical personnel. Others gave terminals to all requesters and provided a "hot line" for user questions. Still others adopted the full information center concept as originally described by IBM.

The information center and end user computing experienced other dramatic changes, in the early 1980s, when powerful personal or microcomputers were introduced into the office. Some data processing managers viewed them as a threat and all but banned their use. Others considered them to be a toy and chose to ignore them. Still others actively incorporated microcomputers into the information center.

As a result of these changes, the auditor today will see end user computing implemented in a variety of manners in the modern organization. No matter how implemented, this is an area that the auditor should include in both general and application data processing controls reviews, because it accomplishes little for the auditor to determine that data processing has adequate controls when significant applications are developed by end users outside of the data processing function.

While information centers and end user computing can be implemented in a variety of ways, usually they can be classified into one of the following:

Information centers with computing facilities

Information centers with support facilities only

Departmental microcomputers

Freestanding end user computers

Although there are some overlaps from one classification to another, we will consider general controls separately for each.

Information Centers with Computing Facilities

The classic information center is a separate room equipped with user terminals. A support staff would be available to train users on the retrieval tools available and help with technical problems. Over time, some information centers have also introduced microcomputers into the facility which may be either free-standing or tied into mainframe machines for file uploading and downloading.

Usually, the information center with computing facilities is part of the overall data processing organization. Chapter 2 described how the information center might report within the structure of the larger data processing department. In some organizations, however, this information center evolved separate from the traditional data processing department. Organizations sometimes placed their information centers under a group such as office automation, reasoning that such office automation personnel are closer to the day to day computing needs of end users than the technologists within the data processing department. Other organizations made their information centers a separate organization, with no organizational connection to data processing even though it used data processing's services.

Before embarking on any review of the information center, the auditor first should determine how it fits into the total organization. Certainly, if the information center is part of the data processing organization, the auditor should consider reviewing general controls here as part of an overall review of data processing department general controls. If organizationally separate, it probably should be treated as a separate audit review project.

The information center with computing facilities usually makes the auditor's review task easier than other end user computing implementations because all computing facilities are located under one roof. The information center with computing facilities generally has the following attributes:

A separate information center facility. Terminals and microcomputers used by the center's end users generally will be located in a separate room. This makes the task of physical security over equipment and software easier, and creates a central library location for documentation manuals and micro-computer program diskettes.

Information center support staff. The typical information center will be staffed with specialists to train and assist end users. They provide the necessary support to ensure that end user applications are built with proper controls and are cost effective. The support staff can also act as an interface between users and data processing department specialists, such as data base administrators.

Policies and procedures. A major advantage of the formal information center is a set of policies and procedures which help to establish better end user system controls. Through the information center support staff, these policies and procedures can be explained and implemented to information center users.

Controlled equipment environment. As will be discussed in a later paragraph, some data processing organizations have difficulty supporting the various types of microcomputer equipment which may be used by end users. Without formal policies, various user departments may use differing hardware and software and may seek support for them from the data processing department. With an information center computing facility, only supported equipment need be installed.

The information center with computing facilities provides the auditor with a central focal point to perform the review. However, it is not the most efficient way to organize the end user computing function in all organizations. It tends to work best when most potential users perform homogeneous functions such as those of accountants and other financial analysts. It does not work as well when end users have differing needs, such as those of engineers, manufacturing materials specialists, and statisticians. These users require different types of equipment and may be located at geographically diverse areas.

Information Centers with Support Facilities Only

This type of information center has no terminals or microcomputers but operates with a central support staff to help end users. Such an information center support staff generally will be responsible for the following functions:

Approving requests for terminal connection to the mainframe computer and arranging for access rights through the data security function

Maintaining a list of approved hardware, software, and accessories

Establishing and promoting the use of information center policies and procedures, including guidance for developing and documenting new applications

Maintaining a telephone "hot line" or help desk to assist end users with computing problems

Conducting training classes

Monitoring use of the mainframe machine by end users, discussing problems with users, and reporting security violations or other matters to appropriate members of management

The information center with only support facilities may not be called an "Information Center" in some organizations. Whatever the name, however, it is a central coordinating function to help and provide guidance for end user computing activities. The auditor should review the controls and procedures within this function and then test those controls within end user areas.

Departmental Microcomputers

Many organizations have not established a formal information center function but make extensive use of microcomputers for various purposes. These machines may be freestanding departmental processors tied into local area networks or to mainframe computers for the downloading of files. Their use is limited only by the creativity of the microcomputer users.

From an auditor's perspective, departmental microcomputers do not have the same level of control and support to establish standards or train and assist users as an information center. There may be some levels of controls over departmental microcomputers but these often are limited to such matters as standards for approved types of hardware or software. While the information center help desk may not exist, microcomputer users in many larger organizations have often formed informal, self help user groups.

The auditor's concerns with this type of environment depend upon the level of decision dependence that is placed upon the various end user applications. For example, one organization may use a microcomputer spreadsheet application to track and report sales statistical data. The particular data may not be available on mainframe applications, and the generated reports may be used for historical reporting purposes only. The level of decision dependence here is probably quite low.

A second organization may develop a manufacturing cost accounting model using a microcomputer database or spreadsheet package. To develop the model, it captures data sent to a normal mainframe application. Even though this microcomputer model may have been built on a somewhat ad hoc basis, there may be a high level of departmental or even total organizational decision dependence placed on the application. The auditor obviously should have a far greater concern about the controls surrounding the latter application. However, it still may be appropriate to review the procedures over backups and application documentation for both situations.

As microcomputers become less expensive, more powerful, and easier to use, the auditor will encounter increasing numbers in most organizations. In addition, as microcomputer users increase their level of sophistication and as more powerful software packages become available, the level of decision dependence on departmental microcomputer applications will increase.

Freestanding End User Computers

Although this method of end user computing is much less common than information centers or departmental microcomputers, some organizations have established specialized mini or mainframe computers dedicated to end user computing purposes. Generally these machines are used for specialized departmental functions. Although established for such purposes as computer aided design, they often are also used for end user business data processing purposes.

The auditor will have a difficult task in identifying controls in this type of a specialized computer environment. The users of such machines often are technical personnel who have a high level of data processing sophistication but little knowledge in application controls concepts. General controls in such a freestanding computer environment are discussed in Chapter 3.

AUDITING END USER COMPUTING GENERAL CONTROLS

Auditors provide an important service to management by reviewing general controls in the end user computing area. As discussed at the beginning of this chapter, audits of end user computing controls often are missed in overall audit plans. They do not quite fit within normal data processing controls reviews but are considered to be beyond the scope of departmental financial or operational audits. Nevertheless, with the growth of end user computing, auditors should give attention to controls in this area.

As a first step, the auditor should understand how end user computing is implemented in the organization. Probably it will follow one of the four forms discussed above. However, it may require inquiries to determine how end user computing actually is being accomplished. For example, a data processing department may have established terminals in a mainframe based information center for end user computing. However, the auditor may then discover that most end user computing takes place on departmental microcomputers outside the information center. Audit procedures then should be oriented to the departmental microcomputer type of review.

The next step for the auditor is to understand relevant end user computing policies and procedures. These might include overall management policies regarding end user computing, data processing organization procedures, or detailed procedures developed by the information center. A review of such procedures might indicate areas where they are too limited and can be improved. It also provides the auditor with a basis for a compliance oriented controls review of end user computing.

Although end user computing and information centers can take a variety of forms, most auditor reviews of general controls can be performed on the basis of whether end user computing is an information center function or a departmental responsibility. Usually, there will be better controls and procedures in the former than the latter implementation.

Information Center Policies and Procedures

The first step for the auditor in reviewing information center based, end user computing is to understand the existing information center policies and procedures. These may have been implemented by the information center management, overall data processing management, general management, or often some combination of all three. General management may have policies requiring that all significant financial systems be the responsibility of the data processing department; the data processing department may have policies on which brands of microcomputer software will be supported; and information center management may have procedures covering requirements for end user application documentation.

The auditor should review the various standards covering end user computing. If these appear to be too fragmented, a valid recommendation to management might be to combine them into a single set of end user standards. The end user information center policies should cover the following general areas:

Security and data integrity. It is important that end users understand the need for protection over data which they may access. If tied into mainframe systems, the end user should follow the information security rules established by the data security administration function. If working on microcomputers, consideration should be given to establishing policies requiring a security system, such as a "DOS shell" type of system. These are microcomputer applications which restrict access to the operating system and file directories through a password screening system.

Documentation. Due to the informal, ad hoc nature of many end user computing applications, user documentation often is not prepared. There should be policies requiring that all regular applications be documented as well as procedures outlining the minimum contents of that documentation. Where practicable, end users should be advised to make use of automated documentation tools such as the various "Spreadsheet Auditor" types of packages, which analyze the templates developed through spreadsheet software and document the formulas built into them.

Reporting standards. It is quite easy for end users to develop reports which look much like the reports produced by "official" mainframe production systems. To avoid confusion regarding the source of such reports, there should be policies requiring that all end user produced reports be identified as to source and originator.

Testing and error control. End users frequently are unfamiliar with the extensive testing and error control requirements that go into larger system production applications. While it is difficult to establish a procedure requiring microcomputer application testing and error controls, there should be an ongoing educational program to provide end users with guidance in this area.

Systems backup and recovery. End users, particularly microcomputer users, often feel that their systems are totally reliable with little need for backups. Procedures should require that users back up all applications on a regular basis. In addition, equipment standards should call for all microcomputer systems with high capacity fixed disc to be equipped with cassette tape units or other devices for producing fast backups.

Acquisition and control of hardware. In the early days of microcomputers, there were many incompatible machines on the market. At present, business oriented machines generally use only one of several basic operating system versions based on a limited number of integrated chip designs. There are numerous "clone" machines with claims of full compatibility. There is a need for a policy for recommended machine types and configurations and recommended accessories such as graphics boards or modems.

Acquisition and control of software. Information center functions often are called upon to support end user software. This support is difficult if the information center attempts to support any package a user group decides to use. Although users have their own reasons for special packages, there should be standards listing software applications which will be supported.

Information Center General Controls

The above outlines the types of end user computing policies and procedures that the auditor should expect to find. Many of these emphasize departmental microcomputers. It is anticipated that there are already procedures, such as data access and data set retention, for end user applications accessed on the mainframe. The auditor should plan to review information center general controls as part of the auditor's review and understanding of internal acccounting controls within the overall data processing function. The objective of this review should be to assess the controls and procedures within the information center organization. The purpose of these controls, of course, is to assure that the information center provides the necessary guidance to produce well controlled and cost effective user developed information systems.

This objective may raise the question in the auditor's mind as to whether the information center or the actual end user is responsible for end user developed systems. The information center should be responsible for providing the necessary guidance for well controlled user developed applications. However, it is the responsibility of the actual end user to build those controls into applications. This is a fine but important distinction.

Figure 11.1 describes an approach for reviewing information center general controls. These procedures allow the auditor to review information center general controls whether the information center has its own terminal and computing equipment or whether it operates solely as a support facility. If the information center uses microcomputers designed for end users in a laboratory environment, the auditor also may want to consider using some of the control procedures for departmental microcomputers outlined in Figure 11.2 below.

There are two steps to this type of general controls review. First, the auditor must assess the overall information center control procedures for their adequacy. The second step is to perform control risk assessment tests to determine whether these procedures are enforced by the information center and followed by the end user community. This can be accomplished by testing several representative applications. Any end user developed application selected for a detailed applications review would be a candidate for this assessment test.

An information center general controls review represents a new direction for

**Control Objectives and Audit Procedures for
Reviewing Information Center General Controls**

Objective 11.1.1. The information center organization should follow the same general organizational controls and procedures found in the overall information systems organization.

   Procedure 11.1.1.1. Review the findings and recommendations from the last general controls review of the information systems function; determine whether any of these findings also are applicable to the information center organization.

   Procedure 11.1.1.2. Review the overall reporting relationship of the information center organization to determine whether it reports at the proper level both to be accessible to end users and to enforce control procedures.

   Procedure 11.1.1.3. Observe information center activities and document its activities and structure.

   Procedure 11.1.1.4. Interview selected members of the end user community to determine their satisfaction with information center services and activities.

Objective 11.1.2. Information center procedures should exist to allow users to access central computer data processing files only on a need to know basis.

   Procedure 11.1.2.1. Review procedures for requesting access to central computer files and determine that such requests are approved by persons responsible for those data files.

   Procedure 11.1.2.2. Select several end user applications which either access or download data from central files and determine that these requests have been approved properly.

   Procedure 11.1.2.3. Using the same selected end user applications, determine that information security levels are appropriate given the level of security established over the same files or databases in production applications.

   Procedure 11.1.2.4. Interview several selected end users with information center applications to determine that they understand the importance of information security procedures.

Figure 11.1. Control objectives and audit procedures for reviewing information center general controls

Objective 11.1.3. The information center should provide adequate training and support services to end users to assure that applications are developed following both information center and overall control procedures.

Procedures 11.1.3.1. Through interviews, class materials, and published schedules, assess the level and extent of training support provided by the information center to the end user community.

Procedure 11.1.3.2. Determine that the information center uses a newsletter or some other means of communication to inform the user community of both classes and new information center processing techniques.

Procedure 11.1.3.3. Select several recent end user developed applications and determine whether it might have been more appropriate to process them through the formal information systems function. If so, consider the reasons why they were developed as end user controlled applications.

Procedure 11.1.3.4. Review published information center guidance materials on application controls to determine that adequate attention is given to this area for end user applications.

Procedure 11.1.3.5. Review information center records to determine that problem resolution requests and other activities are documented properly.

Objective 11.1.4. Reports produced through end user developed procedures should be identified properly and should follow good control procedures.

Procedure 11.1.4.1. Determine that standards exist to balance end user developed applications back to source production data files or applications.

Procedure 11.1.4.2. Select several end user developed applications and determine that their control and balancing procedures are documented and are being followed.

Procedure 11.1.4.3. Determine that there are reporting standards over end user developed applications, covering such

Figure 11.1. Control objectives and audit procedures for reviewing information center general controls *(continued)*

areas as report source or originator identification, and determine that these standards are being followed.

Objective 11.1.5. Information center activities should make proper use of information systems computer resources.

> **Procedure 11.1.5.1.** Determine that a chargeback or cost allocation method exists for tracking and allocating the costs of end user computing activities.

> **Procedure 11.1.5.2.** Review procedures for requesting new information center hardware and software, and determine whether the procedures appear adequate.

> **Procedure 11.1.5.3.** Review physical security procedures surrounding information center operations to determine whether they appear appropriate; such procedures might include:
>> Sign-out procedures for using documentation binders or microcomputer software products
>> Policies prohibiting microcomputer software from leaving the information center facility to prevent illegal copies
>> Sign-in procedures for individuals using the information center facility
>> Restricted access to the information center during non-prime shift hours

Figure 11.1. Control objectives and audit procedures for reviewing information center general controls *(continued)*

many audits. All too often, auditors review general controls within much of the rest of the data processing organization but ignore the information center. This inattention is often caused by such factors as the physical remoteness of the information center or the less technical personnel heading up the function. In many organizations, however, the information center should be included in the auditor's overall controls review procedures. The information center is becoming an important element in many modern data processing departments and should be subject to periodic audit review just as other functions of the department.

Departmental Microcomputer General Controls Reviews

Microcomputers for business analysis applications first came into use in the early 1980s. Although several machines, such as the Apple II, were used for some business analysis purposes, the introduction of the IBM Personal Computer in 1981 suddenly made microcomputers "legitimate" for business data processing purposes. Shortly after the introduction of the IBM PC, the Lotus 1-2-3 electronic spreadsheet software product was introduced. This was a powerful extension of the VISICALC package which was then available on the Apple II microcomputer. It helped to introduce many users to the power and ease of use of microcomputers.

Traditional data processing departments initially viewed these new business microcomputers with a level of distrust if not hostility. Users who previously had to come to the data processing department to request new report applications now could develop their own microcomputer applications. In addition, some of the electronic spreadsheet software available on microcomputers was, in some respects, more powerful and flexible than the mainframe programming tools.

In many larger organizations, these new microcomputers were introduced in an uncontrolled manner. An IBM PC or some of the similar machines that came soon thereafter were inexpensive. Departmental managers could authorize their purchase within their own approval limits. Thus, departmental managers could purchase whatever hardware or software they thought to be appropriate. Because many different microcomputer hardware and software products were introduced in those early days, organizations found themselves with an assortment of incompatible equipment. Microcomputers frequently were installed in departments and placed on desks with no clear objectives on how they were to be used, and personnel were not properly trained to use them.

Over time, many organizations introduced controls and standards into this departmental microcomputer environment. Standards have been established over which brands and types of hardware and software are authorized for purchase. Data processing departments offered training and technical support to microcomputer users. In some organizations, "company stores" have been established for the purchase of department microcomputer equipment. Other data processing functions allowed departmental microcomputers to be connected to their mainframe machines for accessing or downloading data. It took some time, but departmental microcomputers now are viewed as part of the greater information systems capability of many organizations.

Just as data processing departments had misgivings about departmental microcomputers, auditors also initially viewed them with a level of distrust. Traditional audit functions, both internal and external, were composed of financial auditors who did their work in operational areas and computer auditors who traditionally stayed within the data processing department. The departmental microcomputer upset this balance.

Financial auditors perhaps reacted best to the introduction of these departmental machines. Just as end users developed special analysis spreadsheets, many financial auditors began to use those same tools for audit analysis purposes. Traditional computer auditors often had greater difficulty in accepting these new mainframe controls on departmental microcomputers. Auditors sometimes suggested that these desktop machines be kept in locked, secure areas for access control. However, while it generally is appropriate to recommend that the mainframe machine be secured in a locked, environmentally controlled facility, this type of control usually is not necessary for the typical departmental microcomputer.

Departmental microcomputers provide the auditor with some unique control risks and concerns. Because they are not under the protective umbrella of the information center, it is possible that users will develop applications on them without giving proper attention to controls, backups, or system security. Whenever departmental microcomputers are used for significant data processing applications, the auditor should consider reviewing the general controls surrounding their use. It should not be necessary, however, to call in the services of the computer audit technical specialist to perform such reviews.

Departmental Microcomputer Control Characteristics

The auditor should understand some of the unique characteristics of departmental microcomputers and their control implications. These include the following:

Powerful data processing capabilities. While the first business microcomputers, introduced in the early 1980s, had limited internal memory and data storage capabilities, newer ones are very powerful. It is not unusual for a newer desktop machine to have 5 megabytes of internal memory and 40 or more megabytes of disc storage. This means significant applications can be processed on such a departmental microcomputer.

User friendly equipment. Despite their growing processing power, microcomputers often are easy to access and use. Software is built around tutorial help screens, and selection menus allow even the unfamiliar user access to various software applications. In addition, there is not yet a high level of microcomputer security software commonly installed, so that unauthorized persons can access and process many microcomputer applications with little difficulty.

Many microcomputer "experts." Any organization with a large number of departmental microcomputers probably will have a certain number of employees with personal machines at home. Many with home machines become hobbyists and microcomputer "experts" who can perform such functions as accessing protected files or changing programs with little difficulty. While these persons can be a valuable asset to a department because they can help others, they also present a risk because of their ability to access or alter data or programs.

Limited systems development procedures. Chapter 7 discussed the importance of Systems Development Methodology (SDM) procedures for the design, development, testing, and documentation of new data processing applications. Such an SDM typically will not be used in a departmental microcomputer setting because of the limited number of users and the relative informality of their applications development procedures. Departmental microcomputer applications may not be tested or documented properly, or have the necessary application controls.

Easily transportable and fragile media. Microcomputer programs and data are stored on or can be copied easily to diskettes. Because of their small size, they can be taken from the machine area without detection. They also are relatively fragile and can be damaged easily.

Vulnerability to "can't happen" situations. Microcomputer systems can be subject to power surges which destroy data files, to storage media which unexpectedly become unreadable, or to fatal operator errors. Because of the generally friendly nature of the hardware and software, these are viewed as "can't happen" problems, and controls are not installed to protect against them.

Powerful communications capabilities. Equipped with a modem and a telephone line, the departmental microcomputer can gain access to many other computers equipped with dial-in capabilities. Equipped with special emulator boards, the microcomputer can look like another type of terminal to a host computer, with opportunities for security abuses through terminal emulation.

The above characteristics exist whether a microcomputer is installed within an information center organization or is located in a remote user department. The difference, however, is that there probably will be some level of information center management and a set of policies and procedures to monitor microcomputer use. When a machine is located in a separate department, control procedures often will not be as strong.

Departmental Microcomputer Management Policies

Having an understanding of the general control characteristics of departmental microcomputers, the auditor next should gain an understanding of any overall management policies covering the use of departmental microcomputers. In many instances, these have been formulated by general management with the assistance of the data processing function. A good set of management policies covering departmental microcomputers should include the following areas:

Hardware and software acquisition standards. There should be policies regarding which brands or generic types of hardware and software are authorized for purchase and technical support. However, any policies should be flexible enough to be responsive to technological improvements, new product introductions in the marketplace, and unique user requirements.

Report and data identification standards. It is easy in a microcomputer environment for users to reinput data from mainframe systems and modify report procedures to produce similar-looking reports with different results. There should be standards requiring that all such reports be identified as to the data source and who created the report.

Program and data copy policies. In a microcomputer environment, it is easy to make unauthorized copies of software to give to other employees, to use on one's own machine at home, or to even sell for gain. Since such practices are wrong and open the organization to potential legal action, there should be strict policies against unauthorized copying. Employees should be asked to sign an agreement acknowledging this.

Documentation and backup guidance materials. While not a policy or standard, the organization should have some guidance materials available to users of departmental microcomputers outlining good practices for documenting, testing, and backing up systems. Such materials probably should be prepared by the data processing department in a manner similar to the materials prepared for an information center.

In reviewing these overall organizational policies and procedures covering departmental microcomputers, the auditor may find them deficient in one or another area. If so, the auditor should make recommendations where appropriate to upper management. In addition, where a given policy appears weak or poorly defined and unenforceable, the auditor may want to note this for follow-up during a general controls review of departmental microcomputers.

Performing the Departmental General Controls Review

The general controls surrounding departmental microcomputers usually can be reviewed on one of two levels. If the emphasis of the review is to determine the extent to which various departments with microcomputers are following organization standards and procedures as well as other good control methods, the auditor could plan a comprehensive review of all departmental microcomputer systems. This can be an onerous audit task if there are many microcomputers distributed throughout the organization. However, the auditor can develop some general compliance statistics as part of the audit report.

The second approach is to review the general controls of selected departmental microcomputers. These can be scheduled when performing an operational or financial audit within that department. A computer audit specialist need be called in only when the microcomputer application appears to be of particular technical complexity. The disadvantage with this approach is that some departments, because of materiality or other considerations, are seldom subject to audit. As a result, the microcomputer systems in those departments, although possibly significant themselves, are never audited.

The auditor also might select a review approach that is a hybrid of the above two. Some microcomputer systems would be included in periodic departmental audits while others would be reviewed separately. In any event, the overall objective of these reviews is to determine if good general controls have been installed surrounding the departmental microcomputers used in the organization.

Figure 11.2 is a table of control objectives and audit procedures for reviewing departmental microcomputer general controls. The emphasis of these control objectives is on security, including physical and environmental protection, and

Control Objectives and Audit Procedures for
Reviewing Departmental Microcomputer General Controls

NOTE: These objectives and procedures are for reviews of departmental microcomputer systems when there is also a centralized information systems function. The objectives and procedures set out in Figure 3.6 might be more appropriate when the microcomputer system is the main data processing resource for the organization.

Objective 11.2.1. The microcomputer system should be implemented such that its applications support but do not duplicate other organization information systems facilities.

> **Procedure 11.2.1.1.** Review the microcomputer systems configuration in light of central data processing facilities as well as other department computers, and assess whether it is appropriate for the overall information needs of the organization.

> **Procedure 11.2.1.2.** Review procedures for organization approval of microcomputer hardware and software, and determine whether these standards were followed for the departmental microcomputer system being reviewed.

> **Procedure 11.2.1.3.** Determine that the hardware and software installed follows overall organization standards for types of hardware and software.

> **Procedure 11.2.1.4.** Obtain an inventory of hardware and software purchased for the departmental system, and determine the physical location of the inventory.

> **Procedure 11.2.1.5.** Determine that one individual or function is responsible for monitoring and maintaining the microcomputer system including problem solving and installing software upgrades.

Objective 11.2.2. The microcomputer system should be configured such that adequate attention is given to security and integrity.

> **Procedure 11.2.2.1.** Determine if "DOS Shell" password security software has been installed, and interview the individual responsible for maintaining that security system to determine the adequacy of security controls.

Figure 11.2. Control objectives and audit procedures for reviewing departmental microcomputer general controls

Procedure 11.2.2.2. Review available equipment and procedures for backing up key files and determine whether these appear adequate.

Procedure 11.2.2.3. If the microcomputer system accesses the central computer facility or if it accesses an organizational local area network, determine whether procedures for network information security appear to be adequate.

Procedure 11.2.2.4. Determine whether there is an organization or departmental policy prohibiting the illegal copying of microcomputer software, and determine whether the policy has been formally acknowledged by appropriate microcomputer users in the department.

Procedure 11.2.2.5. Determine that policies and procedures exist to limit the introduction of any "non-official software" to the microcomputer system to limit the risk of introducing computer viruses.

Procedure 11.2.2.6. Determine that original copies of microcomputer software are kept in a locked, secure location.

Objective 11.2.3. All end users of the microcomputer system should have adequate training and supervision.

Procedure 11.2.3.1. Review the adequacy of procedures used to train microcomputer systems users including the use of tutorial software.

Procedure 11.2.3.2. Interview selected microcomputer users to assess their understanding of the security and control procedures for applications in use.

Procedure 11.2.3.3. Interview the person or function responsible for end user microcomputer training to determine the person's level of understanding of computer systems integrity and control issues.

Objective 11.2.4. Applications developed on the system should follow good control procedures and should achieve planned objectives.

Procedure 11.2.4.1. Review the library of applications installed on the departmental microcomputer in light of "official"

Figure 11.2. Control objectives and audit procedures for reviewing departmental microcomputer general controls *(continued)*

production applications from the information systems organization; assess whether those departmental applications appear appropriate.

Procedure 11.2.4.2. When an approved departmental microcomputer application uses data from an information systems file, determine that procedures exist to download that data rather than causing users to rekey it.

Procedure 11.2.4.3. Review procedures for designing, implementing, and documenting new departmental developed microcomputer applications such as databases or spreadsheet macro applications.

Procedure 11.2.4.4. Assess whether "Spreadsheet Auditor" or other automated documentation tools are used to document end user developed microcomputer applications.

Procedure 11.2.4.5. Select several end user developed microcomputer applications from over the past year and review them to determine:

> The appropriateness of the application in light of any similar information systems applications
>
> The adequacy of applications documentation, including a description of controls procedures
>
> Procedures for backing up applications files and programs
>
> The adequacy of any output reports produced by the application, with attention given to the source and date of the data
>
> The level of end user training and understanding in the use of the application

Figure 11.2. Control objectives and audit procedures for reviewing departmental microcomputer general controls *(continued)*

information and program security. Applications controls will be discussed later in this chapter.

Many objectives and procedures listed in Figure 11.2 are appropriate also for general controls reviews when microcomputers are located within or under the control of an information center organization. As microcomputer hardware and software mature and as more users become familiar with their use, microcomputers will tend to be located within user departments rather than inside information centers.

In a growing number of organizations, microcomputers are linked into local area networks. This type of configuration has the advantage that microcomputers can communicate with one another and share certain common resources. For example, a single microcomputer can act as a server with common files and programs to all systems attached to the network. In addition, a single special peripheral device, such as a laser printer or a plotter, can be connected to the network for access by all systems on the network.

Many of the general controls discussed in Figure 11.2 also will be appropriate for local area networks of microcomputers. In addition, some of the unique network controls used in such microcomputer systems are discussed in Chapter 4.

Freestanding End User Computers

Previous sections of this chapter discussed controls when the end user develops applications through a formal information center or through departmental microcomputer systems. A third area where the auditor may find end user computing control concerns is where there are freestanding, mini or mainframe computers controlled by end users.

Specialized departments in many larger organizations have installed their own departmental mini or mainframe computers for various specialized purposes. A manufacturing department may have a shop floor minicomputer for automated testing procedures; an engineering department may have a computer for automated design and manufacturing; or a marketing research department may have its own mainframe for data regression analysis. These machines normally are installed because the regular data processing department could not support these specialized requirements adequately. Once installed and running for the unique purpose, other end user applications often are developed on such machines.

An example of such a freestanding computer system might be a machine installed for computer aided design in the engineering department of a larger organization. A group of engineers, dissatisfied with the data processing department's project management system, might elect to write its own project management system and process it on the engineering design computer during off shifts. Other engineering groups also might tie into this project management system and then add some of their own applications. Soon the engineering department might be adding more capacity to its machine to support these additional, unrelated business data processing applications.

This mixed purpose processing environment provides a difficult controls environment for the auditor. Upper management may proscribe that all business data processing take place on the data processing department's machine while only specialized processing can take place on the departmental machines. However, often it is difficult to separate the business applications from the technical ones. It is difficult for management, for example, to determine if a request for additional disc drives is to support a greater level of automated design or to support additional end user applications.

The above comments are not meant to imply that freestanding, departmental computers are always used improperly for business applications. In many instances, these freestanding mini or mainframe computers serve similar functions as the departmental microcomputers discussed previously in this chapter. However, often these types of computer systems are ignored by auditors, because in many organizations the internal audit department has avoided operational reviews of such highly technical departments. Computer auditors have also tended to ignore them as they were outside the regular data processing function.

General Controls on a Freestanding System

Many of the same controls objectives and audit procedures as discussed in Chapters 2 and 3 are applicable here. In many respects, there is a greater control concern in these specialized computer installations. A formal data processing department is experienced and familiar with the importance of such things as systems backups; a technical department controlling its own system may not have an appreciation for control considerations. However, it might be that the automated design records maintained on a freestanding system are more critical than the organization's general ledger file.

The auditor should develop a program to perform controls reviews of all significant freestanding systems controlled by end users. The auditor should first discuss these systems with functional management to understand their purpose and the types of applications developed and run on them. This may require some independent research by the auditor to gain knowledge of the specialized nature of the processing. For example, if a consumer products marketing organization has a separate machine devoted to regression analysis using market research data, the auditor should understand the process and why it may consume such large amounts of machine capacity.

The control objectives and audit procedures for these reviews essentially will be the same as outlined in Chapters 2 and 3. The only difference is that the auditor may want to give consideration to the criticality of the function. A machine used for statistical processing for market analysis may not have the same level of criticality as the prime business data processing computer center. Conversely, a computer system that plays a role in controlling the factory may have an even greater level of criticality.

Using Freestanding Systems for Business Data Processing

As discussed, freestanding mini and mainframe computers often are used for business data processing applications in addition to their intended purposes. Usually this is of major concern only if the departmental machine has the power to duplicate data processing applications. As part of any review of these machines, the auditor should review the controls surrounding any end user business data processing activity.

A first step is to determine managment policies covering such activities. As discussed earlier, management may have stated that all business related data processing must take place through the data processing department's computer center or through the formal information center. If the auditor finds that significant amounts of business data processing are taking place on free-standing departmental machines, there may be a major audit finding.

Many of the control objectives and audit procedures for reviews of free-standing systems are similar to those discussed in Figure 11.1 for information centers and Figure 11.2 for departmental microcomputers. However, because of the mini or mainframe nature of this type of computing coupled with the less formal department controls environment, there also are some unique control considerations as outlined in Figure 11.3.

As larger organizations computerize more specialized functions, the auditor can expect to find a greater number of specialized departmental free-standing mini and mainframe computer centers. Although often ignored because of their unique, specialized data processing tasks, the auditor should include these computer centers as part of a regular general controls review program.

Control Objectives and Audit Procedures for Reviewing Freestanding Systems

NOTE: These objectives and procedures are designed for specialized, freestanding computer systems, such as for engineering research, which may be used also for business data processing by end users. These objectives and procedures should be used in conjunction with Figures 11.1 and 11.2, for information centers and departmental microcomputers, as well as with Figure 3.9 for specialized process computer systems.

Objective 11.3.1. Business data processing on end user systems should take place only with management's specific authorization.

> **Procedure 11.3.1.1.** Discuss the role of the specialized free-standing computer system with management and review any published organization charter to determine whether end user computing is authorized and appears appropriate on the specialized computer system.
>
> **Procedure 11.3.1.2.** Discuss with management the types of applications processed on the specialized computer system to access the level of end user computing activities.
>
> **Procedure 11.3.1.3.** Develop an understanding of the normal activities of the specialized computer system and document findings.
>
> **Procedure 11.3.1.4.** Based on reviews, observations, and discussions, document the type and extent of end user computing activity on the specialized computer system.
>
> **Procedure 11.3.1.5.** Assess whether any end user activity diminishes the overall effectiveness or resource utilization of the specialized processor to perform its designated function.

Objective 11.3.2. Business data processing end user applications developed on specialized, freestanding systems should complement and not conflict with normal applications controlled by the information systems function.

> **Procedure 11.3.2.1.** Review and compare any end user computing applications maintained on the specialized machine

Figure 11.3. Control objectives and audit procedures for reviewing freestanding systems

with normal, information systems applications; determine that those end user applications support and do not conflict with the information systems applications.

Procedure 11.3.2.2. Review procedures for importing data from information systems applications to the specialized system, and determine that adequate control procedures are being followed.

Procedure 11.3.2.3. If data for end user applications comes from the end user area and eventually is processed and exported to a main computer system, assess whether adequate control procedures are followed in the processing and export of data.

Objective 11.3.3. Software tools used on specialized computer systems should be appropriate for any authorized end user business applications.

Procedure 11.3.3.1. Survey the types of software tools used for the specialized end user activity, and assess the appropriateness of these tools for end user business computing.

Procedure 11.3.3.2. Assess whether computer assisted audit tools can be used, if required, on any specialized program file structures.

Procedure 11.3.3.3. Review procedures for testing, documenting, and implementing any end user applications developed on the specialized processor.

Objective 11.3.4. End user data processing applications, regardless of their processing environment, should follow good control and documentation standards.

Procedure 11.3.4.1. Select several applications developed by end users on the specialized computer system and assess the adequacy of their documentation and control procedures.

Procedure 11.3.4.2. Perform a detailed review of the control balancing procedures for one of the applications selected above.

Procedure 11.3.4.3. Determine that specialized end users are aware of and use the application guidance materials devel-

Figure 11.3. Control objectives and audit procedures for reviewing free standing systems *(continued)*

oped by the information systems organization or the information center, as appropriate.

Objective 11.3.5.   Adequate procedures should be followed to maintain the security and integrity of end user applications run on specialized computer systems.

> **Procedure 11.3.5.1.**   Review file and library control procedures on the specialized processor, and determine that controls are adequate to prevent unauthorized access.
>
> **Procedure 11.3.5.2.**   Determine that key systems files are backed up on a periodic basis.
>
> **Procedure 11.3.5.3.**   Determine that one person or function is responsible for maintaining the security and integrity of end user applications on the specialized processor, and determine that these procedures are documented.

Figure 11.3. Control objectives and audit procedures for reviewing free standing systems *(continued)*

AUDITING APPLICATIONS CONTROLS OF END USER SYSTEMS

The auditor's role in reviewing data processing applications controls was discussed in Chapters 5 and 6. Many of these same control objectives and audit procedures are applicable for applications developed by end users. However, there are several unique aspects to applications developed by end users, as follows:

Use of application generators or package software. Although many data processing developed applications today use software packages or fourth generation languages (4GLs), some still use languages such as COBOL. End users tend to use fourth generation languages almost exclusively. As a result, the auditor may be less concerned about local changes to a software application but more concerned about the controls that the vendor has built into the application generator package.

Emphasis on retrieval or reporting applications. End user applications typically are used for special retrievals or reports rather than for file updating. This is contrasted with data processing production applications where transactions are collected and databases updated. This somewhat limits the application control risk for such end user applications.

Emphasis on ad hoc, non-production applications. Many end user applications are initially developed as "quick and dirty" one time applications to solve special needs. Often, there is very little testing and logic checking associated with this development process. However, those same applications, if they achieve their initial objectives, often become regular end user processed production applications. There can be a greater controls risk with such applications.

Limited applications documentation. The importance of end user computing standards covering such areas as applications documentation was discussed earlier in this chapter. However, despite such standards, the auditor often finds that documentation is either lacking or inadequate.

The above illustrate why the auditor may have a more difficult time reviewing and gathering evidence for an end user developed application than for a normal, data processing production application. However, as the number of end user developed applications increases in organizations, the auditor will be asked to review more of them.

Chapter 12, The Auditor and Fourth Generation Languages, contains a section on reviewing end user developed applications. The emphasis of these review procedures is on applications developed with the increasingly popular fourth generation language products. However, these same objectives and procedures are applicable to significant end user developed applications whether developed through a 4GL or other end user application development tool.

END USER COMPUTING AND THE AUDITOR

Applications developed by end users are the result of readily available software tools and computer literate users. With more individuals entering the job market with strong computer skills and the availability of increasingly powerful microcomputers and software tools, the auditor can expect that end user activities will increase in future years. Classic data processing departments will be responsible for strategic applications, databases, and certain critical applications such as accounts receivables and payables, but increasing numbers of retrieval and analysis applications will be developed and controlled by end users.

The auditor should be aware of the unique risks associated with end user computing. In addition, the auditor probably will be more effective in evaluating end user computing controls if the auditor gains a familiarity with the various end user computing packages used within the organization. Some of these can also be used as tools for the audit, as will be discussed in Chapter 12.

CHAPTER 12

The Auditor and Fourth Generation Languages

INTRODUCTION

New programming languages, called fourth generation languages (4GLs), are rapidly being introduced into the world of data processing. These programming languages allow data processing applications to be developed much faster than those created with traditional business languages such as COBOL. They also introduce new applications control considerations for the auditor and provide a powerful new audit tool.

This chapter introduces the use of fourth generation languages as tools for developing applications, discusses the control implications of fourth generation languages, and discusses approaches to reviewing applications developed with them.

Many data processing departments are introducing 4GLs into their organizations. There are many 4GL products, each with somewhat different characteristics. These languages often appear foreign to the auditor who has learned only COBOL or BASIC in college or in prior work experiences. However, auditors will see more 4GLs in the future and should understand them and develop the ability to write simple programs with them.

CHARACTERISTICS AND CLASSIFICATIONS OF 4GLS

The first generation of computer languages was the binary based machine language which gave instructions directly to the machine. Programmers had to enter detailed instruction codes to perform machine functions. To add two numbers, a first generation language program would perform the following steps:

1. Clear the machine's accumulator and add the contents of the storage location where the first number is located to the cleared accumulator.
2. Add the contents of a second storage location to the accumulator.
3. Move the contents of the accumulator to a designated storage location.

The above three step process illustrates the difficulty associated with programming in first generation languages. The first of the early languages required the programmer to code each of the above steps in a binary code of ones and zeroes. In addition, the programmer had to keep track of the actual address of the storage location where the data was located.

The second generation introduced mnenomics, or codes for machine instructions, as well as symbolic names for actual physical locations. For example, the first program step above, written in a first generation language, might be as follows:

011011 000000 000000 000001 110101

The first six binary digits were a symbol for the action to be taken and the remaining digits represented a storage address. A second generation language would reduce that same instuction to something like:

CLA ARBAL

Here, CLA is a mnemonic to clear the accumulator and add contents. ARBAL is a symbolic name for the field to be added.

Second generation languages greatly simplified the programming process by introducing symbolic notations. However, it was still necessary for the programmer to write each programming step. Second generation languages were machine dependent and went by names such as SPRINT for certain UNIVAC machines and AUTOCODER for various IBM machines.

The 1960s saw the introduction of what were called third generation, or high-level, languages. They went by names such as FORTRAN, ALGOL, and COBOL. Although actual language processors were unique to a given computer, the symbols and structure of the programs became more universal. For example, the above three program steps might be written in COBOL as follows:

> MOVE ACCTS-REC-BALANCE TO BATCH-TOTAL
> ADD CURRENT-TRANS-AMOUNT TO BATCH-TOTAL

Third generation languages greatly simplified the programmer's task of developing information systems. However, they still required skilled programmers to develop applications. It took time and effort to understand the unique rules of a given programming language and to implement the application properly.

Since the early 1980s, there has been a steady introduction of what has been called fourth generation languages or applications generators. While these initially appeared in the early 1980s as specialized languages used to help developers or users in limited areas, 4GLs have become common development tools for all areas of business data processing. Figure 12.1 shows the historical progression of these generations of programming languages. The modern auditor should develop a good understanding of both the use and control risks associated with 4GLs.

4GL Characteristics

It is difficult to give a standard definition of a fourth generation language. However, most 4GLs exhibit one or more of the following characteristics:

Non-procedural language. Third generation programming languages generally require a programmer to follow a fixed set of procedures to accomplish a given task. For example, a COBOL programmer who wants to produce a report sorted in a given sequence must first open and read the file of data, second sort the file, and third produce the report. A typical 4GL can ignore this fixed sequence. For example, it may only be necessary to give a command like *"LIST DATA SORTED BY . . . ."* Figure 12.2 shows a sample of a retrieval program written with a 4GL. It is an English-like program which results in a complete output report.

Environmental independence. Many 4GLs are portable regarding computer hardware, operating systems, and telecommunications monitors.

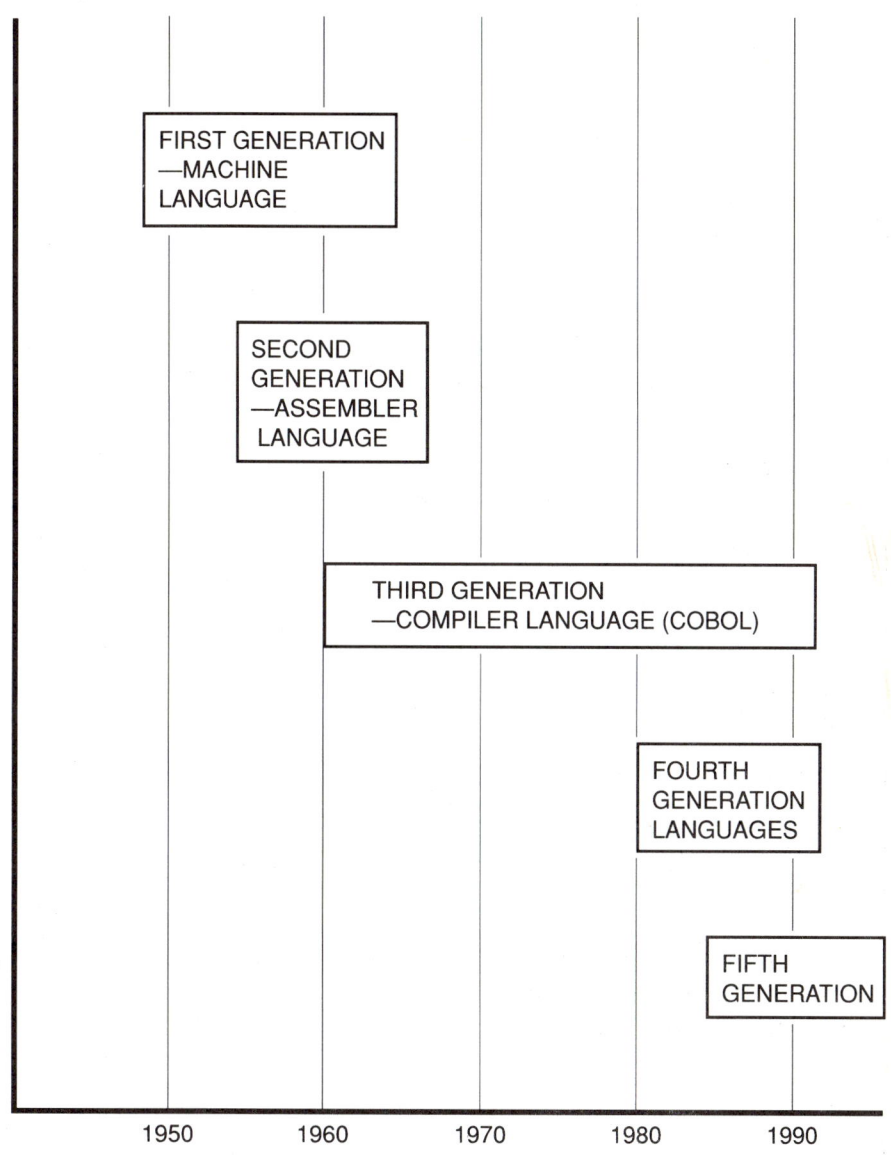

Figure 12.1. Historical progression of programming language

An Example of a 4GL Retrieval Program

This example describes a simple retrieval program that might be constructed to list names and salaries of all employees on a master file. The language syntax shown is based loosely on the FOCUS 4GL product. Syntax for other 4GLs would be similar.

The first step would be to define the data fields in the file of interest. As shown below, it is necessary to define both the field name and size, such as "A6" for six alphanumeric characters. For the payroll file example, this might be described as follows:

```
FIELDNAME = EMP-NO,A6        $
FIELDNAME = L-NAME,A15       $
FIELDNAME = F-NAME,A15       $
FIELDNAME = JOB-CLASS,A3     $
FIELDNAME = SALARY,D6        $
FIELDNAME = STREET,A15       $
FIELDNAME = CITY,A12         $
FIELDNAME = STATE,A2         $
FIELDNAME = ZIP,D5           $
```

Using the FOCUS language syntax as a 4GL example, it is then possible to program a report showing all employees by their number, last name, and salary, as follows:

```
TABLE FILE STAFF
LIST EMP-NO AND L-NAME AND SALARY
COUNT EMP-NO
SUM SALARY
END
```

A program written in COBOL to produce the same report might require several hundred lines of program code. More complex operations can be performed also using a simple 4GL syntax.

Figure 12.2. An example of a 4GL retrieval program

Some 4GLs have been implemented on both mainframe processors and microcomputers.

Powerful software facilities. In addition to being used to develop application programs faster, the typical 4GL has other powerful software facilities, such as ones to design or "paint" retrieval screen formats, develop computer aided training routines, and produce graphical outputs.

Programmer workbench concepts. Most 4GLs support the concept of a "programmer workbench" where the programmer does not need paper, pencils, and filing drawers. The programmer has access through the terminal to easy filing facilities, temporary storage, text editing, and operating system commands. This type of a workbench approach is associated closely with the CASE computer aided systems engineering application development approach discussed in Chapter 7.

Simple language subsets. While 4GLs often have powerful facilities, they generally also have simple language subsets which can be used by less skilled end users through an information center. For the auditor, this is perhaps the most significant characteristic of a typical 4GL. End users and auditors can develop their own retrieval applications rather than rely on the data processing function.

Because of these characteristics, a 4GL often will have a powerful impact on data processing when one is implemented in an organization. While the 4GL generally offers productivity improvements, it often is resisted by some data processing personnel. They may complain, for example, that the 4GL introduces hardware processing inefficiencies, limits programmer flexibility, or gives end users too powerful a tool. Nevertheless, the auditor will probably see an increasing use of 4GLs in data processing organizations.

4GL Classifications

A typical 4GL does not necessarily have all of the characteristics listed in the preceding section. The auditor can gain a better understanding of what a 4GL is by considering the various 4GL types of products. Most 4GLs fit into one of the following classifications:

Query and report generators. Specialized languages that can extract records and produce reports have been around for years. Computer audit software is an example of such generators. Recent years have seen much more powerful languages, however, that can access database records, produce complex on-line outputs, and be developed in an almost natural language. Some of these report generators are tied directly to commercial application software packages.

Embedded database 4GLs. Some 4GLs depend upon self-contained database management systems (DBMS). While they can link to other DBMS facilities, they often operate as self-contained facilities. This characteristic often makes them more user friendly but may also lead to applications which are not integrated well with other production applications. Product names associated with this type of 4GL include FOCUS, RAMIS II, or NOMAD 2.

Related database 4GLs. Other 4GLs are high level language products that usually are an optional feature on a vendor's DBMS product line. These allow the applications developer to make better use of the DBMS product, but they often are not end user oriented. Product names here include SQL + , MANTIS, and NATURAL.

Application generators. These are development tools for generating and maintaining normal transaction processing for business applications. Application generators are tools for data processing development personnel and not end users. Many of these products generate third generation COBOL code which can be further tailored and customized. Product names here include GAMMA, PACBASE, and TELON.

The auditor who encounters the use of a 4GL in a data processing organization under review should inquire as to the type and characteristics of the package being used. This will help the auditor to better understand any control implications associated with that 4GL. Some product names have been mentioned here. However, there are many others. New ones will be introduced, and some existing products will become obsolete, as was the fate of many third generation languages. In the late 1960s, an auditor might have encountered applications written in such languages as ALGOL, FLOW-MATIC, or JOVIAL, which have now all disappeared.

PRE-IMPLEMENTATION REVIEWS OF 4GL DEVELOPMENT PROJECTS

4GLs have introduced many changes to the SDM-oriented methods used to develop computer applications. These changes require the systems developer to rethink the manner in which new systems are developed and implemented.

They also require the auditor to rethink the manner in which pre-implementation reviews of these applications should be performed. Chapter 7 discussed procedures the auditor might normally use for pre-implementation audit reviews. These reviews of new applications under development usually are tied to the classic systems development methodology, or SDM, which uses the phases:

Project initiation and feasibility determination

Systems requirements definition

Detailed design and program development

Applications testing and implementation

Post-implementation reviews

Because they permit processing applications to be developed much more rapidly, 4GLs have modified this systems development life cycle or SDM approach. In many instances, the organization will follow a formal SDM through the project initiation and requirements definition phases. However, there may be little need for the detailed design and applications testing phases. For example if a code generator 4GL is used, the auditor may not find a formal detailed design document or testing plans. These procedures often will be performed more informally. The data modeling activities performed during the initial phases of the applications design will become more significant.

Another approach is to develop a prototype version with the 4GL and then reprogram it with a conventional programming language after everyone has agreed on the application's functions. This approach sometimes is used when the data processing organization is using an embedded DBMS type of 4GL and wishes to rebuild it eventually into a production version using a language such as COBOL and the organization's standard DBMS. Data processing organizations sometimes take this approach because the 4GL-developed application may introduce processing inefficiencies.

A third approach to using a 4GL is to design, build, and implement the complete production application with the 4GL product. This means that many of the normal SDM forms and procedures probably will not be followed. However, as data processing professionals become more accustomed to the use of these new 4GL products, they may follow this route.

Many data processing organizations also take a mixed approach to their use of a 4GL. For example, a 4GL application generator is sometimes used for such things as input screens or output reports. The remainder of the application then is written using conventional methods. This can work well if the 4GL can link easily to the other data processing tools being used. All to often, however, that linkage may not be smooth.

What are the auditor's concerns in this type of a development environment? Chapter 7 outlined three broad objectives for auditing new systems under development:

1. To evaluate application controls prior to implementation
2. To evaluate project definition and justification
3. To evaluate project development controls

The auditor's concerns with respect to the first of these objectives, evaluating application controls, is the same regardless of the type of development technique being used. Any data processing application should be constructed with adequate controls, regardless of whether it is written in COBOL or with a 4GL. However, the use of a 4GL can raise special concerns with respect to the other two objectives. Because developers often view 4GL-developed applications as "quick and dirty" solutions which are soon discarded, the normal SDM procedures which might detect control problems are not used, and then, rather than being discarded, these ad hoc applications often move to permanent production status.

Figure 12.3 contains control objectives and audit procedures for an evaluation of controls for an application developed with a 4GL. The auditor first determines to what extent the 4GL is being used, and then develops procedures to perform a pre-implementation review of that application. Of course, as a first step, the auditor should perform a criticality analysis, as discussed in Chapter 7, to determine which applications to consider for review.

The second broad pre-implementation objective, to evaluate project definition and justification, may be of less audit concern when a 4GL is used, because prototype versions of the application can be built easily and cheaply. If they do not meet user requirements or are otherwise unacceptable, the data processing organization can discard them. Nevertheless, the auditor should look for a good definition of the project to be constructed as well as a user or management justification for the project.

Under this second pre-implementation objective, the auditor should examine the existing SDM in use and consider making recommendations for improvement. Often, the auditor finds that existing SDM procedures have been ignored rather than modified to reflect the changes introduced by a 4GL. However, a 4GL-developed application often increases the importance of the preliminary, project definition phases of application design. Through data modeling and other procedures, many important control and design decisions

**Control Objectives and Audit Procedures for
Reviewing Production Applications Developed with 4GL Tools**

*NOTE: These objectives and procedures are designed to assist the
auditor when reviewing project initiation controls for new appli-
cations developed with fourth generation language tools. This
Figure should be used in conjunction with Figure 7.5,* Control
Objectives and Audit Procedures for a Pre-implementation
Review of the Project Initiation Phase.

Objective 12.3.1. The fourth generation language (4GL) being considered for end user application development should be capable
of operating in the organization's normal business data processing
environment with adequate controls.

   Procedure 12.3.1.1. Through the use of tutorial programs or
   demonstrations, develop a general understanding of the 4GL
   product being used and assess its control strengths and
   weaknesses.

   Procedure 12.3.1.2. Determine whether the 4GL will operate
   within the information system's operations environment,
   including its ability to interface with standard database
   products, information security packages, and normal file
   structures.

   Procedure 12.3.1.3. Through discussions with systems software and operations specialists, assess whether the 4GL tool
   may impose any significant resource constraints on the computer system.

Objective 12.3.2. Applications developed by the information systems organization with 4GLs should follow the same general
request and authorization procedures as conventional application
development projects.

   Procedure 12.3.2.1. Determine that all requests for new or prototype applications using 4GL tools follow the same preliminary systems development methodology (SDM) procedures
   used for conventional development projects.

   Procedure 12.3.2.2. If the information systems organization

Figure 12.3. Control objectives and audit procedures for reviewing production applications developed with 4GL tools

does extensive development work using prototyping, determine that their SDM has been modified to describe a prototype development life cycle process.

Procedure 12.3.2.3. Review procedures for determining when an application will be developed with a 4GL rather than conventional programming methods and determine whether those selection procedures appear appropriate.

Procedure 12.3.2.4. Select several applications that have been developed using a 4GL tool and determine that SDM procedures were followed, including the preparation of documentation.

Objective 12.3.3. Applications developed by end user functions with 4GLs should have the approval of management and should be developed following general information systems SDM standards.

Procedure 12.3.3.1. Determine that there has been a review of all significant end user applications projects by either the end user information systems steering committee, the information center, or the information systems function to determine that the project does not conflict with other approved or implemented applications.

Procedure 12.3.3.2. Select several recently implemented end user 4GL applications and determine that end user management has approved, explicitly or implicitly, these efforts and is aware of the development costs and expected benefits.

Procedure 12.3.3.3. Interview the end users responsible for a selected 4GL-development project in process and assess whether they are following good general SDM techniques.

Objective 12.3.4. Users should be aware of the capabilities and limitations of 4GL-development tools when applying them to new applications.

Procedure 12.3.4.1. Determine that the information systems function has documented properly the strengths and weaknesses of its 4GL-development tools, such as an inability to communicate with certain database structures, and that it has communicated these strengths and weaknesses to end users.

Figure 12.3. Control objectives and audit procedures for reviewing production applications developed with 4GL tools *(continued)*

Procedure 12.3.4.2. If 4GL applications are being developed as prototypes, select several such systems and interview key users to determine that they understand procedures for moving the applications from prototype to full production.

Procedure 12.3.4.3. Interview a sample of end users involved in the development of applications through prototyping and determine that they understand their responsibility to test, review, and suggest changes to prototypes.

Objective 12.3.5. Application controls when using 4GLs should be the same as applications developed through conventional methods.

Procedure 12.3.5.1. Review documentation for several recently implemented 4GL applications and determine that the documentation contains a proper description of application controls.

Procedure 12.3.5.2. Select a sample application recently developed and implemented using 4GL tools and assess its controls including:

Good report formats that include titles, dates, and end of report control totals

Audit trails that allow tracing the processing and disposition of all inputted transactions

Run to run and application to application controls to assure that processing is complete and correct

Error reports or screens which describe the cause of errors and suggest corrective actions.

Figure 12.3. Control objectives and audit procedures for reviewing production applications developed with 4GL tools *(continued)*

are made in the earliest phases of a 4GL project. The actual development requires far less effort.

In reviewing an application developed with 4GL procedures, the auditor should not insist that certain SDM procedures be followed just because they are always used for conventional systems projects. For example, most SDMs require extensive formal printed program documentation. Many 4GLs provide a facility to generate on-line program documentation which may be in the form of diagrams rather than descriptive text. This approach to documentation can be adequate.

The last broad objective for pre-implementation reviews is to review project development controls. This has been a particularly important audit task for traditional, large data processing projects because of the tendency for such projects to run out of control. Auditors often were able to advise both data processing and general management that a given large project lacked proper project controls and might be late or require additional, unplanned resources.

Projects developed with 4GLs often bring about significant productivity improvements. Some users have reported that a small to medium sized 4GL application can be developed with one-tenth the resources required for a conventional, COBOL based project. Figure 12.4 shows this comparative difference in resource requirements.

Figure 12.5 contains control objectives and audit procedures for pre-implementation reviews of project definition and project development controls for 4GL developed applications. The emphasis of these objectives and procedures is on the unique control concerns that 4GLs introduce into the new applications implementation process. This set of objectives and procedures, as well as the ones discussed in Figure 12.3, are not meant to replace but to supplement the ones in similar tables in Chapter 7.

CONTROL REVIEWS OF 4GL APPLICATIONS

Applications Developed by Data Processing Departments

The controls that are built into an application being developed with the use of a 4GL depend upon what is available within the software product. There are dozens of fourth generation language products on the market today. While most of these products are for large scale IBM 360/370 architecture machines such as the 43xx and 30xx series, others are available for different vendor processors. It is important that the auditor understand the controls available in the 4GL tool being used.

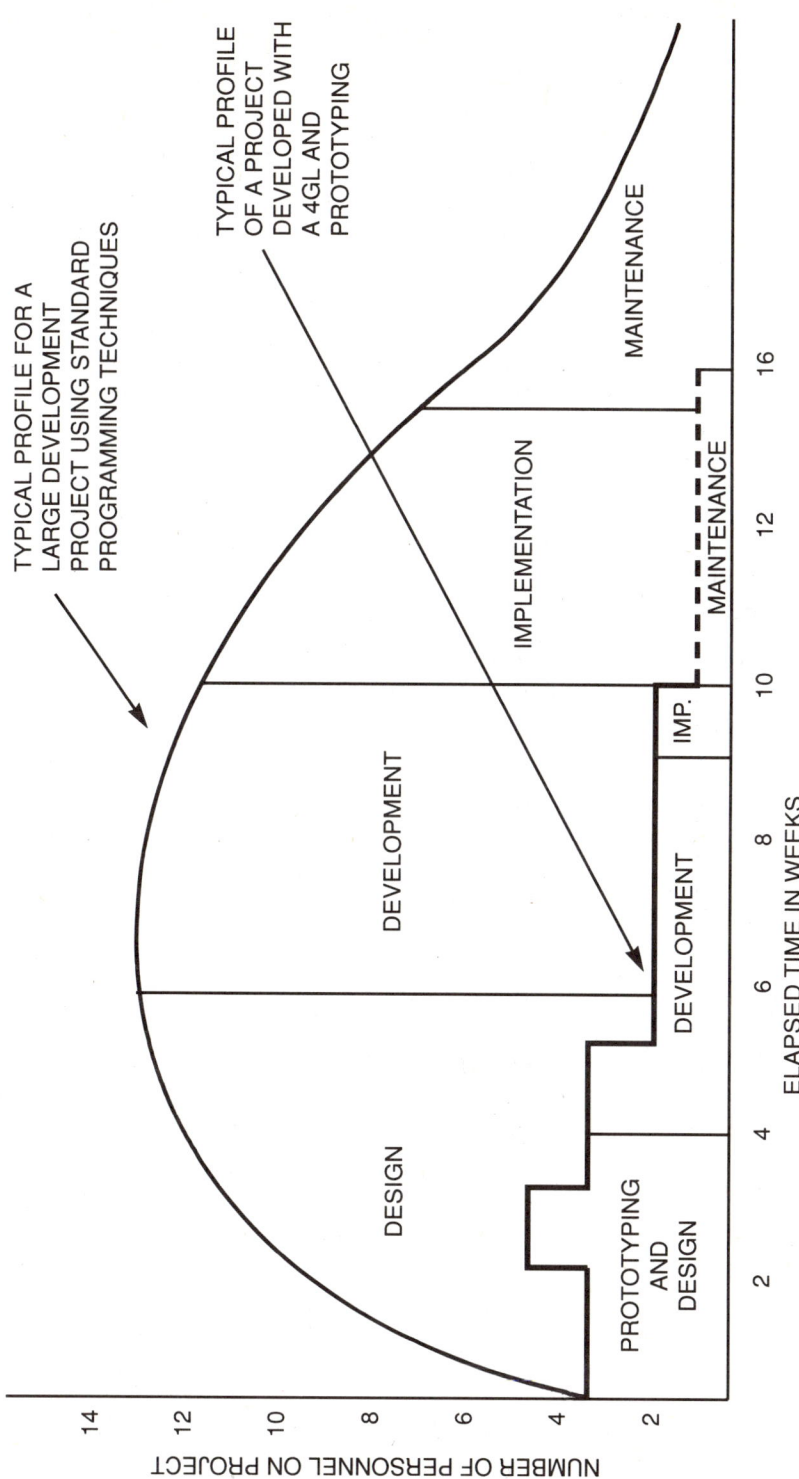

Figure 12.4. Resource requirements for application development utilizing 4GL tools as compared to using conventional SDM methods

Control Objectives and Audit Procedures for a Pre-implementation Review of an Application Developed with 4GL Tools

NOTE: These objectives and procedures are designed to assist the auditor when performing a pre-implementation review of an application being developed with a fourth generation language. These may be used in conjunction with the objectives and procedures contained in Figure 7.6.

Objective 12.5.1. The prototype design of the 4GL application should provide sufficient detail to replace the written documents associated with conventional SDM procedures.

Procedure 12.5.1.1. Determine that the information systems and end user group developing the 4GL application plan to prepare documentation consistent with the features of the 4GL tool.

Procedure 12.5.1.2. Select a recently implemented 4GL application and review the technical systems documentation to determine that it accurately describes the implemented application and is sufficiently complete to allow ongoing maintenance.

Objective 12.5.2. End users who will be responsible for the 4GL application should be involved actively in the development process.

Procedure 12.5.2.1. Interview principle users of the 4GL application development project and determine that they understand their responsibilities in the 4GL design and development process.

Procedure 12.5.2.2. Review user plans for testing and review the 4GL application under development to determine that user objectives have been established, a test plan has been developed, and testing is following that plan.

Objective 12.5.3. The application being developed with a 4GL should give proper attention to data security and integrity.

Procedure 12.5.3.1. Review data and information security plans for the 4GL application and determine that they are consistent with other security procedures within the information systems organization.

Figure 12.5. Control objectives and audit procedures for a pre-implementation review of an application developed with 4GL tools

Procedure 12.5.3.2. Review the results of interim application tests to determine whether the application is being built with proper attention to data integrity and controls.

Procedure 12.5.3.3. If the application uses database facilities, determine that the data administration function has reviewed application plans.

Objective 12.5.4. The application being developed with a 4GL should take proper advantage of the features included in the 4GL software product.

Procedure 12.5.4.1. Through on-line tutorials, seminars, or hands-on practice, develop a good understanding of the properties and features of the 4GL in order to assess its applicability to the application under review.

Procedure 12.5.4.2. Determine whether the application under development makes use of such typical 4GL software features as:

Semantics, syntax, and data integrity checking
Default options to minimize end user decision making time
On-line documentation and "help" screens
Interfaces or linkages to local microcomputers, if appropriate
Data display features, such as graphics, to simplify management reporting

Objective 12.5.5. Where applicable, proper ongoing audit tools should be developed as part of the pre-implementation review of a 4GL application.

Procedure 12.5.5.1. Develop a detailed audit file describing pre-implementation audit activities for the application, using the same 4GL techniques used by the application developers.

Procedure 12.5.5.2. If appropriate, develop computer assisted audit integrity testing programs using the same 4GL product to test the application during all phases of the project.

Figure 12.5. Control objectives and audit procedures for a pre-implementation review of an application developed with 4GL tools *(continued)*

Many applicable control characteristics were discussed in the objectives and procedures for reviewing controls in new applications under development set out in Figure 12.3. Fourth generation language processors will have differing levels of capability to allow the following control procedures to be built into their applications:

Audit trail logging facilities

Checkpoint, transaction backout, and restart facilities

Database access paths

Exits to and from third generation languages

Use of runtime monitors

The auditor should develop a good understanding of the capabilities of the 4GL product being used. This often can be accomplished by reading documentation and discussing the package with data processing personnel. For example, the auditor may want to determine that a facility to automatically log transactions has been invoked in the application under review, if that feature is available.

If the 4GL being used does not have some of these desirable control characteristics, the auditor may recommend that they be built through the use of linkages to third generation languages. While this type of recommendation is most appropriate for a pre-implementation review, it also could apply to an operational application. In addition, the auditor may suggest to data processing management that such control procedures be incorporated into all new 4GL applications.

Other than understanding the unique characteristics of the 4GL product, much of the process of reviewing a 4GL application is the same as with any data processing application. Chapter 5 discussed the process of reviewing controls, testing, and gathering evidence to attest to the integrity of an application. The objectives and procedures outlined there generally are applicable to 4GL-developed applications.

One aspect of the audit of a 4GL application that often is easier than traditional applications is the process of testing or evidence gathering. Many 4GLs have retrieval languages for use by users. As will be discussed later in this chapter, these can be used as an audit tool in place of conventional audit software. This particularly is convenient for embedded DBMS processors which use an unusual type or format of database. While it may be difficult to access that database with conventional software, the 4GL retrieval language typically will have no problem with data access or retrieval.

Applications Developed by End Users

Because of their ability to create ad hoc reports quickly, the auditor often finds 4GLs being used by end users. Many 4GLs are powerful tools for that purpose. Others, because they may be linked closely to the organization's DBMS, may be too powerful to give to an end user. The prior section discussed how some traditional application controls, such as the use of a formal SDM, may need to be modified when 4GLs are used by a data processing department. This lack of formal development and application controls may be an even greater problem when 4GLs are used by end users for their own application development.

Chapter 11 discussed the need for general controls in the information center and over end user developed applications. In addition to these general controls, the auditor may want to review the specific application controls surrounding critical 4GL end user developed applications. The auditor's review approach here should follow the general guidelines for applications reviews discussed in Chapter 5.

Auditors typically will not find the same type of critical applications under end user development and control as found within the data processing systems and programming function. In many instances, the auditor finds that end users are doing little more than using the 4GL tool to write retrieval reports against data processing controlled databases. However, the auditor first needs to develop an understanding of the types of 4GL applications that are under end user control.

Understanding the 4GL End User Application

Chapter 5 discussed approaches for selecting traditional, data processing developed applications for review. It then suggested the types of documentation that the auditor should review to develop an understanding of the application review candidate. Auditors sometimes will have a more difficult task in developing this understanding for 4GL end user applications, because these applications often are developed on an ad hoc basis and not well documented. However, such applications generally are not as complex as data processing production applications.

Many 4GL applications are of little concern to the auditor or management. Some may be of a personal productivity nature such as personal scheduling calendars. Others may be simple retrieval applications used to improve decision making within one department or function. For example, a cost accounting department might use a 4GL for special analysis reports of certain product costs. Such reports essentially will have no impact on the organization's financial statements and present minimal risk.

Other end user 4GL applications may be used for important strategic business decisions. There could be a significant control or business risk from such applications. However, unless such applications are processed through a formal information center and unless there are good general controls within that information center, it is difficult to even identify such applications.

To identify potential review candidates from the various 4GL end user developed applications, the auditor may want to consider conducting a survey. The auditor should develop an information survey form that asks users to list the types of applications they are running, the databases or files being accessed, the frequency of use, and the control responsibility for the application. A sample of such a survey form is shown in Figure 12.6.

Depending upon the organization, this survey can be used for specific departments or throughout the end user community. Preliminary survey results should be discussed with the information center function as well as with data processing management to resolve any questions. It also may be necessary to contact end users directly to determine that survey results are complete and accurate.

This completed survey should help the auditor classify 4GL end user applications as follows:

1. Applications that perform critical functions in their own areas and might be candidates for audit review.
2. Applications which appear to perform significant internal accounting control functions that might be recommended for conversion to data processing controlled production systems.
3. End user applications presenting minimal audit risk.

The first of the above groups of candidates should become the pool for potential applications reviews. Group two candidates also may require an applications review to gain sufficient evidence to make a recommendation for conversion to a production system. The applications in the third group are the various retrieval reports and other personal productivity applications found in many organizations. There may be little need to go beyond identifying these for statistical purposes.

Auditing 4GL Applications Developed by End Users

Once the auditor has developed background data on the end user 4GL applications in use, some may become candidates for a more detailed audit review.

4GL Application Survey

*NOTE: This survey is designed to gather audit background informa-
tion about 4GL applications developed in user departments. Since an
organization typically will use only one 4GL product, the survey should
refer to the 4GL product by its name. A form should be completed for
each significant 4GL application.*

DEPARTMENT _____ NAME _____

NAME OF APPLICATION _____

APPLICATION TYPE OR FUNCTION:

    ☐ SELF CONTAINED REPORTING APPLICATION

    ☐ SELF CONTAINED DATABASE APPLICATION

    ☐ APPLICATION RECEIVING DOWNLOADED DATA FROM OTHER SYSTEMS

    ☐ APPLICATION UPLOADING DATA TO OTHER SYSTEMS

    ☐ OTHER _____

DESCRIBE GENERAL FUNCTIONS OF THE APPLICATION:

APPROXIMATE TRANSACTION VOLUME _____

OUTPUT REPORTS _____

FREQUENCY OF USE _____ CONTROLS RESPONSIBILITY _____

IS APPLICATION DOCUMENTED? _____ ARE SECURITY CONTROLS USED? _____

LOCATION OF DOCUMENTATION _____

BACKUP OPERATOR _____ DATE IMPLEMENTED _____

Figure 12.6. An information survey to determine appropriate 4GL-devel-
oped applications for review

The actual approach for such a review is similar to that of reviewing normal production applications. Chapter 5 suggested the auditor approach these application reviews as follows:

Understand the purpose of the application
Review the application with key users
Describe the system for audit workpaper purposes
Identify system control points
Test key controls

There is little difference in these review steps when the application is developed through a 4GL rather than through a traditional data processing function. The auditor will want to develop an understanding of the application and then develop a specific set of control objectives and audit procedures for the applications review. The various lists of objectives and procedures contained in Chapter 5 as well as Figure 12.3 should help the auditor create a review approach for such an end user developed application.

When working with an end user developed application, the auditor will want to emphasize areas where there may be a greater control risk. These generally include:

Poor change controls. Because of the ease of developing and subsequently modifying 4GL applications, good techniques for documenting changes to the system often are ignored. There is often little distinction between "test" and "production" versions of a 4GL application.

Poor error controls. End users often are not familiar with data processing procedures for error screening and detection. Being novice developers, they tend to assume that there will be no data errors.

Application logic errors. Because of the "quick and dirty" approach that often is taken to develop 4GL applications, there is a greater chance that the completed application may contain logic errors.

Limited documentation. End user 4GL applications are developed and maintained by only a limited number of key persons. If a certain minimum level of documentation has not been prepared, it is quite possible that the application could not be maintained if those key persons were unavailable.

Figure 12.7 contains a set of control objectives and audit procedures that the auditor should use to perform a review of an end user developed 4GL applica-

Control Objectives and Audit Procedures for
Reviewing Applications Developed by End Users

Objective 12.7.1. As a first step to reviewing the end user developed application controls, the auditor should have a good understanding of the objectives and functions of the application.

> **Procedure 12.7.1.1.** Review existing documentation for the application to gain an understanding of its objectives and functions.

> **Procedure 12.7.1.2.** Review project initiation documentation for the application to gain an understanding of initial objectives and interview key users to assess whether the application is meeting original design objectives.

> **Procedure 12.7.1.3.** Based on the objectives and control implications of the application, assess whether an end user developed 4GL application appears to be appropriate.

Objective 12.7.2. The application should follow good control techniques and should achieve its specified objectives.

> **Procedure 12.7.2.1.** Review application controls and determine that they provide assurance that all transactions have been processed correctly.

> **Procedure 12.7.2.2.** Review user procedures for monitoring the 4GL application and maintaining any external controls; assess the adequacy of these procedures.

> **Procedure 12.7.2.3.** Review project initiation documentation for the application to gain an understanding of initial objectives and also interview key users to assess whether the application is meeting its original design objectives.

Objective 12.7.3. The application should follow good error control techniques for the transactions processed such that errors can be detected and corrected by responsible personnel.

> **Procedure 12.7.3.1.** Review 4GL program logic being developed to screen and correct errors and determine whether this logic appears to be comprehensive.

> **Procedure 12.7.3.2.** If the application receives transactions

Figure 12.7. Control objectives and audit procedures for reviewing applications developed by end users

from the information systems central computer or uploads transactions to it, determine that there are procedures to resolve any differences in transaction error processing between the two systems.

Procedure 12.7.3.3. Determine that the application has sufficient reporting or other procedures to identify all error conditions.

Procedure 12.7.3.4. Determine that the application has sufficient documentation covering the correction of error conditions through either "help screens" or descriptive reports.

Objective 12.7.4. Documentation covering the application should be sufficient such that prime users as well as others can operate and maintain the application.

Procedure 12.7.4.1. Review user documentation for the 4GL application and assess whether it is easy to understand and is comprehensive.

Procedure 12.7.4.2. Review technical documentation covering 4GL programming code and processes to assess whether it is complete.

Procedure 12.7.4.3. Determine that there is more than one end user trained in the programming and technical aspects of the 4GL as used for the application.

Objective 12.7.5. Changes to the application should be made only with management approval and should be documented fully.

Procedure 12.7.5.1. Review procedures for making changes to the application's programming code and assess whether access controls to the code are sufficient.

Procedure 12.7.5.2. Determine that a log file exists for documenting all changes to the 4GL application and for noting management approval of those changes.

Procedure 12.7.5.3. If the 4GL is maintained on the information systems department main computer, determine that program library controls for updating the 4GL application are consistent with those used for normal production applications.

Figure 12.7. Control objectives and audit procedures for reviewing applications developed by end users *(continued)*

tion. These procedures assume that the organization has good general controls within the information center function and over data processing operations.

Auditors should be aware of end user developed applications created through the use of fourth generation languages. While some of the earlier retrieval languages provided users with little more than facilities to develop ad hoc retrieval reports, 4GLs now allow end users to implement complete data processing applications outside the formal data processing function.

4GL SECURITY IMPLICATIONS

Fourth generation languages also introduce some security concerns when they are used as either a data processing department application generator or as a tool for end user development work. Many of these security concerns arise because many 4GLs are unique, specialized products that may not interface easily with standard security software products within the data processing function.

Some of the current 4GL packages cannot operate easily within the rules imposed by program library management software or information security software. Often, the data processing function does not discover this problem until they have purchased and installed the 4GL package and established a large base of enthusiastic users. They must then take steps to develop other procedures for program library updates and access control security.

Auditors play an important role in informing management of the security issues surrounding the use of 4GLs. Their security vulnerabilities fall into the following three areas:

Security over target files and applications. As mentioned, some of the current 4GL products do not interface with existing information security packages. Thus, even when data processing department standards require that "all" data sets be protected by such departmental standard security software, exceptions must be made for those tied to the department's 4GL. In addition, these same 4GLs generally do not have their own mechanism to control and monitor security violations.

Security over the development process. "Source code" in its traditional sense does not exist with 4GLs. Therefore, it is difficult to control programs at a source level and monitor changes to applications. In addition, some 4GL products allow the skilled programmer to write exit routines into the 4GL code which may manipulate other areas of the computer's operating system.

Security over the 4GL development environment. Whether the 4GL is installed on a microcomputer, a specialized development work station, or a normal terminal, security should be a concern. While the security concerns are similar to those over any physical device, they are particularly important here because of the power associated with the 4GL.

Security risks associated with the use of 4GLs should be identified as soon as possible. The best time, of course, is during the software evaluation process. Auditors should request, when possible, to become involved in this review and evaluation process. In this manner, they can ask appropriate security and control questions before the package is purchased and installed.

In many situations, however, the auditor will not have the ability to review a package before installation. When reviewing controls and procedures associated with the installed package, the auditor should attempt to assess its security risks. Figure 12.8 is a checklist which should be helpful in performing this assessment. Of course, any "No" answer from this list should not result only in a controls weakness finding in the auditor's report. Rather, these should suggest areas for corrective action.

For example, the auditor may find that the 4GL installed at the data center allows exits or calls to other processors to be included in the 4GL code. This easily could result in improper manipulation of programs and data. The auditor might suggest that, if possible, only a subset of that language—without the exit facility—be distributed to end users. Data processing departmental policies also should be instituted regarding the use of such exit procedures. In addition, the auditor might suggest that all data processing production applications be subjected to quality assurance review to look for such exits.

There is a danger that significant security concerns surrounding the use of 4GLs will be ignored because of user enthusiasm for the 4GL and for its productivity improvements. The auditor can play an important role in identifying these concerns.

THE 4GL AS AN AUDIT TOOL

Generalized audit software packages have been a significant tool for audit retrieval and analysis since they first became readily available in the early 1970s. While many of these earlier packages used punched cards and were batch oriented, some of the generalized audit software products available today are quite powerful. Fourth generation languages now offer another tool for audit retrieval applications.

Checklist for Assessing the Security Risks of an Installed Fourth Generation Language

SECURITY FEATURES	YES	NO	N/A
1. Is the 4GL compatible with the information security package installed at the data center?			
2. Using the organization's prime information security package, can access rules for users of the 4GL package be identified to specific individuals and particular data sets?			
3. Does the 4GL product have its own security features or functions built around its own database system?			
4. Are the 4GL product security features compatible in form and function with the prime information security software?			
5. Can security over the 4GL database be maintained at a record or data item level?			
6. If a 4GL security function is used, does it have the ability to independently monitor and report security violations?			
7. Is it possible to protect 4GL code from unauthorized changes?			
8. In addition to the overall information security system, does the 4GL have a password system to protect individual programs and report files?			
9. Are unauthorized password access attempts immediately signaled or highlighted?			
10. Can tables or other constant data be hidden within the 4GL code?			
11. Does the 4GL software maintain archival copies of program changes to track change history?			
12. Can files properly assigned to one set of 4GL applications be accessed by other 4GL programs without proper authorization?			

Figure 12.8. Checklist for assessing the security risks of an installed fourth generation language

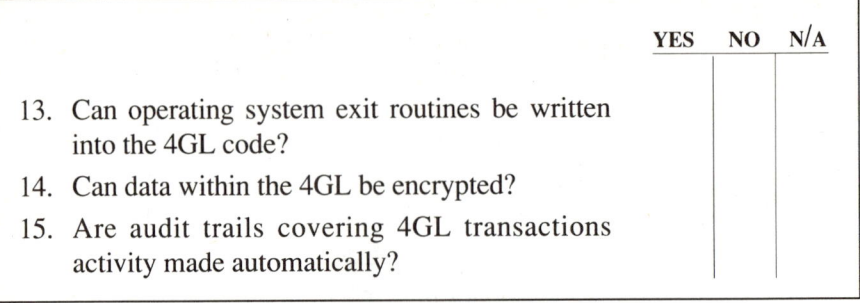

	YES	NO	N/A
13. Can operating system exit routines be written into the 4GL code?			
14. Can data within the 4GL be encrypted?			
15. Are audit trails covering 4GL transactions activity made automatically?			

Figure 12.8. Checklist for assessing the security risks of an installed fourth generation language *(continued)*

The earlier sections of this chapter discussed the power and flexibility of many of the fourth generation languages in use today. In addition to being powerful tools for developing production applications, these 4GLs can also function as powerful audit retrieval languages. They provide much of the report writing and file manipulation capabilities associated with the best of today's generalized audit software packages. If the audit department is still using one of the older audit software packages, a 4GL may provide even more power than the audit software.

Earlier sections of this chapter suggested that when a data processing organization has installed and is making extensive use of a 4GL, the auditor should gain a hands on familiarity with that package to understand its functionality and to test applications developed with it. A logical next step is to use the 4GL for other audit retrieval work. This can be as a substitute or a supplement to the audit department's existing generalized audit software package.

The main advantages to using a typical 4GL as an audit retrieval tool are its power, ease of use, and flexibility in developing retrieval reports. While generalized audit software can produce reports easily, the packages often do not produce well-designed output report formats. Generalized audit software reports are well suited for auditor workpapers while the outputs from 4GLs can often be better used as more formalized management reports.

Some of the other advantages of 4GLs for generalized audit software include the following:

Graphical reporting. The typical 4GL product today has powerful graphical reporting capabilities. This type of reporting is becoming a medium for management communication. Although some generalized audit software packages do provide a limited set of graphical reports, they generally are quite rudimentary compared with the typical 4GL.

Multidimensional data manipulation. Auditors frequently find it necessary to manipulate multidimensional data to derive audit analysis results. This is a capability of 4GLs but not generalized audit software.

Specialized database access. Some 4GLs have the ability to create their own database formats while many others have the ability to access different DBMS formats. While the latter is available in some generalized audit software packages, generally the 4GL must be used to access its own database format.

File building. Several of the 4GLs allow test data files to be built through full screen inputs. This is particularly useful for a "test deck" approach to reviewing applications as discussed in Chapter 6.

Microcomputer to mainframe links. Many 4GL products have powerful and convenient tools for downloading data to a microcomputer for detailed analysis. This capability is limited in generalized audit software.

The above advantages point to the superiority of 4GLs over existing generalized audit software. However, generalized audit software has some definite advantages in other areas. Some specific audit functions that are usually built into generalized audit software must be programmed, often with some effort, when using a 4GL. These audit functions include:

Statistical analysis and sampling. A typical generalized audit software package has the ability to perform audit sampling with minimal auditor extra coding. The mathematics must be coded in most 4GLs.

Data file analysis routines and reports. Audit software can sequence check, age, and duplicate record check with no extra coding.

Specialized audit reporting. Generalized audit software can prepare specialized reports such as audit confirmation letters. These take more work, at least the first time, with a 4GL.

Job accounting log analysis. Several of the generalized audit software packages read and analyze system log files from large scale IBM computers. This can be accomplished using 4GLs only with great difficulty.

The above relative advantages of 4GLs and generalized audit software show that there is an audit purpose for each. If the audit department is concerned primarily with financial analysis testing, a 4GL often proves to be a superior product. If the audit department does extensive data analysis work, they might do well to continue using their generalized audit software package. However, if the data processing function already has a 4GL installed and the audit department has no generalized audit software, the audit function should consider whether the 4GL will meet requirements before starting a search for generalized audit software.

EVOLVING 4GL STANDARDS

Fourth generation languages will become a fixed feature in many data processing organizations as both an end user and as a development tool. In the near future, many different products with different features will be available. However, over time, the weakest of these competing products will tend to disappear. From those remaining, the auditor can expect to see some evolving standards.

From these standards will evolve a common language and data description format for much of the business data processing of the future, just as COBOL evolved as the third generation language standard for business data processing. While it is difficult to predict the exact nature of this future format, it is possible it will follow along the lines of IBM's SQL 4GL database retrieval product.

In data processing, of course, no "standard" remains with us for long. Fourth generation languages can be expected to be replaced soon by what are sometimes called "fifth generation languages." These are known also as "expert systems" and "artificial intelligence languages." They are discussed in Chapter 14.

Section V

FUTURE TRENDS FOR COMPUTER AUDITING

C H A P T E R **13**

Integrating Financial and Computer Auditing

INTRODUCTION

Many topics covered in the preceding chapters of this book are of a technical nature and appear to be directed to the "computer audit specialist." However, that specialist designation is avoided in the book. Rather, the objectives and procedures discussed throughout the book are directed to "the auditor." This choice of terminology has been deliberate. An objective of this book is to provide guidance to the computer audit specialist and the non-computer, financial or operational auditor in understanding modern data processing security, audit, and control procedures.

The typical internal audit function today has separate computer audit and financial/operational audit groups. Similarly, many public accounting firms have separate groups specializing in computer auditing. This separation is because computer auditors were necessary to understand and evaluate data processing controls and to communicate audit results to data processing personnel. The traditional financial/operational auditor typically did not have the technical skills to deal with the data processing function. However, with the increased use of office microcomputers, end user computing, and prototyping procedures for major applications development work, it is now often difficult to separate control responsibilities between users and data processing personnel. To evaluate many types of data processing controls, it is necessary for the traditional computer auditor to spend audit review time in user areas. Similarly, financial/operational auditors often must work closely with data processing personnel to perform their audit procedures.

Figure 13.1 shows how a traditional audit function has been organized. Often, communication between the computer audit group and the financial/operational audit group has been limited, sometimes to a one way communication channel. The only way the two groups have worked together at times has been for the computer auditors to write audit retrieval programs for the other auditors. This organizational separation and the traditional roles played by the two groups have limited the potential joint projects.

With many new information systems oriented toward end users, many audit functions can be integrated so that one auditor can evaluate data processing controls and perform traditional financial/operational audit procedures. This chapter discusses approaches to integrating better the financial/operational and computer functions in a typical internal audit department. The advantages and disadvantages of the approach will be discussed. The chapter also suggests the position descriptions and skill requirements for such an integrated audit department.

ADVANTAGES OF INTEGRATING THE AUDIT FUNCTION

There are many unique audit skill specialties. For example, there are financial auditors who are experts in pension benefit plans; operational auditors who are experts in manufacturing process control functions; and computer auditors who are specialists in telecommunications. The need for so many specialties argues for more, not less, separation of duties within an internal audit department. However, many reviews of financial systems in operational areas require both data processing and basic control evaluation skills. There are advantages and disadvantages to integrating computer auditors with an audit department's other auditors. The decision to integrate depends upon the nature of the organization and management's long range goals for the audit department.

Many audit departments today find that their computer audit function and financial audit function disagree about the scope or objectives of audit projects. For example, an audit department may request that its computer audit specialist review controls of an on-line financial system. The traditional computer auditor might review just general financial system controls within

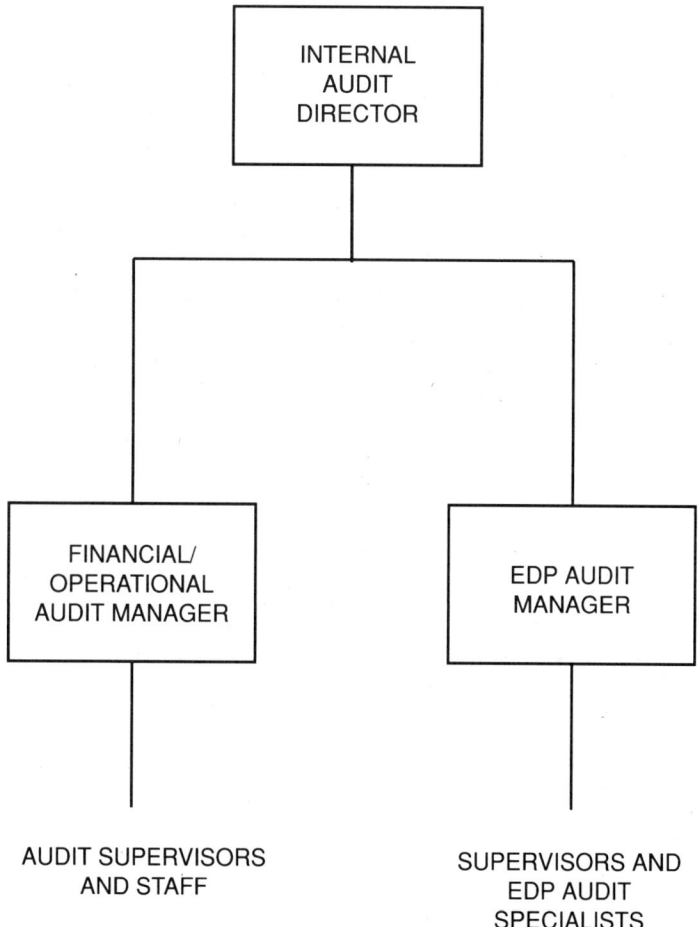

Figure 13.1. Traditional internal audit department organization

the data processing department, such as controls over program library updates, and give little attention to the on-line financial applications controls that should be in place in user areas. The audit report that will result from such a review may point to data processing problems but will provide little help for the financial auditor attempting to understand why there is so much difficulty in balancing control totals. That same financial auditor, however, deserves some of the blame. All too often, the financial auditor will identify as a problem the report from an automated application and decide that it is a "computer audit problem" and take no further action.

An integrated audit team approach to such an automated system might be able better to plan and perform the overall review. The members of the audit team with stronger data processing skills could spend time with the applications developers but also would review the output reports and retrieval screens used in the end user areas. Similarly, the financially oriented auditors might do a more effective job, through their use of available retrieval languages, to sample and survey application files.

The integrated audit team, whether internal or external, often can be very effective in reviewing complex, automated applications. Members of that team can contribute their unique skills to produce a better overall audit effort. While there probably will always be a need for specialists, some of the advantages to having a more integrated audit function include:

The changing nature of audit evidence. Chapter 6 discussed the changing nature of audit evidence as organizations move toward paperless applications. When attempting to develop an understanding of a paperless system, the auditor must have knowledge of data processing controls and procedures. Auditors who have an understanding of the accounting and other control procedures built into such a paperless system, however, also are necessary to evaluate it properly. A joint or integrated review approach is often the best.

The growth of end user computing. Chapter 11 discussed the growth of end user controlled computing activities. In many organizations today, it is not unusual to find, for example, a cost accounting analyst who is responsible for end user computer applications and sounds more like a data processing department programmer than a cost accountant. It is difficult to justify separate computer audit and financial audit functions when dealing in this type of an end user computing environment.

The auditor's use of automated tools. Audit functions are becoming increasingly automated. Auditors use computer assisted audit procedures to test major automated applications and use microcomputers to schedule proposed audit adjustments and generate audit workpapers. Through the use of these procedures, many traditional financial auditors have gained some data processing skills. Similarly, computer auditors helping to set up such automated audit tools often have gained an increased understanding of the needs and concerns of financial auditors.

DISADVANTAGES OF INTEGRATING THE AUDIT FUNCTION

Chapter 2 discussed how computer auditing evolved as a separate profession in the early 1970s. Although financial and operational applications became increasingly dependent upon automation, many organizations elected to retain separate audit functions. They did not consider the advantages of an integrated approach for the following reasons:

The need for specialized technical computer audit skills. While a combined audit approach may be best for the review of complex, automated applications, there are other areas where this combined approach may not work due to specialized auditor skill needs. For example, organizations may require computer audit specialists in areas such as telecommunications or operating system controls. The audit generalist often does not have the appropriate level of technical knowledge to operate effectively in such areas.

Problems with audit "cultural" differences. Many computer auditors have data processing backgrounds while many financial auditors have public accounting audit backgrounds. These two groups tend to differ in their training and education as well as problem solving approaches. It may be difficult to build an effective audit team because of these training and work culture differences.

Career pathing differences between audit specialists. The typical computer audit specialist often plans to move, at some future point in time, from audit back to data processing, or to functions such as quality assurance or data security. The typical financial auditor sees an eventual move to financial management as a more logical career path. If audit staffs are integrated, organization management and the auditors themselves may have difficulty seeing a clear career path from the audit department.

The above illustrate a few of the potential disadvantages to integrating the audit function. However, increased dependence on automated information systems in the modern organization may force greater audit function integration. It is possible that most modern audit functions of the future will not have a separate computer audit function.

APPROACHES TO INTEGRATING THE AUDIT FUNCTION

There are a variety of approaches to combine audit functions to increase effectiveness and efficiency. Many depend upon the overall organization, including its information systems environment. An approach that can be effective today may need to be modified in the future if the organization moves to a greater emphasis on distributed computing with many key systems located in end user areas.

Public Accounting Approaches to Audit Integration

Public accounting auditing standards are now a strong influence encouraging audit integration. However, this has not always been the case. From the middle 1970s through 1984, external auditors performed their reviews of data processing controls following the AICPA Statement on Auditing Standards (SAS) No. 3, *The Effects of EDP on the Auditor's Study and Evaluation of Internal Control*. That standard required external auditors to review data processing controls in accounting systems even though they did not need to rely on the data processing controls.

The requirements of *SAS No. 3* caused many public accounting firms to place the responsibility for reviewing data processing controls on non-auditing personnel. Some firms gave the responsibility to their Management Advisory Services (MAS) consulting groups while other firms set up special computer audit groups. The MAS computer auditors often had minimal appreciation of the application control concerns of financial auditors; and their reviews tended to emphasize general controls issues which, in some instances, were not relevant. For example, computer auditors in this era spent considerable time on physical security access controls and fire protection systems. At the same time, the financial auditor may have been looking for on-line data validation controls as well as procedures for controlling information security.

In 1984, the AICPA superseded *SAS No. 3* with *SAS No. 48, The Effects of Computer Processing on the Examination of Financial Statements*. This standard directed external auditors to consider data processing controls as part of the overall internal controls environment. While the auditor should consider the unique characteristics of data processing control procedures, they should be considered as part of the overall system of internal accounting controls. Although the audit procedures used to gather evidence may differ, *SAS No. 48* pointed out that specific audit objectives do not change whether accounting data is processed manually or by computer.

SAS No. 48 also advised that auditors should consider whether persons with specialized data processing skills (computer audit specialists) may be needed to help understand the flow of transactions and help design control procedures for computerized applications. In addition, *SAS No. 48* stated that even when such specialists are used, the auditor responsible for the work "should have sufficient computer-related knowledge to communicate the objectives of the other professional's work; to evaluate whether the specified procedures will meet the auditor's objectives; and to evaluate the results of the procedures applied as they relate to the nature, timing, and extent of other planned audit procedures." In effect, this guidance material required the non-computer auditor to have a level of understanding of computer control techniques and procedures.

Subsequent changes to auditing standards for public accountants have reinforced the need for all auditors to have an understanding of data processing controls. Current standards follow *SAS No. 55* and require auditors to evaluate the level of what is now called "control risk" in complex data processing environments. While specialists can be used to perform a review or assessment of control risk, the financial auditor is expected to have a basic level of data processing understanding.

Public accounting firms have adapted to these newer standards in a variety of ways. However, the standards do establish that data processing controls should be considered as part of the overall controls environment and the non-computer auditor should have a good understanding of the work of the computer audit specialist. This suggests that external financial auditors and computer audit specialists should work closer together. In effect, the AICPA auditing standards are requiring greater integration between the financial audit and the computer audit functions.

Internal Audit Approaches to Audit Integration

Internal audit departments follow a somewhat different set of rules and standards than external auditors. While many internal auditors come from public accounting backgrounds and are familiar with AICPA auditing standards, they also follow the guidance of their professional organizations, The Institute of Internal Auditors (IIA) and the EDP Auditors Association (EDPAA). In addition, internal audit departments must follow the direction of their own organization's management and board of directors' audit committee.

The internal audit professional groups provide mixed guidance to audit department integration. The statements in the IIA's *Statements on Internal Auditing Standards* are general and designed to apply to both financial/operational and computer audit specialists. However, IIA also has separate publications and conferences for the computer audit specialist. The EDPAA, on the other hand, is a professional organization just for computer audit specialists and tends to promote the specialty as a separate profession. The EDPAA also provides support and guidance for highly technical areas of computer auditing which probably would never be part of an integrated audit function.

Many internal audit departments today have separate computer audit and financial functions as illustrated in Figure 13.1. Even though business automation and technological trends point to the advantage of integrating audit department functions, an effective integration often is difficult to implement. The following sections describe some approaches which have worked in achieving an integration of the internal audit functions.

Joint Audit Assignments

There always will be some areas where computer audit specialists are best suited or where financial or operational auditors have unique skills. Normally, these are narrow areas such as a controls review of the data processing telecommunications function or an operational review of compliance with regulatory rules. There also are many other areas where auditors of both technical backgrounds could contribute to achieving an understanding of controls and making effective management recommendations.

A successful approach taken by some internal audit departments is to make joint assignments. Depending upon the specific project, the audit team could be supervised by either computer or financial auditors and staffed with an appropriate mix. Care should be taken, of course, to define the roles of all members of such a joint audit team. They should understand the overall objectives of the work and carefully plan their individual roles.

The most obvious area of opportunity for joint audit projects is in the review of automated financial and operational application systems. This type of review is particularly well suited for a joint review by auditors with various specialties. Many of the references to "the auditor" in Chapter 6 were made with such a joint team in mind. With such a joint project, the computer specialists might concentrate on the following areas:

Technical systems and programming documentation, controls, and procedures

End user controls over information center retrieval systems

Information security rules and procedures in both user and data processing departments

The design and implementation of continuous audit monitors or other computer assisted audit techniques

The financial auditors might contribute to the project in the following areas:

Reviews of procedures and user documentation, whether developed by data processing or user departments

Financial and accounting related control procedures

Human factor issues associated with forms, output reports, or on-line screens

Compliance with organization policies and regulatory requirements

There can and should be a role for a computer audit specialist in reviews of non-automated applications to help identify areas for potential automation opportunities. In many instances, the computer audit specialist may be aware of other automated applications within the overall organization which might be useful to the non-automated area under review.

Although often considered solely the responsibility of the computer audit specialist, data processing operations controls reviews can also benefit from the contribution of the non-computer audit specialist on the review team. Some of the areas where non-computer personnel might contribute to a joint data processing controls review include:

Input and output controls, including the submission of documents for batch systems and the distribution of output reports

Procedures and controls for monitoring the costs of data processing operations and pricing its services to end users

Procedures and controls for authorizing remote terminal access rights to the main computer

Procedures and controls for assigning equipment to end users

Administrative procedures and controls for the computer security function

Joint audit projects may be the best way to integrate computer and financial/operational auditing functions in the typical internal audit department.

This approach is particularly appropriate if the organization is sufficiently large to justify multi-participant audit projects. Joint audit projects also are sometimes appropriate for a smaller audit department with one computer auditor and two or three financial auditors.

A joint project audit organization is illustrated in Figure 13.2. The typical problem with this organization is that the supervisors remain within their specialty areas. Also, conflicts arise when staff members from one group are loaned to the other for joint projects. Loaned staff members may feel they are being asked to report to two bosses.

Cross Training Approaches

Another approach to organizing an integrated audit function is to cross train some of the staff members. Financial/operational auditors can be trained in computer audit techniques and data processing controls to allow them to perform some types of data processing reviews, and the computer audit staff can be trained in financial and operational auditing techniques. This approach works best with staff members who have a strong interest in learning other audit procedures. While some staff members always are eager to learn, there must be an incentive to persuade others to develop these skills. Sometimes, an appropriate incentive is to let it be known that promotions in the department require that candidates have skills in both disciplines.

Locating good sources of training can be a real problem with this strategy. There are numerous seminars and classes offered in both computers and computer auditing techniques. There also are more limited offerings in financial and operational auditing courses. Many of these programs, particularly the computer courses, often are too elementary, too oriented to narrow technical areas, or out of date. Before attending or having an employee attend such a class, a detailed syllabus should be reviewed to identify course offerings that are out of date or otherwise incomplete.

Textbook or seminar training will accomplish little, however, if the persons taking the training do not use it. After a financial auditor has completed a computer audit class, that individual should be assigned to a computer audit project as soon as possible. If the audit department has sufficient resources, peer members of the staff with computer audit skills can review the work of this person and make suggestions for improvement.

COORDINATION WITH THE QUALITY ASSURANCE FUNCTION

Many larger organizations have a separate data processing quality assurance function with the responsibility of reviewing data processing activities and

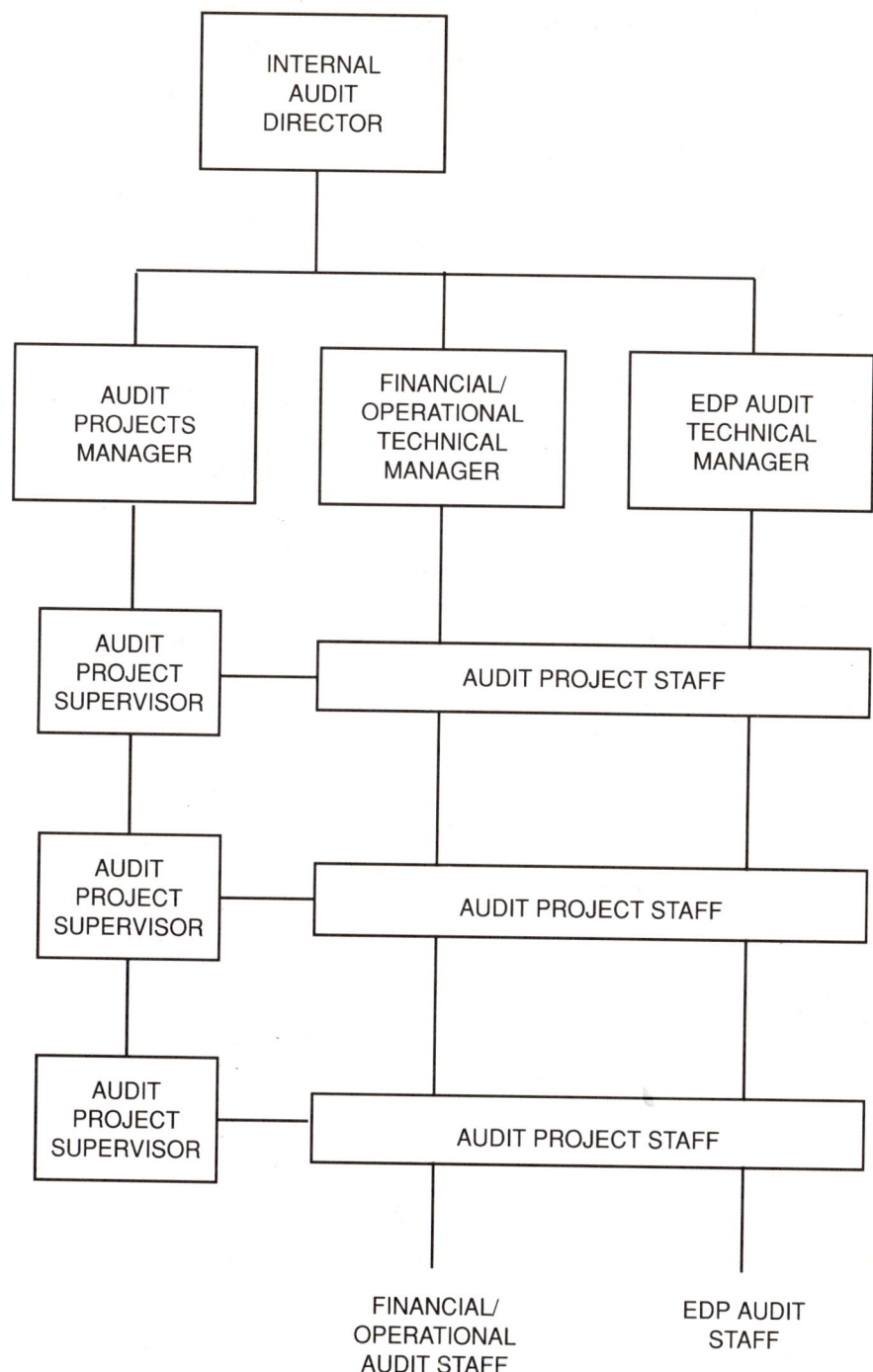

Figure 13.2. Joint project audit organization

suggesting corrective action on an ongoing basis. Such functions are similar to classic computer audit functions; the essential difference is that computer auditors determine whether things have been done correctly after the fact while quality assurance personnel attempt to do so before final implementation. Internal audit departments should have close relationships with the quality assurance function.

As discussed in this chapter, one of the impediments to an integrated audit function is the need for certain specialized audit skills in the larger department. If personnel with these skills exist in the quality assurance function, however, there may be less need to have these skills within internal audit. Such a transfer of responsibilities, however, takes both negotation and management approval.

Reliance on quality assurance will work when quality assurance agrees to monitor and suggest improvements in technical areas, such as telecommunications controls or operating systems maintenance. The more generalist internal auditors, then, can review the work of the quality assurance function on an ongoing basis. They would expect the quality assurance department to have workpaper documentation and evidence of reports suggesting corrective action. Internal audit would report to data processing and to general management on the overall performance of this quality assurance group. Internal audit could perform this function without the same level of technical skills as if they were doing the actual technical reviews.

The strategy to rely on a quality assurance function, though, will not be acceptable to some organization managements. They will want an independent function to review data processing technical areas, and quality assurance is generally part of data processing management. If this is the case, internal audit management may want to retain some highly skilled specialists on staff to perform such technical work along with a more generalist, integrated audit function performing most normal audit projects.

THE SUCCESSFUL MODERN INTERNAL AUDIT FUNCTION

This chapter discussed how the modern audit function might better integrate its computer and financial/operational audit functions. Most modern systems use data processing technologies too much, and modern user departments are too involved in the applications development process to justify a totally separate computer audit and financial/operational audit department. The successful audit function of the future will integrate these skills with some additional technical specialists to help.

How might such a modern internal audit function be organized? Figure 13.3

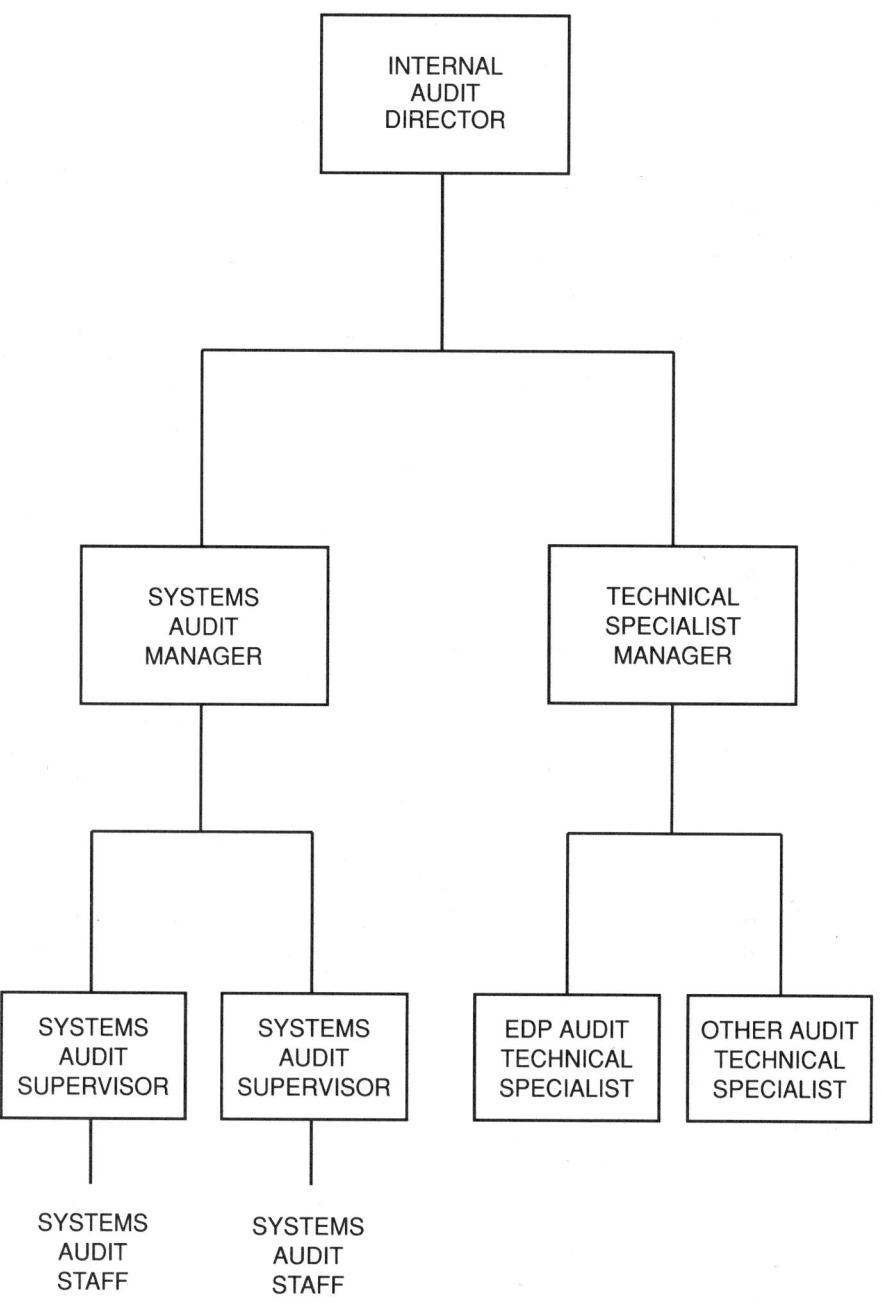

Figure 13.3. Potential audit organization of the future

shows an organization chart for a potential, medium sized audit department of the future. There are three main operating groups in such a department: systems auditors, computer audit technical specialists, and other audit technical specialists. The sizes of the groups of technical specialists would depend upon the needs and requirements of the organization. An organization with only a medium sized, minicomputer based data processing department using primarily packaged software might not require any technical specialists. However, if that same organization were performing complex manufacturing procedures, it might want a technical operational auditor with manufacturing or industrial engineering skills.

The Systems Auditor

We have described a new position in the modern, integrated internal audit department—the systems auditor. This is an audit professional with strong skills in both financial/operational auditing and computer auditing. This probably will be the audit professional of the future, and certainly will be the internal audit professional of the future.

Figure 13.4 is a position description for such a systems auditor. It requires that the audit professional have skills in both financial auditing and computer auditing. Such an audit professional does not need specialized technical skills in such narrow technical areas as database management systems or telecommunications controls. However, auditors of the future probably will be trained with these skills. The students coming out of colleges and universities today have far greater data processing knowledge and skills than those of just a few years ago. Similarly, professional qualification examinations, such as the CPA examination or the IIA's Certified Internal Auditor test, have increased their emphasis on data processing concepts and skills.

Internal audit management should recognize, of course, that the position described in Figure 13.4 represents an ideal. An advertisement on the company bulletin board today seeking someone with those exact skills may not elicit many responses. However, that description represents the auditor of the future in the modern organization, and it should be an audit organization's goal to build personnel with these skills.

The Computer Audit Technical Specialist

There will continue to be a need for the computer audit technical specialist in many organizations because of the technical environment of the organization.

Systems Auditor Position Description

Job Requirements: A generalist internal auditor with a strong knowledge of audit and internal control techniques including operational auditing, cost and financial accounting, and data processing technologies. Job requires the review, testing, and risk assessment of various manual and automated systems. The Systems Auditor will report to the Systems Auditor Supervisor and will be able to use the skills of technical audit specialists for in-depth help with various audit technical issues.

Education Requirements: Bachelor's degree in accounting or computer information systems. An MBA is preferred.

Professional Certification: CIA, CISA, and/or CPA.

Required Knowledge Skills:
   Internal audit procedures following the IIA's *Standards for the Practice of Internal Auditing*
   Financial and cost accounting principles and concepts
   Information systems design and organization concepts
   Familiarity with programming data retrieval languages such as fourth generation languages or a computer audit retrieval language
   Good understanding of information systems physical and information security concepts
   General understanding of data processing concepts such as database technologies, telecommunications, and CASE
   Strong microcomputer skills in the use of spreadsheets, wordprocessing, telecommunications
   General understanding of risk analysis concepts and statistical sampling

Figure 13.4. Systems auditor position description

For example, the internal audit department working with the larger data processing organization, as described in Chapter 2, will require the services of one or more computer audit technical specialists. Such persons may also be required in an organization with a large distributed processing telecommunications network as described in Chapter 4.

A position description for a computer audit technical specialist is shown in Figure 13.5. However, the particular requirements of such an individual depend upon the data processing department's technical environment and the nature of its various applications. This person or persons will be responsible for reviewing controls in specialized areas as well as providing technical consultation and support to other members of the audit function.

THE AUDIT DEPARTMENT OF THE FUTURE

This chapter described an integrated audit function where many of the tasks and functions now performed separately by computer audit specialists and by financial/operational auditors are integrated in a single function. Such an audit function generally does not exist at the present time. However, the aggressive, forward thinking audit function probably will evolve into this type of organization.

The audit professional also should consider a career move to become either more of a generalist, a systems auditor, or a technical audit specialist. The computer audit specialist of today who spends much time looking at general controls within the computer operations area and does not get into user areas to evaluate applications controls, risks the danger of becoming obsolete in the era of modern data processing procedures. That auditor should develop financial and/or operational audit skills as well as computer auditing skills to operate as the organizations auditor of the future.

Computer Audit Technical Specialist Position Description

Job Requirements: A strong data processing specialist with an interest in the audit, security, and internal control aspects of information systems. Job requires an understanding of computer systems activities within the organization as well as the requirements of the internal Systems Auditor group. The Computer Audit Technical Specialist will serve as an advisor to the Systems Auditor group and will work on special projects as required. Position reports to the Audit Specialty Manager who may also have technical specialists in other disciplines on the staff.

Education Requirements: Bachelor's degree in computer sciences or related area. An advanced technical degree or MBA is preferred.

Professional Certification: CISA, CDP, and/or CIA.

Required Knowledge Skills:
> Strong knowledge of information systems design including the use of CASE tools and structured design methodologies
> Good knowledge of programming principles and practices, including the ability to write code in one procedural language, such as COBOL, and a fourth generation language
> Strong knowledge of database procedures including an understanding of audit related risks and exposures
> Knowledge of operating systems concepts including the use of log file and other diagnostic tools
> Familiarity with computer security tools and techniques for mainframe, microcomputer, and other processors used in the organization
> Knowledge of telecommunications general concepts and local area networks
> Knowledge of audit procedures following the IIA's *Standards for the Practice of Internal Auditing*
> General understanding of risk analysis concepts and statistical sampling

Figure 13.5. Computer audit technical specialist position description

CHAPTER **14**

Evolving Technologies and the Auditor

THE IMPORTANCE OF KEEPING CURRENT

This book has tried to emphasize current data processing and auditing procedures. It covers many of the new computer and information systems technologies that the modern auditor encounters. Technology, however, moves fast, and the auditor should keep up with current issues. It is not necessary to be an "expert" in these new and evolving technologies, but the auditor with up-to-date knowledge can better adapt standard audit procedures to new technological environments as they appear.

The basic concepts of internal control have changed little over time. However, advances in technology often force the auditor to adapt and rethink internal control concepts. The auditor who refuses to budge from old concepts will become obsolete rapidly and of little use to management.

An example may help to explain this need to adapt to changing technologies. In the early 1980s, microcomputers were beginning to be introduced into the office setting as a serious, user managed business data processing tool. However, many computer audit "experts" at that time questioned whether these machines could ever be used for "serious" data processing because of their limited capabilities and because they were not subject to classic data processing general controls.

In 1984, the author of this book participated in a panel on microcomputer controls at a major computer audit conference. One of the panelists, from a major public accounting firm, *insisted* that to ensure a proper controls environment all office microcomputers must be kept in separate, locked rooms accessible by authorized personnel only! That panelist came from an era of mainframe machines when computers always were kept in secure facilities. The panelist was unable to rethink classic control concepts in light of changing technologies.

The effective, modern auditor should continuously adapt classic concepts of internal control to new, evolving technologies. This chapter introduces several of these evolving technologies which the auditor may well encounter in the near future.

NEW TECHNOLOGIES—LOOKING INTO THE CRYSTAL BALL

It is difficult to predict the future. Prognosticators tend to look at today's most recent advances and project them to future years in the same manner in which unscrupulous securities salespersons take the past few years of a stock's performance and make a straight line projection into the future. Another example of this type of faulty prediction is the work of economists in the mid-1970s who looked at the rise in the price of crude oil and predicted it would cost several hundred dollars per barrel by the year 2000. Other market conditions soon changed that trend.

Another crystal ball trap is to grasp one new idea and predict that everyone soon will be using it. One only need look at some of the projections published in the 1950s. One would have had us all commuting to work today in our personal helicopters!

Despite these concerns about predicting the future, there are several trends which will impact data processing technology. Some of these trends are new while others have been under development for some time and are not yet commercial successes. The forward thinking auditor should be aware of these trends.

Artificial Intelligence

There have been attempts since the earliest days of computers to make computers "think." Much of the early work in this area involved programming computers to play games such as chess or translate foreign languages. However in recent years, this entire field, which is known as "aritificial intelligence," has exploded with many applications moving out of the research laboratory and into commercial applications.

Artificial intelligence (or AI) is that portion of computer science that is concerned with designing systems that exhibit characteristics normally associated with human intelligence—such as learning, reasoning, and understanding problems. Some of the major branches of AI include:

Natural language applications. These are systems which can accept and respond to data and queries in a natural language, such as English, as opposed to a traditional computer language, such as COBOL. Many rudimentary natural language applications exist at present. As their use and sophistication grows, the auditor will see a decrease in the level of traditional, paper oriented documentary evidence. For example, if it is possible to give an information system a verbal authorization through a microphone attached to a terminal, there certainly will be a reluctance to document that authorization for audit trail purposes.

Expert systems. Many decisions that are made in decision and policy areas require "rule of thumb" judgments of experts in those decision areas. Experienced personnel or experts in a subject area often make a decision without going through an extensive analysis of analytical data. A computer system that captures all of the expert's knowledge and then makes best guess decisions based upon this knowledge is known as an expert system. Of the artificial intelligence applications discussed here, these probably will have the most immediate impact on auditors. Expert systems are discussed in more detail later in this chapter.

Machine vision and robotics. Cartoon and movie characterizations of intelligent, human-like robot machines have been with us for some time. However, practical examples based on artificial intelligence techniques are being introduced onto the factory floor. Such machines, for example, can select the correct part, properly orient it, and add it to a product under assembly.

Machine learning and self programming applications. Artificial intelligence techniques are used to perform such tasks as writing programs to prove mathematical theorems. A similar theorem is input and the system develops a program to prove a new one. In the commercial area, this is known as programming by induction.

The auditor is likely to see an explosion in the use of artificial intelligence applications in the near future. These applications will change many concepts of applications controls, systems development, and our reliance on computerized procedures.

Highly Parallel Processors

The largest and most powerful supercomputers still work in a serial manner. The machines must perform one instruction before they can do the next. The human brain, however, works in a parallel manner. Our brains can process numerous thoughts or tasks simultaneously. Many experts feel that a computer system which could act more like a human brain and process many instructions simultaneously could solve complex problems much more effectively.

Considerable work is taking place now in combining many small computers to process data and make decisions in a parallel manner. The result of this probably will be even more powerful supercomputers to solve such complex problems as global weather predictions. It could result in different approaches to programming and in more powerful microcomputers.

There will be some audit and control implications in the future use of such parallel processors. Because of the theoretical capacity of these processors, more complex tasks can be performed which may not lend themselves to traditional audit testing and evidence gathering.

The introduction of parallel processor computers should increase greatly the complexity of application systems. These machines will be used initially for specialized purposes such as engineering modeling and design. However, the auditor can expect also to see highly complex business modeling or statistical decision making systems in the future.

Telecommunications Advances

Many new technologies are being developed for the commercial marketplace. These are increasing the speed and accuracy of data communications. These technologies also are merging voice, data, and even video imaging. While these advances may not have a direct effect on the auditor's internal controls oriented review work, they will increase the numbers of networked or distributed systems.

Another telecommunications related advance will be an increased ability to interconnect various computer systems. Computer systems, large and small, will be able to treat one another as peers and share data and programs easily. Thus, a user may be able to access an application through a local terminal which accesses and shares data from a worldwide network of databases. The locations of these distributed databases will be transparent to users.

Microcomputer Technology Advances

When microcomputers first arrived in the workplace, they had 64k bytes or less of random access memory and operated at relatively slow speeds. New microcomputers today, based on the Intel 80386 chip for example, are capable of accessing up to 4 gigabytes of random access memory and almost unlimited virtual memory. They can operate at five or six times the speed of earlier microcomputers. These new machines definitely will change the way information systems are developed and designed. These advances are discussed in more detail later in this chapter.

Even though the capability of the Intel 80386 chip has not yet been realized, an even more powerful 80486 chip has already been announced. A similar path of advanced development is taking place with Motorola's 68040 chip. The power of the microcomputer can be expected to increase, and future business systems will be designed to take advantage of those increases.

ROM or WORM Disc Storage Devices

Compact discs (or CDs) have revolutionized the way music is recorded and replayed. Music is recorded digitally on a smooth disc and reread through a laser device. These small metal plated discs carry high volumes of recordings, are convenient, and are virtually indestructible. ROM or WORM discs may cause some of the same changes for information systems.

ROMs and WORMS refer, respectively, to "Read Only Memory" and "Write Once [then] Read-only Memory." They use essentially the same technology as music CDs to store vast amounts of computer data on a small platter. For example, an entire encyclopedia or set of reference data can be put on a single ROM disc. WORMs can be used to record archival data which may be reread at a later date but can never be erased.

This is not an extremely new technology, but inexpensive computer peripheral devices are just beginning to take advantage of these tools. When widely available, they will impact information systems design. They also will become a powerful audit productivity tool.

Digital Image Processing

Optical scanners and bar code readers have been part of data processing technology for some time. However, a similar and much improved technology, called digital image processing, is evolving rapidly. This technology uses scanners which can read almost any document, including special forms and photographs, and write that document to a WORM disc. An exact image of the document can be retrieved later, and document data can be ported over to other database files.

Digital image processing offers major savings in productivity since manual filing systems can be replaced. The technology also presents a challenge to the auditor since traditional paper trails might be eliminated. Once a document has been captured, however, it cannot be erased from a WORM disc. It therefore will be necessary for applications based on this technology to have good database retrieval tools.

Superconductivity Advances

While ROMs and WORMS represent a technology waiting for implementation, computers based on superconductive materials represent the more distant future. If the temperature of an electrical conductor is reduced to a very low point, below $-400°F$, electrical current flows with essentially no resistance and therefore at very high speeds. A computer system designed to operate at these low temperatures would be extremely fast but is not now practical due to the extreme requirements for chilling the processor.

Superconductive materials are being developed now that do not require extremely low temperatures. These materials have not moved outside the laboratory but probably will be incorporated into computer systems in the future that will be faster and more powerful than anything in existence today.

The above list is not intended to be all inclusive of the newer technologies which will impact the data center and the auditor in future years. However, they represent some short and long range directions in which computers and information systems may evolve. Some, such as processors based on superconductivity, do not have an immediate direct effect upon the auditor. Others, such as expert systems, are beginning to be used in the business environment and will have a more immediate effect.

EXPERT SYSTEMS AND THE AUDITOR

For many years, the Campbell Soup Company had one expert within its organization to solve mechanical problems with certain models of its plant soup cookers. Local plants would call this person and explain the symptoms, and the expert would suggest a solution. That expert's advice was based on some formal rules and also on informal rules of thumb. This problem solving technique is similar to that found in many organizations where there is a very small group of experts in some unique, specialized area. Campbell Soup had a problem several years ago, however, when their expert was ready to retire.

Prior to that retirement, a group of expert systems developers attempted to capture all of the formal rules as well as the informal rules of thumb that the soup machine expert used in problem diagnosis. These were loaded into a computer system that was able to scan through the various problem symptoms and rules to solve the problem. Now, when a plant calls in with a problem, the symptoms are entered into the "expert" computer system, and it suggests possible corrective actions. This is an example of an expert system.

Characteristics of an Expert System

Most computer programs today are based upon numerical rules or algorithms. As long as the decision variables can be stated numerically, programs can process through these rules and generate correct results. However, human decisions tend to be based on human knowledge and heuristic rules of thumb rather than predefined rules that can be coded into a typical computer program. For example, an auditor probably could not easily develop a conventional computer program to help decide when there is a controls weakness in a general controls review. An analysis of the many other factors and just good auditor judgment are necessary to make controls descisions.

The concept behind an expert system is to capture the knowledge of a human expert and place it into a computer system such that other users can access that system and take advantage of the expertise to make better decisions based upon these programmed expert rules. Expert systems usually consist of three components: a knowledge base that contains the expert's rules, an inference engine which can develop relationships from the knowledge base, and a user interface to allow communication with the system. These three components are illustrated in Figure 14.1. Software packages which combine the three components are called expert systems shells.

Expert systems are based on "If-Then" rules which are used to build the knowledge base. These "If-Then" rules are of the same type of rules that an

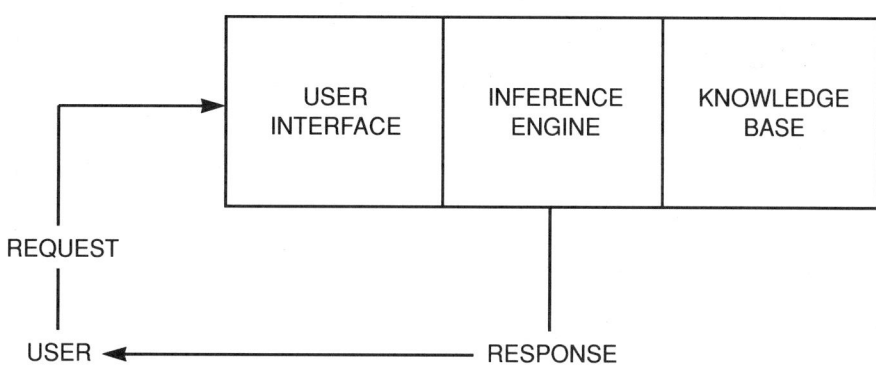

Figure 14.1 Components of an expert system

auditor uses when arriving at a conclusion. For example, a simple set of rules that the auditor could use to determine if physical security is good in a large data center might be as follows:

> *If* the computer is located in a physically secure room, *and*
> *If* the door to the room has an access locking device, *and*
> *If* only restricted personnel are given access keys,
> *Then* computer room physical security probably is adequate.

The above represents a simplified set of rules. Normally, hundreds of such rules are built into a knowledge base to help the expert system's inference engine look at the various conditions and make a best guess recommendation.

Currently, considerable development work is successfully taking place in expert systems. Applications have been developed and implemented for such activities as medical diagnosis, manufacturing parts analysis, and computer systems configuration control. Several major public accounting firms have developed expert systems for corporate tax accrual accounting, computer controls evaluation, and auditing risk analysis. The auditor should expect to encounter more expert systems in the future.

Reviewing Expert Systems Controls

In the future, auditors routinely will encounter expert systems being purchased or developed within their organizations. Because of the "If-Then" heuristic rules built into these systems, they will be developed and implemented in a different manner than conventional information systems. Nevertheless, it will be particularly important that the auditor review such expert systems applications for controls.

The most difficult problem confronting the auditor with such a review, however, is that there is often no one "correct" output or system answer. In addition, such systems are not developed through a single set of requirements or specifications. Rather, knowledge rules and inference relationships are added, changed, or deleted until the system is fine tuned and generally gives correct results.

Expert systems usually are developed by following a procedure similar to prototyping as discussed in Chapter 7. However, expert systems require a more iterative approach of collecting knowledge, translating it into rules, and testing those rules against the expert's judgment.

When reviewing such a system, the auditor should look for a strong set of testing procedures which predefine conditions and expected results. When the system's conclusions are different than expected, there should be detailed test evaluation procedures to determine the reason. In many instances a "test deck" technique, as discussed in Chapter 6, is a good method for testing expert systems. There should be a standard set of "test deck" conditions to be tested along with a standard set of documented conclusions.

The auditor should determine that the systems developers of an expert system use this "test deck" every time there is a change to the system. In addition, the auditor may want to maintain a separate version of this same "test deck" to test the application on a regular, periodic basis. Because expert systems developers and users will add or change systems rules from time to time, the auditor should ascertain that such comprehensive testing takes place whenever there is a change. Because of the way "If-Then" rules are linked in an expert system, a small change in one portion of the application may cause the system to yield very different results elsewhere.

Expert Systems as Audit Tools

Expert systems also have strong potential applicability to the audit function. Some areas where expert systems might be useful include:

Risk analysis decision tools. Auditors often are asked to evaluate the degree of relative audit risk associated with audit decisions, such as the risk associated with the controls being evaluated or the risk of reaching the wrong conclusion from a sample result. An expert system could ask all the questions that an expert auditor would ask and then provide the auditor with some suggested risk assessment conclusions. Such a system might help the auditor to select applications to be tested from among a large population, and also could provide documentation to support the selection.

Expert systems for specialized areas. Many organizations have special areas of operation where only a few persons in the audit department have any strong knowledge. For example, a food packaging company might have a single agricultural commodity trading division that is quite different from the balance of the organization. If there is only one expert within the audit department to review such a specialized division, an expert system might capture that expert's knowledge so that other auditors also could perform reviews as backup help. Also, an expert system could supply audit department support in certain highly technical specialized areas. For example, if an audit of the systems programming function required special computer audit expertise, an expert system could capture the rules, procedures, and conclusions necessary to perform such a review.

Expert systems for audit decision support will become commercially available in the not too distant future. These may be offered by public accounting firms or other commercial organizations. They generally will be used to solve limited domain problems, such as risk analysis or statistical sampling problems. While they are costly to develop, expert systems can be powerful tools to increase the overall efficiency and effectiveness of the audit function.

Audit Implications of Expert Systems

Expert systems almost certainly will be a technology that the auditor will encounter in future years. In addition to representing a different approach to programming computer applications, expert systems also present auditors with a new audit decision challenge.

Due to the manner in which expert systems are constructed, auditors may encounter problems relying upon them. Although auditors typically look for controls over access to data and programs, these controls may not exist in typical expert systems. Due to their nature, expert systems developers or users constantly add and change rules in the knowledge base or alter relationships in the inference engine.

Auditors have standard methods of testing and validating conventional data processing systems. Such testing methods, however, usually are not appropriate for the typical expert system. The conclusions developed by expert systems cannot be tested by matching expert system output data to the type of formula found in conventional audit software.

As we become accustomed to using expert systems, there is an additional danger that too much reliance may be placed on them. Due to their nature, expert systems cannot be expected to give the best answer all of the time, only most of the time. When a systems user has received a good answer from an expert system on many successive occasions, there is a danger that users will follow the results of an expert system even though it may be clearly wrong in a given situation.

REENGINEERING EXISTING COMPUTER APPLICATIONS

Microcomputer memory capacity improvements, disc capacity improvements, interconnection improvements, and more powerful microcomputer operating systems—How will this rapidly evolving technology impact existing applications? These trends all point to microcomputers in the near future which have power comparable to some of today's mainframe computers. This type of computing power will definitely alter the manner in which information systems are designed and data processing departments are organized.

The typical organization has a large inventory of existing, older computer applications which were developed over time using personnel and programming approaches no longer available. These applications, along with their revisions, represent major investments to the organization. Because of that investment, management generally is reluctant to throw them away and start fresh.

New software tools are being developed to help organizations reengineer these older applications to make them more maintainable and to allow for the introduction of any needed application changes. These software tools will read old program code and restructure it into a more maintainable form. They also will generate current documentation from the reengineered programs. These changes often are transparent to the application user and result in better documented and more maintainable programs.

The auditor should become aware of these evolving software tools. When an organization has a large inventory of old computer programs, it might be appropriate to recommend their use. If a data processing organization is embarking upon such a reengineering process, the auditor should follow these activities to determine that audit documentation is current and that no controls related changes are introduced during the reengineering process.

EVOLVING TECHNOLOGY AND KEEPING CURRENT

This chapter suggested a possible series of new information systems technologies which may impact the auditor. However, the technologies mentioned may represent only a portion of what the auditor may see in the next five to ten years. One can see how true this is if one steps back and examines what was predicted only five or ten years ago for today. Some of those predictions were right on target while others missed the mark.

This all points to the importance of keeping current in future technological developments. In addition to their roles as the internal control experts, auditors served and will continue to serve as advisors to management. Faced with rapidly evolving technology, however, the effective auditor must keep current with a good general knowledge of new technology developments and related control implications.

How does one keep current? Seminars and evening courses are available. However, due to the work required to construct such course material, often it is not current enough. The best sources are technical and trade periodicals. These help the auditor gain a general understanding of some of the evolving technological trends.

There are numerous data processing publications which provide the auditor with background information. The weekly newspaper, *Computerworld*, has industry, technical, and other related news. While the busy auditor may not have the time or interest to read through the entire publication, it provides information of new product announcements. There are many good data processing periodicals. Some are available free if the reader is involved in recommending or approving products of the publication's advertisers. There is a subscription fee for others.

Another source of information on new technical developments are trade publications for the auditor's industry. Whether involved in banking, food distribution, or government services, there will be some specialized trade publication available. While these generally are devoted to general, industry related news, they often have descriptions of new technical developments or stories about how industry leaders are solving problems. Such publications ordinarily are available within the auditor's organization.

Data processing and trade publications provide information on overall trends. The next step is to understand how these trends impact audit activities. Here, the professional organizations can help. Both the EDP Auditors Association and The Institute of Internal Auditors publish articles in their journals and hold periodic conferences covering evolving technology, "state of the art" topics.

The creative modern auditor should spend a certain portion of time keeping current on technological developments. Only this way will the auditor be able effectively to audit and review modern computer security and controls.

AUDITS OF THE FUTURE: MANUAL SYSTEMS

This book discussed many of the audit, control, and security implications of modern computer systems. These modern computer systems are found in virtually all organizations. It is essential that the modern auditor have some knowledge of the data processing audit, control, and security concepts surrounding those systems.

While this chapter discussed new technologies, many of them may become commonplace in the near future. In fact, the unusual system of the future may be today's traditional manual, paper oriented system. As tomorrow's auditors become familiar with various advanced technologies, those auditors may need to seek out a specialized book such as this for guidance in reviewing manual systems. This may be solace to today's auditor attempting to understand computer audit, control, and security concepts.

Appendix: How to Use the Supplied Diskette

Included with this book is a diskette containing copies of the tables of control objectives and audit procedures published as figures in this book. These are in ASCII format so that the reader can load them into an IBM-PC or compatible computer for printing or modification. This will allow the auditor to tailor these audit programs.

Each figure included on the diskette is presented as a separate file, following the same numbering system as in the book. For example, Figure 2.4, Control Objectives and Audit Procedures for Reviewing General Organization Controls, is carried on the diskette as the ASCII file, FIG2-4. The contents of the diskette are listed on pages ix and x at the front of the book.

This diskette requires an IBM-PC or compatible computer with MS DOS 2.0 or later. A standard 5¼" floppy disc drive is required, and a printer is recommended. Files can be printed using standard DOS functions or can be imported into a word processing software package for modification. Virtually all standard word processors have the ability to import ASCII files for editing and printing.

The procedure listed below describes how to make a backup copy of the original diskette. Since this is always a good idea, please take the time now to make a backup copy before doing anything else.

1. Start your computer either from the hard disc or with a disc containing the DOS operating system in drive A.

2. In response to the DOS prompt, A> (for floppy disc systems) or C> (for hard disc systems), enter:

DISKCOPY A: B:

3. After pressing the ENTER or RETURN key, place the original diskette in drive A and a formatted, blank diskette in drive B. Follow the prompts on the screen to make a backup copy. When complete, remove the original diskette and store it in a safe place.

4. If you are using a floppy disc system place your DOS diskette back in drive A and leave the new backup copy in drive B. Hard disc users can leave the diskette in drive B.

You do not need a word processor to see, edit, or print your files. By using simple DOS commands, you can have the files sent to the screen or to the printer. You may also customize or change the files by using a DOS text editor such as EDLIN. Please consult your DOS manuals for further information about EDLIN.

To print one of the files, use the DOS PRINT command. For example, if you want to print Figure 2.4, type the following:

PRINT B:FIG2-4   (press ENTER)

Since you may not always want to have the file sent to the printer, DOS also has a command which will display the file on the screen. Use the DOS TYPE command:

TYPE B:FIG2-4   (press ENTER)

For displaying or printing a file with word processing software, consult that software documentation. Usually the applicable procedures are discussed in sections describing how to import ASCII files or files produced on other work processing software. However, remember to make a backup copy of this diskette before starting any other processing.

Readers are encouraged to tailor or modify these audit programs according to their own requirements and needs. The author would be interested in seeing copies of any modified audit programs or suggestions for improvement for potential inclusion in the next edition of this book. Send any such materials, in either diskette or paper format, to:

Robert R. Moeller
Senior Corporate EDP Audit Manager
Sears, Roebuck & Co.
Dept. 968A — BSC 61-03
Sears Tower
Chicago, Illinois 60684

Index